MW01448052

Information Warfare and Organizational Decision-Making

For a listing of recent titles in the *Artech House Information Warfare Series,* turn to the back of this book.

Information Warfare and Organizational Decision-Making

Alexander Kott
Editor

ARTECH HOUSE
BOSTON | LONDON
artechhouse.com

Library of Congress Cataloging-in-Publication Data
A catalog record for this book is available from the U.S. Library of Congress.

British Library Cataloguing in Publication Data
A catalogue record for this book is available from the British Library.

ISBN-10: 1-59693-079-9
ISBN-13: 978-1-59693-079-7

Cover design by Yekaterina Ratner

© 2007 ARTECH HOUSE, INC.
685 Canton Street
Norwood, MA 02062

All rights reserved. Printed and bound in the United States of America. No part of this book may be reproduced or utilized in any form or by any means, electronic or mechanical, including photocopying, recording, or by any information storage and retrieval system, without permission in writing from the publisher.
   All terms mentioned in this book that are known to be trademarks or service marks have been appropriately capitalized. Artech House cannot attest to the accuracy of this information. Use of a term in this book should not be regarded as affecting the validity of any trademark or service mark.

10 9 8 7 6 5 4 3 2 1

Contents

Introduction ix

CHAPTER 1
Know Thy Enemy: Acquisition, Representation, and Management of Knowledge About Adversary Organizations   1

Introduction to Organization Warfare   2
Organizational Intelligence   6
Leadership Analysis   12
Social Network Modeling and Analysis   14
Representative Analytic Process and Tools   19
Application Considerations   23
    Endnotes   25

CHAPTER 2
Learning from the Enemy: Approaches to Identifying and Modeling the Hidden Enemy Organization   29

An Elusive Enemy   29
Domain of Organization Identification   31
Adversary Identification Research   33
Identification Focus: Definition of Command and Control Organization   36
From Structure to Strategy: Characterizing Behavior of a C2 Organization   38
Looking Beyond the Smoke Screen: Definition of Observations   42
Discovering the Dots: Finding Identities of Actors   45
Connecting the Dots: Discovering Patterns of Interactions and Activities   48
Behavior Learning and Simulating Enemy Activities   53
The Overall Process of Identifying Enemy Command and Control   55
Experimental Validation of the Organization Identification Process   57
    References   59

CHAPTER 3
Who's Calling? Deriving Organization Structure from Communication Records   63

The Tasks   64
CCR Structures and Their Graphical Representation   64
The Ali Baba Scenarios   66
Social Network Analysis   67
Previous Work Exploiting Time in Social Networks   71
The Windowing Down the Lines (WDL): The Algorithm   73
    The Scenario   74

The Data	74
Formatting the Data	74
Time Windowing	75
Time Overlapping	77
The Algorithm	79
Simple Simulation Tests	81
Evaluation of WDL with the Ali Baba Datasets	82
Without Topics	82
With Topics	84
The RSF Algorithm	86
References	88

CHAPTER 4
Means and Ways: Practical Approaches to Impact Adversary Decision-Making Processes — 89

Planning Operations for Organization Impact	90
Targeting Decision-Making	93
Targeting the Entire Organization	94
Effects-Based Targeting	95
Inducing Effects for Defeat	100
Denial or Destruction	101
Disruption	103
Deception	104
Direction and Reflexion	105
Targeting for Effects	107
Vulnerability Assessment	108
Plan Development	108
Strategic Analysis	109
Organization Behavior Analysis	109
Application Considerations	110
Endnotes	111

CHAPTER 5
Breakdown of Control: Common Malfunctions of Organizational Decision-Making — 115

Tardy Decision	117
Low and High Threshold	120
Excess of Timidity or Aggressiveness	121
Self-Reinforcing Error	122
Overload	124
Cascading Collapse	125
Misallocation of Authority	127
Lack of Synchronization and Coordination	128
Deadlock	131
Thrashing and Livelock	132
Acknowledgments	133
References	134

CHAPTER 6
Propagation of Defeat: Inducing and Mitigating a Self-Reinforcing Degradation   135

A Simple Model for Self-Reinforcing Decision Overload	136
Propagation of Disruptions in Organizations	141
Active Compensation	144
Dynamic Reorganization to Mitigate Malfunctions	146
Modeling Team Decision-Making: Decision Responsibility and Information Structure	147
Measuring Decision-Making Performance	150
Forecasting Decision Requirements	151
A Simulation of the Firefighting Example	151
Mitigating and Inducing Malfunctions	153
Acknowledgments	156
References	156

CHAPTER 7
Gossip Matters: Destabilization of an Organization by Injecting Suspicion   159

Is Gossip Good or Bad?	160
What Is Gossip?	162
Anonymous Information Exchange Networks	163
Hypothetical Organization, Part I: It's All Bits	164
Hypothetical Organization, Part II: Miscreant Markets	165
TAG Model Overview	167
The Mechanisms of Trust	169
Trust Model	169
Honesty Models	171
Gossip Model	171
Gossip and Disruption	172
Setting the Baselines	172
A Virtual Experiment in Disruption	173
Results of TAG Experiments	174
Interpretation	175
Giving and Taking	177
Acknowledgments	181
Endnotes	181

CHAPTER 8
Crystal Ball: Quantitatively Estimating Impacts of Probes and Interventions on an Enemy Organization   191

Organizational Analysis Approach	192
Organization's Options for Dealing with Mission Complexity	192
Enemy Organization Dynamics and Counteraction Strategies	193
Actionable Implications	194
Main Methods	195
Developing Strategies Against Enemy Organizations	195
Probe Identification	196

Intervention Planning	196
Test-Case Scenario: A Human-Guided Torpedo Attack on a U.S. Military Vessel in a Foreign Port	197
Enemy Organization Model	199
Enemy Mission Complexity Characteristics	200
Estimating Impacts of an Example Probe	203
Estimating Impacts of Example Interventions	205
Analytical and Numerical Methods Underlying VDT and Related Approaches	207
Organization Modeling and Strategy Development with SimVision	208
Implications for Detecting and Defeating Enemy Organizations	211
References	211

CHAPTER 9

Organizational Armor: Design of Attack-Resistant Organizations	213
How Organizations Cope with Disruptions	214
Attack-Resistant Design Solutions	218
Organizational Design Formalism	221
Precursors for Superior Organizational Performance	222
A Computational Approach for Predicting Organizational Performance	225
Normative Design of Robust and Adaptive Organizations	228
Empirical Validation of Normative Design Methodology	230
Reverse-Engineering Organizational Vulnerabilities	233
Robust and Adaptive Designs of Attack-Resistant Organizations	236
Illustrative Example—Redesigning an Organization to Enhance Attack Resistance	242
Example Scenario	242
Enemy Attacks	245
Redesign Principles	246
Attack-Specific Courses of Action	247
Engineering for Congruence with Mission in the Face of Attacks	248
Adaptation	250
Analyze Organizational Design	251
References	252
About the Authors	257
Index	265

Introduction

Organizations are among the most valuable and yet most fragile structures of our civilization. We rely on organizations to obtain information, to comprehend and process information, to accumulate and internalize information, to make decisions based in part on that information, and to execute those decisions. Organizations are ubiquitous because they are indispensable. From the most primitive to the most complex societies, organizations of various types have constituted the backbone of societal decision-making.

This book explores recent developments in computational solutions to problems of exploiting or mitigating vulnerabilities within organizational decision-making processes. It describes a range of computational techniques that can help to guide attacks on an adversary's organization or the defense of one's own.

From an engineering perspective, the applications of the techniques described in this book cover a broad range of practical problems. They include planning and command of military operations against an enemy command organization, military and foreign intelligence, antiterrorism and domestic security, information security, organizational design, military psychology and training, management practice, and organizational computing. In particular, one discipline—information warfare [1]—traditionally paid special attention to issues related to attacking and defending military command organizations.

Computational approaches applicable to such problems originate in several scientific disciplines: computational organization theory, organizational and management sciences, artificial intelligence planning, cognitive modeling, and game theory. A key objective of this work is to demonstrate important close relations between ideas coming from such diverse areas.

There are several reasons why the authors believe a book on this topic is particularly timely now. Earlier, in the 1990s, researchers were attracted to related topics in the context of information warfare, particularly in the subfield of command and control warfare [2], and in techniques for simulating command and control organizations for simulation-based wargaming [3]. More recently, the breadth and depth of related research have extended dramatically. The number of publications and conferences related to organizational modeling and analysis has grown by an order of magnitude in the last few years. This phenomenal growth was largely due to the aftermath of the September 11, 2001, terrorist attack [4] and the resulting attention to (1) ways to understand and influence terrorist and insurgent organizations, and (2) concerns about the vulnerability of domestic organizations.

Additional interest in these issues is spurred by massive (and expensive) transformational processes in the U.S. military. It is accompanied by the rethinking of approaches to organizing the command structures and processes in the age of information dominance and massive pervasive networking [5]. It is also accompanied by newly heightened appreciation of the complexity, nonlinearity, and vulnerability of the highly dynamic processes that occur in decision-making organizations.

Cases of large-scale organizational manipulation, such as Enron [6] and the United Nations (UN) Oil-for-Food program [7], also engender interest in means to detect and prevent such sophisticated attacks on organizational safeguards.

Furthermore, it has been only within the last 5–10 years that significant progress has been made in practical computational methods relevant to quantitative understanding of organizational performance [8, 9].

An Organization's Vulnerability to Attacks

Vulnerabilities of organizations, as well as their strengths, originate in the very purpose and nature of an organization. The reason an organization exists is to process a much greater volume of information (and, correspondingly, make and execute more decisions) than could be done by a single decision-maker, while at least maintaining and hopefully even increasing decision quality. This increase in information processing and decision-making capacity is accomplished by dividing a large volume of information processing and decision-making into smaller tasks and distributing those tasks among multiple entities—individuals or groups.

To integrate and coordinate these tasks, additional collaboration and supervision tasks are created. The result is an opportunity for each decision-maker to work on a manageable volume of assignments, to specialize in a particular class of tasks, to develop and apply significant skills, to have decisions reviewed and vetted by multiple specialists with a variety of perspectives and backgrounds, and to have the decisions effectively coordinated and combined into coherent higher level conclusions and decisions.

However, even a well-designed and established organization, with strong structure and culture, with properly trained and motivated (or designed, if artificial) decisions-makers of sufficient capabilities, can experience debilitating failures. Many of them originate from within the organization, while others may be induced by external influences, including hostile actors.

One major source of such failures is the limits on the rationality of human decision makers [10]. There are multiple aspects of such limitations: constraints on the amount and complexity of the information that a human can process or acquire in a given time period; biases in decision-making; and the rules, norms, and procedures that an organization or a broader society and culture impose on the decision makers. For example, time pressure is a well-recognized source of errors in human decision-making—as the number of decision tasks per unit time grows, the average quality of decisions deteriorates (see [11] and Chapter 6).

Lack of information and uncertainty about the likelihood of a decision's consequences, or the utilities of the respective alternatives, also have major impacts on decision-making. For decades, expected utility theory has been accepted as a normative model of rational choice and served as the dominant theme in the analysis of decision-making under uncertainty and risk. Yet many experiments demonstrate that real human decision-making exhibits consistent and pervasive deviations (often termed *paradoxes*) from the expected utility theory. For example, humans tend to prefer those outcomes that have greater certainty, even if their expected utility is lower than those of alternative outcomes. Indeed, a large volume of literature shows that bounded rationality is a more accurate characterization of human decision-making than is the rationality described by expected utility theory [12, 13]. Cognitive limitations lead to the use of decision-making heuristics that produce biases, which underlie the so-called paradoxes of expected utility theory. The anchoring and adjustment heuristic, for example, is particularly relevant to naturalistic decision-making situations [14]. In such situations, decision-makers, particularly highly experienced ones, tend to base decisions on decisions made in similar situations in the past.

Although such heuristics can serve as valuable cognitive shortcuts, they also are rich sources of potential vulnerabilities. For example, deception techniques are often based on the tendency of human decision-makers to look for familiar patterns, to interpret the available information in light of their past experiences. Deceivers also benefit from *confirmation bias*, the tendency to discount evidence that contradicts an accepted hypothesis [15].

Naturally, some of the human limitations can be alleviated by computational aids. For example, computational approaches can help in detecting a deception and minimizing its impact [16]. Artificial decision-making agents (expert systems, intelligent agents, planning and decision aids, control systems, and the like) can greatly improve the speed and accuracy of decision-making, especially when the information volume is large. Yet they also add complexity to the system, leading to new and often more drastic types of errors, especially when interacting with humans. In effect, instead of acting as error-reducing systems, they can be error-inducing systems [17] and themselves become chinks in organizational armor.

Although limitations of individual decision-makers are important sources of failures, an organization is more than an aggregation of its individual members. One important tradition in organizational science holds that an organization is primarily an information-processing system [18]. Organizations consist of multiple information-processing and decision-making entities—individual humans, cohesive groups of humans, and computer-based information processing and decision support systems—all of which can contribute to failures of the organization. Some of these sources of failure are rooted in the limitations of each individual decision-making entity. Others have to do with the relations between the entities.

Galbraith, for example, argued that an organization's design and its ability to produce successful performance are largely driven by the need to avoid information-processing overload of the organization's decision-makers, which depends on the information-processing requirements within an organization, which in turn are driven by the degree of uncertainty in its environment and tasks [19].

More generally, the effectiveness of an organization depends on how well its structure fits the organization's mission and environment. Factors such as the complexity and uncertainty of the organization's environment play a critical role in determining how well a given organizational structure and process will perform in that environment [20, 21]. Thus, by manipulating the organization's environment and tasks, an adversary can disrupt and degrade the performance of an organization (Chapter 9). In designing a way to attack an information-processing organization, an adversary can make a conservative assumption that the decision-makers within the organization are sufficiently competent, adequately trained, and given appropriate incentives. Yet, such decision-makers are still subject to bounded rationality and will inevitably make mistakes, especially under the conditions of information overload and imperfect information. These limitations create opportunities for the adversary to degrade the information processing that occurs in an organization of such decision-makers (Chapters 6 and 8).

To minimize its exposure to such manipulations, an organization may want to prevent an adversary from learning about the organization's internal workings. However, the very nature of an information-processing organization makes such prevention very difficult.

Because the organization has to coordinate its tasks among multiple decision-makers, its members have to exchange information, engendering a significant volume of communications. This volume can increase as the decision-making units become more distributed, the relative decision-making capability in any unit declines, or the environment becomes less predictable. Communications can be intercepted and exploited by an adversary. Further, the communications reflect the structure of the tasks assigned to the communicators. This helps the adversary to infer the structure of the organization and its processes, as we discuss in Chapters 2 and 3.

Still, many vulnerabilities of an organization stem from other factors, such as social forces within an organization, which go beyond the purely information-processing perspectives. Information flows in organizations are constrained not only by information-processing limitations, but also by such psychological and social factors as the unwillingness to bring bad news to a superior, or the reluctance to contradict a widely accepted truth. For example, groupthink—the tendency of decision makers within a cohesive group to pressure each other toward uniformity and against voicing dissenting opinions [22]—can produce catastrophic failures of decision-making. Groupthink interacts with information-processing activities to cause feedback loops [22, p. 196] (e.g., information overload contributes to groupthink tendencies).

Organizations are also vulnerable to failures induced by internal power struggles, by incongruent and often competing objectives of the decision-makers within the organizations, by effects of incentives and punishments within an organization, and other factors [23] that may not necessarily support the overall goals of the organization and often cause a significant degradation of the organization's effectiveness. Organizational decision-making is often driven by the formation of coalitions between decision-makers, which involves complex exchanges of resources and favors [24]. Such coalitions are fragile and can be manipulated (Chapter 7). Protection of an established position, power struggles, preservation of a legacy, following a rule or a tradition, personal "chemistry" between the decision-makers all have a dramatic impact on how a real organization functions, and all constitute potential sources of vulnerabilities exploitable by an adversary of the organization.

The Road Map of the Book

Although new theoretical contributions constitute the core of each chapter, a key goal of the book is to appeal to practitioners and developers of applied systems. To this end, each chapter of the book includes a discussion and recommendations regarding at least one application actually developed or potentially enabled by the class of techniques described in the chapter.

The structure of the book is built around several key questions that our readers are likely to ask:

- What do I need to know about an adversary's organization, and how can this knowledge be obtained? Conversely, what information about my organization would be particularly valuable to an adversary?
- What practical approaches exist to affect the performance of an adversary's organization, and what happens when such approaches are applied?
- How can I quantitatively estimate the impact that I can effect on an adversary's organization and, conversely, that an adversary might impose on my organization?
- How can I increase the ability of my organization to cope with an adversary's attack? Conversely, what countermeasures is an adversary likely to take in order to reduce the effectiveness of my attack?

Throughout this volume, key ideas are explored and illustrated in the context of examples, hypothetical and real, that cover a broad range of organizations: military command and control organizations (Chapters 2, 4, 5, 6, and 9), terrorist organizations (Chapters 3, 4, 5, and 8), business corporations (Chapter 5, 6), and Internet-based criminal organizations (Chapter 7).

We begin in Chapter 1 with an outline of the challenges and methods of collecting and analyzing information on adversary organizations to support operations that influence, degrade, or defeat them. We introduce the information requirements for such operations and approaches to collect, represent, and analyze intelligence on targeted organizations. We then describe the relatively well-understood analytic methods applied to leadership analysis in hierarchical organizations, the more general approaches applied to understanding networks, the technical tools applicable to these problems, and the plethora of challenges associated with making progress in this area.

Clearly, in order to impact an adversary's organization, one needs a significant amount of information about that organization—particularly about its structure and processes. Yet, every competent organization operating in an adversarial environment will make strong efforts to protect such sensitive information. With only limited and imperfect observations of the adversary's organization, one wishes to find ways to infer the hidden organizational structure and processes. Therefore, Chapter 2 discusses how one can learn the structure (e.g., decision hierarchy, allocation of resources and roles to decision-makers, and communication and information structures) and processes (allocation of tasks to decision-makers, task execution schedule, decision policies, and so forth) of an enemy organization based on partially observed sequences of communication transactions between the enemy actors and activity events. The available observations may be noisy and incomplete. The key idea of this approach is to generate multiple hypotheses of the structure and processes of the adversary's organization, to estimate the likely characteristics of the observable events and communications based on each hypothesis, and then compare the actual observables with the predictions.

In order to generate the hypotheses as discussed in Chapter 2, one needs a library of organizational structures and missions. However, it is a very nontrivial task to create such a library and to validate and maintain it. Also, the creators of the library are likely to populate it with known organizational and procedural patterns, and this may produce misleading results when an organization in question is designed in a novel, unconventional fashion. Clearly, it would be desirable to avoid the burden of creating, validating, and trusting such a library. In fact, there exist approaches that have the advantage of not requiring the library (although not without their own drawbacks), and we discuss some of them in Chapter 3. This chapter is also a good place to discuss the key ideas of the popular *social network analysis*, and to point out its limitations. The analysis of a hypothetical organization called Ali Baba illustrates the key points and results.

There are a number of ways one can utilize the information about an organization, such as what has been discovered about the Ali Baba organization. One could, for example, use it to collect more complete intercepts of communications among the Ali Baba members, to determine their upcoming actions, and to take defensive measures against those actions. In this book, however, we are interested in ways of affecting the performance of such organizations. To this end, Chapter 4 outlines a range of measures, most of them falling into a class commonly called *information operations,* which are used in practice to produce a variety of effects on organizational performance and behavior.

The chapter introduces fundamental approaches to influencing the adversary's decisions, such as denial and deception, as well as the range of related approaches to modeling decision-making. To illustrate these principles, the chapter offers two case studies—one deals with targeting an integrated, structured, hierarchical air defense system and another applies to an unstructured network organization.

When one uses such disruptions against an adversary organization, the intent generally is to cause the target organization to malfunction, to perform in a way undesirable for the organization. To this end, it would be useful to know what these malfunctions are. What occurs within an organization when it is disrupted by an external agent? Is it true that even a properly designed and trained organization cannot avoid the disruption?

To answer these questions, Chapter 5 pursues two directions. First, it explores the possible mechanisms and behaviors associated with organizational malfunctions. Second, it considers how such behaviors can be intentionally caused or exacerbated by means identified in the previous chapters (e.g., by deception or denial). The chapter attempts to reach beyond the organizational and management science literature by looking at what can be learned from discrete event system theory, as well as classical and model-predictive control literature focused on chemical and power plant control. It also makes use of several historical examples where actual organizations exhibited such malfunctions.

Beginning with Chapter 6, we abandon qualitative discussions of organizational malfunctions and take a more quantitative approach based on computational models. We explain the differences between intellective and emulative models and then introduce intellective models that exhibit elements of several malfunctions: self-reinforcing error due to positive feedback, overload and saturation of decision-making elements, and cascading collapse due to propagation of overload and erroneous decision-making through the organization. A simple dynamic model shows how organization enters into a self-reinforcing vicious cycle of increasing decision workload until the demand for decision-making exceeds the capacity of the organization; it also offers us useful insights into the conditions under which an organization can experience rapid increases in the fraction of erroneous decisions and even a classical form of instability. We then extend the model to more complex networked organizations and show that they also experience a form of self-reinforcing degradation. These computational experiments suggest several strategies for mitigating this type of malfunction, such as a suitable compensating component (e.g., a brokering mechanism that dynamically redistributes responsibilities within the organization as it begins to malfunction). A key lesson is that correctness of the information flowing through an organization is critical to its ability to function. We see that even a modest increase in the error rate produced in one node of an organization can lead to a profound, far-ranging degradation of the entire organization's performance.

Another characteristic of the organization's information flows—trustworthiness—can be at least as critical as information accuracy. Even with perfectly correct and useful information circulating through the organization's communication channel, suspicion or lack of trust with regard to the information can induce a dramatic collapse of the organization. Therefore, in Chapter 7 we explore such phenomena by focusing on a particular type of organization, where the effects of trust and suspicion are relatively easy to understand and to model. We examine how certain types of organizations that exist on the Internet can be disrupted through loss of trust induced by deception or other means. A key mechanism that controls trust is gossip. We learn that gossip is a fundamental component of who we are as social beings. It is part of our species-culture, and it is, in a literal sense, also a part of us. What might this tell us about organizational decision processes and potential vulnerabilities? And what might this also tell us about insulating an organization against these vulnerabilities? To answer these questions, we take a simple, but surprisingly ubiquitous, type of organization—anonymous online information networks—and build a computational simulation to explore the sensitivities of this type of organization to gossip.

In Chapter 8, we further strengthen our emphasis on quantitative analysis of organizational performance and adversarial effects on such performance. Here our objective is not only to determine the direction of organizational behavior but also its quantitative characteristics. Consider that an effective way to learn more about an organization is to use probes—actions that cause observable and measurable effects that confirm the organization's structure and behavior. However, it is difficult to design an effective probe without the means to estimate how the structure and processes of an organization relate to the quantitative characteristics of observable reactions to probes. Also, a disruptive intervention is most beneficial to the attacker when it achieves effects with certain desirable quantities and avoids undesirable side effects. Again, one needs the means to estimate the quantitative characteristics of impacts that interventions are likely to produce.

The approach discussed in Chapter 8 is based largely on the computational *virtual design team* (VDT) system for organizational modeling, extensively tested in practical applications to multiple industrial and other corporate organizations. Main methods include symbolic representation and modeling of organizational actors, tasks, and workflows; probabilistic simulations of candidate organizational structures to identify information flows, determine required coordination, and rework problems; and operational mapping and analysis linking the necessary "real-world" aspects of the problem to their computational representations. To illustrate the application of these methods to the enemy-disruption problem, the chapter analyzes a specific test-case scenario of a suicide-bombing attack by an enemy organization utilizing a human torpedo to target a military vessel abroad.

Finally, Chapter 9 takes a primarily defensive perspective. Instead of finding ways to attack vulnerabilities of organizations, we now explore how to design organizations that can successfully resist attacks. The ability to engineer organizations that can withstand hostile interventions, or rapidly adapt to them, is of great importance in highly competitive or adversarial environments. Some organizations are less susceptible to the kinds of attacks discussed so far in this book. We wish to identify the features of these organizations that make them less susceptible to a given attack. Knowing such features, we would like to design an organization that incorporates these defensive features. Chapter 9 presents a methodology for designing attack-resistant organizations.

The methodology is grounded in experimental research conducted within the Adaptive Architectures for Command and Control (A2C2) program. The A2C2 research used distributed dynamic decision-making team-in-the-loop real-time wargaming to compare the performance of different organizations for the simulated mission scenarios. The key point is that the congruence between an organization and its missions allows analysts to predict how the organization would perform under different situations, including those imposed by the adversary's attacks.

References

[1] Libicki, M. C., *What Is Information Warfare*, Washington, D.C.: National Defense University, 1995.

[2] Waltz, E., *Information Warfare Principles and Operations*, Norwood, MA: Artech House, 1998.

[3] Mavor, A. S., and R. W. Pew, (eds.), *Modeling Human and Organizational Behavior: Application to Military Simulations*, Washington, D.C.: National Academy Press, 1998.

[4] National Commission on Terrorist Attacks, *The 9/11 Commission Report: Final Report of the National Commission on Terrorist Attacks Upon the United States*, New York: W. W. Norton & Company, 2004.

[5] Hayes, R. E., and D. S. Albert, *Power to the Edge: Command and Control in the Information Age*, Washington, D.C.: CCRP Publications, 2003.

[6] McLean, B., and P. Elkind, *Smartest Guys in the Room: The Amazing Rise and Scandalous Fall of Enron*, New York: Portfolio Trade, 2004.

[7] Barton, J., (ed.), "United Nations Oil for Food Program: Hearing before the Committee on Energy and Commerce," U.S. House of Representatives, Diane Pub. Co., 2003.

[8] Carley, K., and M. Prietula, (eds.), *Computational Organization Theory*, Hillsdale, NJ: Erlbaum, 1994.

[9] Prietula, M., K. Carley, and L. Gasser, (eds.), *Simulating Organizations: Computational Models of Institutions and Groups*, Menlo Park, CA: AAAI/MIT Press, 1998.

[10] Simon, H., *Models of My Life*, New York: Basic Books, 1991.

[11] Louvet, A. -C., J. T. Casey, and A. H. Levis, "Experimental Investigation of the Bounded Rationality Constraint," in S. E. Johnson and A. H. Levis, (eds.), *Science of Command and Control: Coping with Uncertainty*, Washington, D.C.: AFCEA, 1988, pp. 73–82.

[12] Tversky, A., and D. Kahneman, "Judgment Under Uncertainty: Heuristics and Biases," *Science*, Vol. 185, 1974, pp. 1124–1131.

[13] Kahneman, D., and A. Tversky, "Prospect Theory: An Analysis of Decision Under Risk," *Econometrica*, Vol. 47, No. 2, March 1979, pp. 263–292.

[14] Klein, G., *Sources of Power: How People Make Decisions*, Cambridge, MA: MIT Press, 1999.

[15] Bell, J. B., and B. Whaley, *Cheating and Deception*, New Brunswick, NJ: Transaction Publishers, 1991.

[16] Kott, A., and W. M. McEneaney, (eds.), *Adversarial Reasoning*, Boca Raton, FL: CRC Press, 2006.

[17] Perrow, C., *Normal Accidents: Living with High-Risk Technologies*, Princeton, NJ: Princeton University Press, 1999.

[18] March, J., and H. Simon, *Organizations*, New York: Wiley, 1958.

[19] Galbraith, J., "Organization Design: An Information Processing View," *Interfaces*, Vol. 4, May 1974, pp. 28–36.

[20] Galbraith, J., *Designing Complex Organizations*, Reading, MA: Addison-Wesley, 1973.

[21] Burton, R., and B. Obel, *Strategic Organizational Diagnosis and Design*, Boston, MA: Kluwer Academic Publishers, 1998.

[22] Janis, I. L., *Groupthink: Psychological Studies of Policy Decisions and Fiascoes*, Boston, MA: Houghton Mifflin Company, 1982.

[23] Shapira, Z., (ed.), *Organizational Decision Making*, Cambridge, U.K.: Cambridge University Press, 1997.

[24] March, J. G., *Primer on Decision Making: How Decisions Happen*, New York: The Free Press, 1994.

CHAPTER 1
Know Thy Enemy: Acquisition, Representation, and Management of Knowledge About Adversary Organizations

Ed Waltz

The *organization*—a formal group of people who share a mission and an arrangement to structure roles, relationships, and activities—is a fundamental unit of society, and the understanding of organizations is critical to understanding cooperation, competition, and conflict in civil society, business, and nation-state or nonstate warfare. The study of organization theory has progressed in the past 40 years to become a dominant element of the business management curricula and a necessary component of business administration. March and Simon's classic text, *Organizations* (1958) introduced the notion that while organizations are rational, they are bounded in their rationality, determining only what methods are sufficient to achieve their current necessary goals (*satisfying* rather than *optimizing* to determine the best means to achieve the highest possible goal) [1]. Furthermore, March and Simon recognized that organizations are adaptive organisms, not machines, and warned that stimuli by management do not produce predictable or repeatable results. Subsequent texts produced ever-refined models of the organization, and in the 1970s many researchers focused on the critical decision-making processes of Cold War leaders. Graham Allison's *Essence of Decision* replaced monolithic models to explain national organizations with a more complex network of actors with internally competing goals and diverse perspectives [2].

Consider a sample of the applications of organization theory in classic texts that address organizations in competition or conflict. The specific issue of how decisions are made in deliberate planning, policy formation, and crisis by international policy makers was developed by Jervis in *Perception and Misperception in International Politics* [3]. Jervis applied cognitive psychology to explain the role of individual actors within states, to explore their methods of developing perceptions of adversarial international situations, and to infer what other actors may do (their intentions). Janis and Mann added the element of psychology to understanding organizational decision-making behavior in *Decision Making: A Psychological Analysis of Conflict, Choice, and Commitment* [4]. Weick introduced approaches to the organizational sociology of decision-making in *The Social Psychology of*

Organizing and described models of the cognitive processes that provide the context for decision-making in *Sensemaking in Organizations* [5]. The archives of *Harvard Business Review* trace the continual quest to describe the social structure and operational behavior of competing organizations from the perspectives of bureaucracy, economy, technology, sociology, and psychology.

The major focus of this organizational research and its application in business, social, and government organizations has been to understand the organization in an effort to accomplish ever-changing competitive missions, increasing efficiency, agility, innovation, and other corporate virtues. Moreover, these efforts are pursued to introduce change by its leadership, from *within the organization,* and with full and open access to the internal data that describes its planning, personnel, operations, and behavior.

The focus of this book, however, can be viewed as the flip side of the foregoing applications that seek to influence organizational behavior *for the good, from the inside*. In an adversarial competition or conflict, organization warfare seeks to understand and then change a target organization's behavior to achieve the goals of the attacker; generally these goals include deterrence, dissuasion, deception, disruption, degradation, or total defeat of the targeted organization. We use the term "warfare" broadly here. In corporate competition, an organization may employ many of the principles and technical methods, herein, to ethically collect, analyze, and understand an adversary to employ ethical methods to influence the market and its competitor. The organization may also employ these methods to detect and deter an adversary's unethical efforts at manipulation.

This chapter introduces the challenges and methods of collecting and analyzing information on adversary organizations to support operations to influence, degrade, or defeat them. In the first section, we introduce the needs for information in such operations and then the approaches to collect, represent, and analyze intelligence on targeted organizations. The "Leadership Analysis" section describes the analytic methods applied to leadership analysis in hierarchical organizations, and the following section describes the more general approaches applied to understand networks. Technical tools that may be applied to these problems are summarized in the "Representative Analytic Process and Tools" section. Finally, the issues associated with making progress in this area are briefly summarized.

Introduction to Organization Warfare

Cooperation, competition, and conflict between organizations, and the operations that support them, are not at all new. The disciplines of foreign policy, diplomacy, and military warfare (another means of diplomacy, as Clausewitz noted) have always strategized about the means to cooperate, coexist, or conquer the organizations of nation-states. Chinese strategist Sun Tzu observed the obvious need to know the adversary in his classic *The Art of War* [6]: "One who knows the enemy and knows himself will not be in danger in a hundred battles. One who does not know the enemy but knows himself will sometimes win, sometimes lose. One who does not know the enemy and does not know himself will be in danger in every battle." Sun Tzu further noted that military leaders must know individuals within the adversary's organization [7]: "Generally, if you want to attack an army, besiege a walled city, assassinate individuals, you must know the identities of the defending generals, assistants, associates, gate guards, and officers."

Whether the target is the leadership of a foreign nation, a terrorist organization, an insurgent group, organized crime, a foreign intelligence service, or a military unit, the principles of organization operations described in this text apply. In recent years, this activity has become formalized, particularly within the U.S. defense community, as a component of *information operations* (IO), defined as [8] "the integrated employment of the core capabilities of electronic warfare (EW), computer network operations (CNO), psychological operations (PSYOP), military deception, and operations security (OPSEC), with specified supporting and related capabilities to influence, disrupt, corrupt, or usurp adversarial human and automated decisionmaking while protecting our own." Air Force Doctrine Document (AFDD) 1 more specifically defines influence operations that focus on individuals and their organizations [9]:

> *Influence Operations* are defined as those operations that focus on affecting the perceptions and behaviors of leaders, groups, or entire populations. Influence operations employ capabilities to affect behaviors, protect operations, communicate commander's intent and project accurate information to achieve desired effects across the cognitive domain. These effects should result in differing objectives. The military capabilities of influence operations are psychological operations, military deception, operations security, counterintelligence operations, counter-propaganda operations and public affairs operations.

While this definition of influence operations and many of the references throughout this chapter are based on U.S. military sources and doctrine, the reader is reminded that the application of organization intelligence and influence is broader than military alone. The principles of organization intelligence may be applied in, for example, law enforcement, political posturing, or corporate competition.

Recognizing the need for improved understanding of adversary organizations and their decision-makers to support such influence operations, the U.S. Information Operations Roadmap identified the need for understanding of organizations as a critical element of intelligence support to IO [10]:

> Better depiction of the attitudes, perceptions and decision-making processes of an adversary. Understanding how and why adversaries make decisions will require improvements in Human Intelligence (HUMINT) and open source exploitation, as well as improved analytic tools and methods.

As a component of IO, we may consider the *organization operations* described in this book as those activities that seek to understand the elements, structure, and dynamics of an organization to perform activities that influence its behavior by processes that influence, shape, and ultimately control its behavior.

The classical approach to engaging opposing organizations to conduct influence operations requires three phases of activity (Figure 1.1):

Network aspect	Access	Exploit	Influence
Social network	• Understand, map social network • Gain access to people: overt and covert	• Expand access across social net and human agent channels • Refine and track organization model	• Influence perception and behavior
Information network	• Gain access to sources, external traffic, and internal information (content)	• Exploit information to support access channels • Expand net breadth and depth of access	• Insert information • Open remote access ports
Physical network	• Gain access to physical media, channels, links, and nodes • Gain access to resources, facilities, channels	• Secure persistent access to additional information channels	• Create, open physical channels to inject information

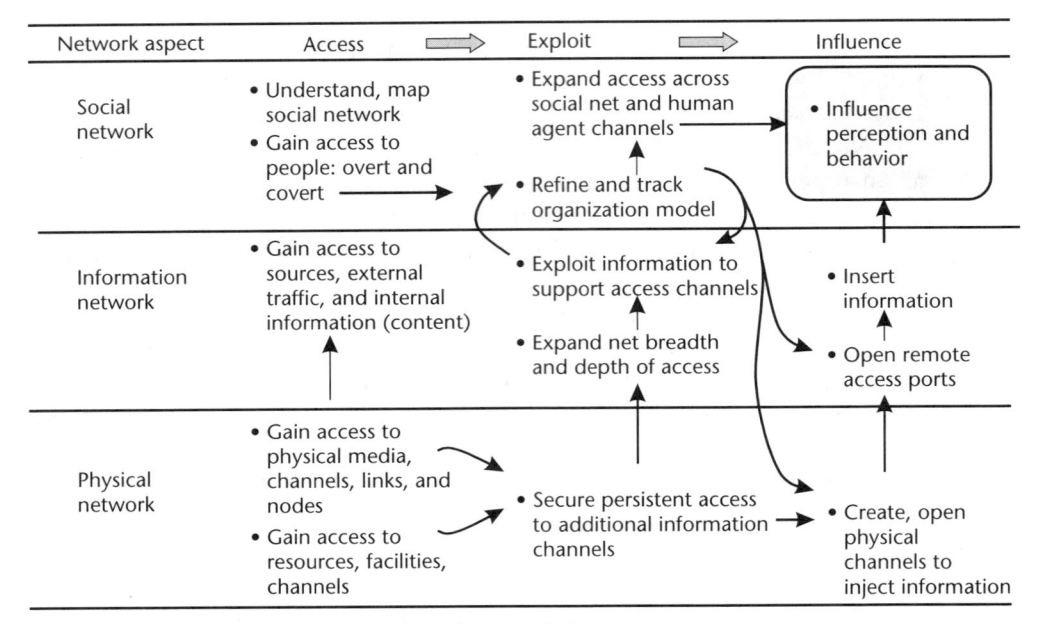

Figure 1.1 Organization access, exploitation, and influence.

- *Access*—The first phase of engagement requires a means to gain access to information about the organization to understand its goals, intentions and plans, resources, participants, and behavior. The access can be first via open and public sources of information that provide information on the goals, personnel, resources, public statements, and activities of the organization. In military conflict, access may also include signals interception and covert sources of access. (Most of this book focuses on the use of open means of access that do not rely on covert sources. Indeed the text focuses on what can be accomplished without covert access—by observing the external manifestations of an organization's behavior and by external influences.)
- *Exploit*—Initial information is exploited to improve the understanding of the targeted organization; this, in turn, is used to refine the collection of information used to further improve understanding. Intelligence operations may seek to escalate the level of access (penetration) to deeper sources of information within the organization, including the recruitment of internal human sources (agents) or the establishment of covert access to information systems or communication processes.
- *Influence*—The next stage is the use of the channels of access to influence the organization through critical leaders or decision-makers or through the entire social process of the organization. This influence, introduced in detail in Chapter 4, includes activities with goals that range from disruption (to reduce efficiency) to control.

In addition, the operators who implement the influence plan should have a means to observe the effects of their actions to enable them to refine both their target models and their actions to the adapting target. This feedback provides the ability to assess the performance of the organizational models, the effectiveness of their influence, and be warned of adverse and unexpected consequences not considered in their models.

Notice in the figure that these escalating stages of knowledge about and influence on the organization occur at three levels or aspects of the organization: the *social network* of people, their formal and informal social roles within the organization; the nonphysical *information networks* that allow information (intents, plans, procedures, commands, finances, reports, and so forth) to flow among participants; and the *physical networks* of organizational resources (facilities, power, communications, computation, transportation, security, and the like) [11]. The figure illustrates the general flow of intelligence that proceeds from access to physical, information, or social sources to exploitation of the sources, before planning influence operations. Consider the following representative sequence of escalating activities conducted against a paramilitary organization:

1. *Physical and information access*—Conduct broad area surveillance to identify, and then observe, organization facilities to assess the size and scope of the organization; identify communication and courier channels to scan for traffic.
2. *Social access*—Conduct surveillance and covertly contact periphery actors (e.g., sympathizers, third parties, and the like) to develop information on the social structure of the organization. Prepare profiles of motives, perspectives, and motivations of the organizers, leaders, and participants.
3. *Physical and information exploitation*—Obtain limited information access by forensic analysis of captured computers and radio interceptions, and by covertly placed sniffers on the organization's networks. Perform quantitative social network analysis to derive organization structure.
4. *Social exploitation*—Task technical intelligence collectors to further exploit the information systems to focus on leaders and planning; interrogate defectors to describe organization members and structures to validate social network analysis; perform a complementary qualitative leadership analysis.
5. *Vulnerability analysis*—Identify potential vulnerabilities in the organization structure, composition, or behavior (behavioral pathologies); identify active probes of the organization that may confirm or disconfirm these vulnerable points.
6. *Influence operations planning*—Establish influence objectives based on mission intent against the paramilitary organization; develop candidate influence actions (using access at the physical, information, and social levels of the organization), and estimate effects and observable indicators that those effects are occurring. Assess the potential behavior of candidate operations by wargaming in behavioral simulation tools.
7. *Physical and information influence*—Perform the planned actions using physical, information, and social channels of access, observing the indictors of performance to measure impacts.
8. *Influence effects*—the desired result is a change in the perception of the situation by the targeted paramilitary organization, its trust in its information flow, and the planned changes in its behavior.

Intelligence has traditionally distinguished between two levels of the organization: organizational physical capabilities (objective, physical measures of combat strength, orders of battle, command and control, and weapons performance) and intentions analyses (e.g., more subjective assessments of individuals' and organizations' aggregate mental makeup, attitudes, and morale). The importance of this understanding remains as vivid today as when a CIA analyst encouraged the continued pursuit for means to understand adversary organizations, personnel, and their intentions 50 years ago [12]:

> The stress on measurable physical facts [capabilities] is justified. While we are making important strides in understanding and measuring motivation and mental processes, we are not yet far long in that field to measure intentions as precisely as we can capabilities and. . . the danger of deception is a very real one. Even so, since decision-making is so inevitably bound up with consideration of the personal element, it is the better part of discretion and of valor as well, to consider intentions. They are so often the sparkplug of human action.

Organizational Intelligence

Organizational intelligence includes a broad range of activities, including the functions of collection of information from multiple sources, search (to identify and track emerging social patterns to discover the potential formation and dissolution of organizations), analysis to detect and discover organizations (including their participants and activities), and then characterization of the organization. The process may also include active, covert penetration to understand the organization's operations and enable internal influence.

The purpose of organizational intelligence includes the development of a description of a targeted organization that includes the following categories of information:

- *Mission*—Background to the formation, strategic intent, mission, and operational objectives of the organization; an understanding of the mission requires information on the organization culture, motives, means available, and intentions. The organizational means include its people, as well as other resources, including economic capital, external relationships, and nontangible political or social capital.
- *Individuals, entities, or nodes*—Identification of key leadership personnel, their positions and roles, and the overall size and makeup of the members of the organization.
- *Structure, relationships, or links*—Description of the relations between individuals and the network formed by these relationships.
- *Behavior*—Description of the dynamics of the organization, including the causal relationships between entities as well as the processes by which transactions and decisions take place.

Organizational intelligence is developed by collection of information from a variety of sources as well as comprehensive analysis to develop a description of the target organization. The full range of intelligence sources is tasked to collect information on the adversary's organization, to assemble a picture of the organization's mission, structure, and behavior. Table 1.1 [13], based on the required intelligence support to IO identified in U.S. Joint Publication 3-13 Information Operations, summarizes the typical categories of information requested from collectors. The table also identifies the typical sources used to collect these categories of information introduced in the following paragraphs:

- *Open-source information* (OSINT) refers to intelligence derived from open sources available to the general public (e.g., the Internet, newspapers, books, scientific journals, radio and television broadcasts, and public records). These sources may provide critical contextual background on an organization's mission and activities, as well as limited information on the size. OSINT has limited detailed value on clandestine organizations (e.g., terrorist, insurgent, and paramilitary) organizations.
- *Human intelligence* (HUMINT) includes intelligence derived from information collected and provided by a wide spectrum of human sources that range from covert controlled agents, defectors to a third country, and informants on domestic criminal gangs, to civilians with knowledge or overt business relationships with target organizations. Controlled agents may be able to penetrate the organization to provide information on actors and relationships or to gain physical access to install means to monitor traffic (e.g., phone taps, computer network sniffers, and so forth) to derive social network models.
- *Signals intelligence* (SIGINT) and *network intelligence* (NETINT) refer to intelligence derived from the interception and analysis of external traffic data and internal content of free-space electromagnetic signals and computer network signals, respectively. Communication sources can include radio and cell phone transmissions, e-mail, instant messages, and telephone calls. Electronic record sources that may also provide key transaction data include expense, banking, and activity records of organization personnel obtained by exploitation of computer network data.
- *Geospatial intelligence* (GEOINT) includes intelligence derived from exploitation and analysis of imagery and geospatial information to describe, assess, and visually depict physical features and geographically referenced activities on the Earth. Observation of the facilities, sites, and other locations used by the organization or frequented by members may provide insight into organization size, activity, or other context-setting information. GEOINT data also provides information to support active collection efforts in the other disciplines.

Collection of such information is among the most difficult due to operational security (OPSEC) and countermeasures used to deny an adversary from obtaining information concerning an organization, its planning, and operations. U.S. intelligence lacked such information in 2003 on Saddam Hussein's core Iraqi organizations; the *Commission on the Intelligence Capabilities of the United States Regarding Weapons of Mass Destruction* noted [14], "the dearth of information made any analysis of Iraqi political calculations largely speculative, and analysts therefore relied on historical information and observed behavior."

While the elements in Table 1.1 focus on the targeted organization, the contextual description of the social environment that shapes the organization is of equal importance. An evaluation of U.S. collection on the social environment surrounding Saddam Hussein's leadership organization described the importance and difficulties of obtaining this critical contextual information to understand organizations [15]:

Table 1.1 Typical Organizational Intelligence Needs

Domain of Analysis	Required Information Elements	Typical Sources of Information
Cognitive Properties of the Organization	*Key individuals*—Identity and psychological profile of key leaders and influencers (e.g., advisors, key associates, and/or family members) affecting attitudes, perceptions, and decision-making environment *Decision-making calculus*—How decision-makers perceive situations, reason about them, plan actions, execute outcomes of their actions, and then assess those outcomes; the cultural basis of their perspective and doctrines applied to decision-making *Organizational and social structure*—Formal organizational structure (roles, responsibilities) and informal social relationships *Cultural and societal factors*—Affecting attitudes and perceptions (e.g., language, education, history, religion, myths, personal experience, and family structure) *Credibility assessments*—Assessments of key individuals or groups; description of their sphere of influence *Historical factors*—Key historical events that affect an individual's or group's attitudes and perceptions of others	Organization statements, reports, news releases, messages Accumulated lists of members and associates, directories, attendance lists HUMINT sources reporting on organization activities, plans, intentions SIGINT and NETINT intercepted communications, message traffic, e-mails, courier messages OSINT historical and cultural information; family and genealogical information
Information Properties of the Organization	*Information infrastructure*—Description of the capabilities of lines of communication, networks, nodes, their capacity, configuration, and the related computation (this includes nonelectronic networks, such as rat lines and dead drops) and other nontechnical forms of communications) *Technical design*—Description of all technical elements of the information infrastructure, equipment, and data models of all layers of the OSI stack. *Social and commercial networks*—These process and share information and influence (kinship and descent linkages, formal and informal social contacts, licit and illicit commercial affiliations and records of ownership and transactions, and so forth) *Information*—External traffic descriptions; internal content and context of information obtained from the information infrastructure	SIGINT and NETINT intercepted communications, message traffic, e-mails, courier messages HUMINT descriptions of technical components, networks, and supporting infrastructure OSINT commercial specifications; operating characteristics
Physical Properties of the Organization	*Geospatial data*—Description of location, geographic coordinates, organizational facilities, infrastructure, physical lines of communication, and so forth *Physical site data*—Description of sites, facilities, power and communications equipment, and critical links between physical locations.	GEOINT imagery, terrain, and supporting MASINT data about physical sites and facilities HUMINT and OSINT descriptions of physical sites, construction, and equipment

Source: [13].

The Intelligence Community knows how to collect secret information, even though in the Iraq situation it did not perform this function well. On the other hand, the acquisition of "softer" intelligence on societal issues, personalities, and elites presents an even greater challenge. This latter information can be found in databases, but they are too often only accessible indirectly and with considerable effort. It may also reside in the minds of groups of people who are accessible but not easily approachable and who do not fall into the category of controlled agents. Although there is a strong argument that the clandestine service should not divert its attention away from collecting "secrets," information on the stresses and strains of society may be equally, if not more, important. This type of information, however, does not fit with the reward system in the collection world and can be difficult to fully assess and to integrate with other information.

The collected data is processed (e.g., translated to a common language and placed in structured databases) and then analyzed to derive a coherent understanding of the organization, with a level of fidelity that will allow influence operations to be conducted. Automated processing is performed to translate languages, convert technical data into an understandable form, or bring multiple data sources into a common reference prior to analysis. The analysis process can include two complementary analytic perspectives:

- *Leadership and nodal analysis* processes focus on identifying and characterizing key decision-makers, subordinate nodes, their immediate relations, decision-making methods and styles, and the propagation of decision effects. This approach, introduced in the "Leadership Analysis" section, focuses on *individual entities* and their leadership styles; it supports those operations that focus influence on these key nodes in the organization. This form of analysis often supports influence operations aimed at having direct and immediate effects form dramatic actions against critical network nodes.
- *Holistic analysis* focuses on characterizing and analyzing the organization as a whole, as a fully integrated organism. One example is the method of *social network analysis* (SNA), introduced in a later section of this chapter. SNA analyzes the static relations between members, based on the proposition that overall organization behavior requires an understanding of the *relationships* among the entire collection of entities in the organization. Dynamic simulation methods, illustrated in this text, are also used to represent and explore the dynamic relationships within the organizations, emphasizing the need to understand causality between influence actions, organization decisions, and effects. Static SNA and analyses and dynamic simulation methods support influence operations that seek to shift the perceptions or behavior of the entire organization in more subtle and protracted manners.

These analytic methods proceed from decomposition (analysis) of the collected raw data to component parts and a construction (synthesis) of hypotheses that describe the organization [16]. The process, illustrated later in this chapter, begins with automated or semiautomated processes that search across data sources to form descriptive and quantitative models of the organization for review and refinement by subject matter experts. The process then proceeds to all-source analysis that marshals all sources of evidence into hypotheses (e.g., organization charts and process diagrams, membership directories, estimated decision processes, budgets, and orders of battle) by subject matter experts.

Consider the fundamental stages of the organization analysis-synthesis process (Figure 1.2) that depict three levels of increasing fidelity of intelligence about the organization [17].

- *Enumeration*—This first stage lists, organizes, and evaluates the composition of the organization, generally in the categories identified earlier in Table 1.1. The focus is on describing the size, scope, and scale of the organization, and identifying the key leaders and influencers. At this level of analysis, the intelligence product may include a report or portal on the organization, with a supporting database of individuals and their profiles, groups within the organization, locations of facilities, estimated resources, and so forth.
- *Relation*—The second stage analyzes the organized data to infer the structure and estimate the properties of relationships between individuals, groups, facilities, equipment, and other resources of the organization. This analysis process extends the enumerated data to describe the roles of individuals (including those known to exist and those presumed to exist but not yet identified) and their operational relationships. The intelligence products of this stage may include a relational database linking the enumerated data, qualitative organization structure charts, and qualitative social network diagrams.
- *Simulation*—The simulation stage further estimates the behavior of the organization, including its decision-making and operational processes. This

Approach Characteristics	1 Enumeration	2 Relation	3 Simulation
Description	Enumerate the entities within the organization; identify, characterize, and index	Model the structure of relationships with the organization; identify and characterize formal roles, responsibilities, and actual, informal roles and relationships	Simulate the behavior of the organization and the dynamics of interaction among entities, as a function of its perception of the environment, its resources, and goals
Analytic focus	Size: Organization scope, size, and scale of personnel, resources, locations, capabilities	Structure: Relations (roles, responsibilities, linkages) between organization entities	Behavior: Function and performance of individuals interacting; operational effectiveness of the functioning organization
Modeling tools	• Organized data sets in databases	• Relational databases • Network graphs • Statistical analysis	• Computational social science; social simulations
Typical elements of organization model	• Lists of nodes (entities) • Personnel profiles • Psychological analyses • Timelines of activities	• Social network graphs (nodes and links relations) • Organizational net analysis (roles)	• Actors and goals • Environments • Resources
Examples	• Modernized Intelligence database (MIDB) • Operational net assessment (ONA) database	• Pathfinder tools • Analysts' Notebook • Clementine	• Large-scale social simulations

Figure 1.2 Three increasing levels of organizational analysis fidelity.

requires an extrapolation from the static relation data to estimated dynamic models based on the organization's prior behavior patterns or known properties of similar organizations. The intelligence product of this final stage is a simulation model that allows the analyst to conduct exploratory analysis of the dynamics of the organization to compare to actual behavior to refine the model, while allowing the planner to evaluate candidate influence strategies to assess their effects using the shared model.

It is important to note three characteristics of this organizational analysis process. First, this analysis process is inherently inferential, because in general the entities and relationships are not directly sensed or reported in source data. While the enumeration process is dominated by sources of information that explicitly represent entities (e.g., extracted names in text), the relation process requires inference from the behavior of entities (e.g., extracted verbs or time-references in text) to relationships. (Even when intercepted conversations may directly imply a relationship, the analyst must bring to bear significant contextual knowledge to infer the kind, strength, and relevance of the relationship between speakers.) The inferential process is generally difficult because of the scale (*dimensionality*, or number of entities and relationships to explain), scope (*domain breadth*, or dynamic range of behaviors possible), and temporal changes in relationships (e.g., changes in organization structure over time or changes in roles and authority).

Second, organizational analysis requires broad contextual knowledge about the environment in which the organization operates. While the enumeration process depends on contextual knowledge of entity properties, the relation and simulation processes require much broader knowledge of context to understand the many manifestations of relationships (e.g., formal and informal interactions, relative power and position, or frequency of transaction) and behaviors.

Third, organizational analysis is traditionally the domain of expert judgment, performed by analysts with limited application of quantitative analysis, such as statistical social science measures. The recent introduction of tools, such as automated entity-relationship extraction, social net analysis, organization simulation and network visualizations, has introduced new capabilities, requiring new methodologies to assure the best use of subject matter expertise, contextual domain knowledge, and quantitative analysis.

The next two sections focus on the analysis of the social or human level of the organization. Similar methods may be used to analyze and synthesize models of the physical and information level of the organization, the other critical components of organizational analysis that focus on networks of physical infrastructure (e.g., electrical power or transportation) or information (e.g., financial, communication, or computation nets). The study of networks at all three layers are elements of more general network science that a recent Board on Army Science and Technology recommended be further developed into a mature discipline. The report noted [18], "Does a science of networks exist? Opinions differ. But if does, network science is in its infancy and still needs to demonstrate its soundness as a science on which to base useful applications."

Later, we will illustrate an analytic workflow that implements these levels of analysis, integrating analytic tools and methods to perform organizational intelligence analysis; first, we introduce the methods of leadership and social network analysis in the next sections.

Leadership Analysis

Leadership analysis focuses on understanding the key decision-makers as critical nodes of an organization—those leaders whose perceptions and decision processes determine the formation of policies and critical decisions. These leaders, whether heads of foreign states, ministers of defense, military officers in highly coordinated integrated air defense units, team leads of loosely knit terrorist organizations, leaders of paramilitary units, or charismatic organizers of ad hoc insurgencies, pose the most difficult and perhaps most important challenge to intelligence organizations. Former U.S. Director of Central Intelligence (DCI) George Tenet articulated this challenge [19]: "To this day, Intelligence is always much better at counting heads than divining what is going on inside them. That is, we are very good at gauging the size and location of militaries and weaponry. But for obvious reasons, we can never be as good at figuring out what leaders will do with them. Leadership analysis remains perhaps the most difficult of analytic specialties."

Leadership analysts assemble profiles of leaders, track their decisions and statements, compare established policies, observe the political-social environments that influence them, and monitor the behavior of the organizations they lead; all of this is performed to infer the leaders' perspectives of situations, their goals, decision-making styles, plans, and intentions. The analyses may include psychological profiles of individuals that describe the elements of background, goals, core personality, ideology, modifying factors, conflicts, and vulnerabilities [20].

Traditionally, high-level leadership analysis within intelligence has focused on the qualitative analysis of nation-state and military leaders, attempting to understand critical policymaking processes, military planning, and decision-making in crises. Foreign policy analysts have applied increasingly sophisticated models to describe the complexities of such processes (Table 1.2). Graham Allison, in *Essence of Decision*, distinguished classical rational actor models that represented the state as a single actor from more complex models that recognize state decision-making as a result of the negotiation among many institutional or bureaucratic actors with differing goals [2]. The rational actor model adopts utility theory, where state decision-making is represented as a choice by optimal selection from among a set of alternatives, chosen to optimize a utility function. Institutional and bureaucratic models recognize the tensions between institutions (e.g., the tension between economic goals to grow and establish bilateral trade relations, and military goals to develop strength and dominate relations) and the complex integration of multiple decision, collaboration, and negotiation processes to arrive at collective policies and decisions. These models seek to explain factions and coalitions that form to arrive at decisions. Irving Janis, Leon Mann, and Robert Jervis added the additional factors of psychology, group dynamics, and complexity to describe the behavior of national leadership as a living organism resulting in emergent decision-making behavior that may not be predictable from the knowledge of all participants [21]. Models at this level are generally narrative explanations of the environments, the principal actors,

Table 1.2 Nation-State Level Organizational Models

Model	Nation-State Organizational Behavior	Alternative Submodels
(Classical) Rational Choice Model	*Basis:* nation-state as *actor* (or team of actors): *Rational actor:* Decision-making is a *choice* by optimal selection from among a set of alternatives. Basis of decision-making is overall utility	*Unitary:* single rational actor *Factional:* internal factions in cooperative-competitive decision-making *Strategic:* power game among multiple international players
Institutional Model *Bureaucratic Government Politics*	*Basis:* nation-state as machine. Government machine or process model: Decision-making is an output or result of integration or bargaining compromises among institutions. Basis of decision-making is integrated process result	*Organizational process:* coordinated decision-making among independent institutions. Bureaucratic politics: collaboration, bargaining, and compromise among competitive institutions of government with different goal sets
Interacting Multiactor Models	*Basis:* nation-state as *organism*. *Mixed-actor models:* Decision-making is a group process. Basis of decision-making is emergent group behavior, dynamics, psychology	Many multiactor model structures interacting; independence of actors and large number of relationships produce emergent decision-making behaviors of the entire complex system

influences, factions, and coalitions, as well as the decision-making dynamics among actors, groups, and institutions.

At the intermediate level of qualitative leadership analysis, individual groups within the larger organization are analyzed to understand the dynamics of a single leader and subordinates. This level of analysis focuses on the relations and interaction between the leader and group members; groups and their dynamics are categorized by their group decision-making styles. Vroom and Yetton have characterized eight specific styles that range from autocratic decision-making by the single leader to unguided group decision-making styles [22]. Each style has implications for the influence planner (Table 1.3), and the organization intelligence analysts seek to identify the preferred operating styles and indicators if the styles are changing (e.g., due to unseen changes in leaders).

A survey of alternative taxonomies of organizational leadership behaviors, performed to identify approaches suitable for organizational modeling, identified additional aspects of leadership characteristics beyond decision-making styles, including approaches to searching for and structuring information for problem solving, using information for decision-making, and managing people and resources [23].

At the more detailed level of group analysis, where more data on the participants and behavior is available, analysts may turn to more quantitative methods of analysis, including the following:

- Statistical analysis of the organization characteristics estimated by different observers and subject matter experts from different perspectives (e.g., political, economic, military) using group dynamics Q-methodology (to assess patterns of subjective perspectives secured by questionnaires) and R-methodology (trait clustering among objective variables estimated by the experts);

Table 1.3 Vroom-Yetton Organizational Decision-Making Style Categories

Group Leader Style	Decision-Making Process Styles	Description of Small Group Decision-Making Process Sequence and Roles	Representative Influence Approaches
Autocratic	A1: Autocratic or directive	Leader defines issues; diagnoses problems; generates, evaluates, and chooses alternatives	Focus influence operation on key leaders and advisors
	A2: Autocratic with group input	Leader defines and diagnoses the problem; elicits inputs from group members	
	A3: Autocratic with group review and feedback	Leader defines and diagnoses the problem; elicits inputs from group members; chooses solution; and elicits feedback from group	
Consultative	C1: Individual consultative	Leader defines problem; elicits ideas regarding the problem and solution alternatives; chooses among alternatives	Operations timing must target the consultative process and key contributors
	C2: Group consultative	Same as C1 but leader shares problem definition process with the group	
Group	G1: Group decision	Leader generates and shares the problem definition; group performs diagnosis, alternatives generation, and evaluation; group chooses among alternatives	Operations must address many, if not most, participants; may confuse, delay, or render the group indecisive, depending on makeup
	G-2: Participative	Leader guides (facilitates) the group as a whole through the entire process	
	G-3: Leaderless team	The group has no designated leader; process leader or facilitator emerges	

- Comparative statistical analyses of the behavior of a target group's behavior to that of a better-understood reference group (e.g., a different country or nongovernment organization that behaves similarly);
- Quantitative estimate of measures of collaboration and information sharing within the organization [24].

In the next section, we introduce social network analysis, as an initial quantitative approach, representative of the static methods to analyze the structure of the targeted organization described earlier in the section on organizational intelligence as the second level of organizational analysis, or *relation* methods. In subsequent chapters of this text, *simulation* methods are introduced to study the dynamics of organizations and the effects of decisions within the organization on its behavior.

Social Network Modeling and Analysis

The quantitative approach to the analysis of organizations requires data on the organization members, or actors, and relationships between the actors. Social network analysis (SNA) is one formal analytic method that quantifies the static, topological properties of the organization as a whole, rather than any individual leader-actor. Indeed the premise of SNA is that relationships are more important than individual actors and that the behavior of the network can best be described by the entire collection of interacting actors [25]. We introduce SNA here to illustrate one of the possible approaches to explicitly model organizations; the approach translates real-world organizations into abstract network graphs, applying graph theory to quantify two categories of network elements:

- *Actors* are the interdependent representations of *entities* within the organization. In graph theory, an actor is a *vertex* of the graph or *nodes* of the network. Note that an entity may represent an individual person or the aggregate of a group of people (e.g., entire communities, organizations, social populations, or even nation-states).

- *Relationships* or *ties* describe the type and character of the relative roles, interactions, associations, transactions, or other interdependency between actors; relationships can describe the transactions by which material or nonmaterial resources (e.g., capital, communications, or commands) are passed between actors and are a property of a *pair* of actors. In graph theory, a relationship is an *edge* of the graph or *links* between nodes of the network. Notice that many events may be described as relationships, the nature of association between actors.

The purpose of SNA is to identify and understand regular patterns, or *structures*, in network graphs that represent known behaviors in real organizations; these structures are represented by structural variables or metrics. These metrics allow analysts to understand behavioral characteristics of an organization as a whole, identify critical actors, locate potential vulnerabilities for exploitation, or compare one organization to another. In order to apply SNA to the analysis of an adversary organization, of course, information must be collected (or estimated) about the actors and relationships within the targeted organization. This data can be derived from the sources described in previous sections, and the SNA process can help guide further collection to refine network models.

Consider two representative military organizational networks illustrated in Figure 1.3 to distinguish the actors and relationships in a nondirected graph of a paramilitary organization derived from radio intercepts [Figure 1.3(a)] and a directed graph that describes the direction of information flows within an integrated air defense systems' (IADS) command and control (C2) network [Figure 1.3(b)].

The undirected graph of a paramilitary organization—Figure 1.3(a)—represents actors and relationships by dots and lines, spatially arranged to distinguish the relational structure of leaders, operational commanders, and operational units. The leader subnet at the top is an identifiable "star" structure with a central figure (strategic leader); this subnet is characterized by high *centrality*. Operational leaders that translate strategy to operational plans for the three distinguishable units at the bottom have a high degree of *betweenness*, because of their linear, series connection between leaders and units. The highly interconnected operating units at the bottom are recognizable as dense *clusters* characterized by neither centrality nor betweenness. The undirected graph, derived solely from *empirical* radio intercept data does not distinguish the flow of transactions between members; however, from these structural properties and the *contextual* knowledge of organizations, we can infer the flow of commands and relative importance of the entities.

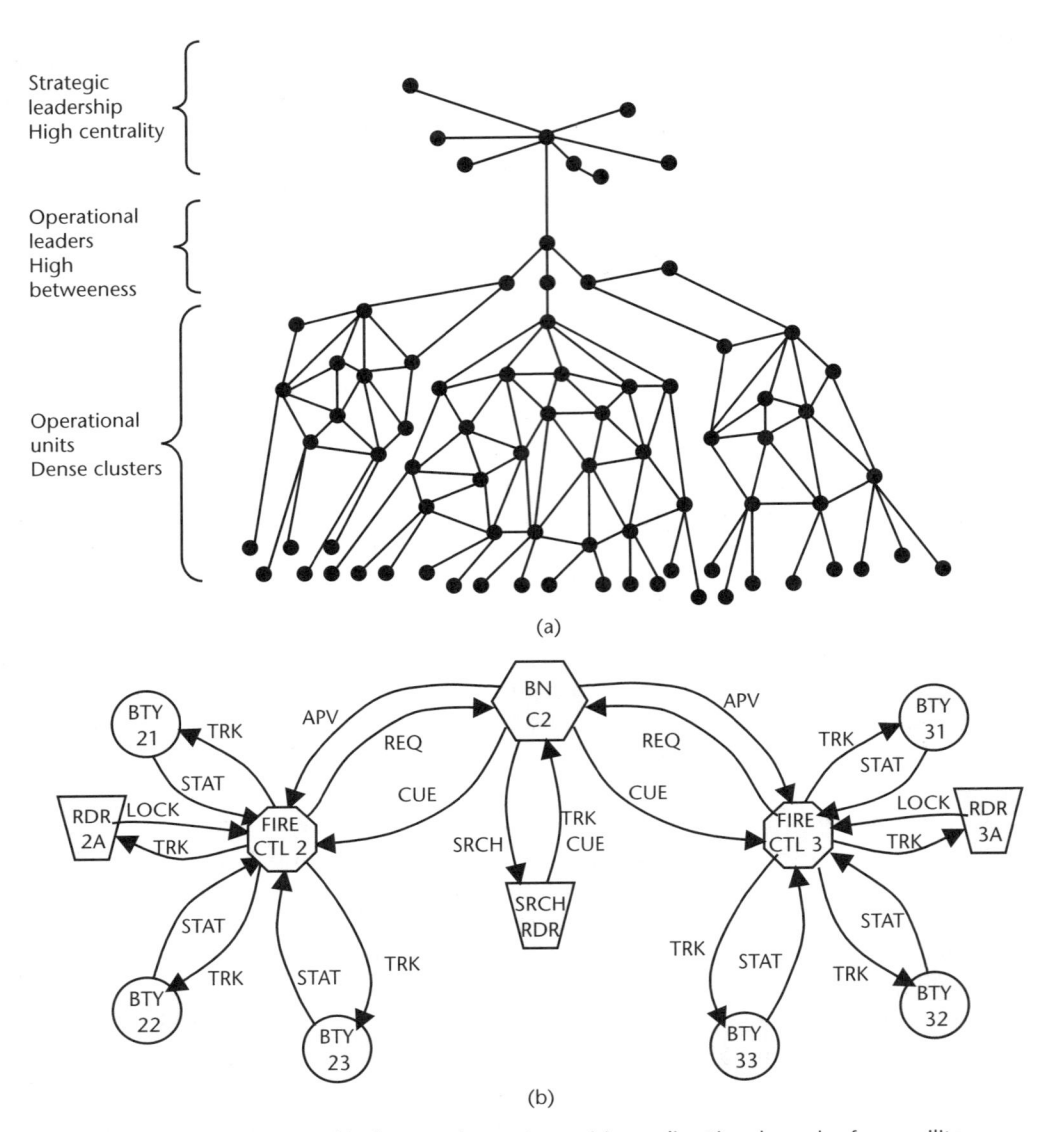

Figure 1.3 Examples of graphical network structures: (a) nondirectional graph of paramilitary network, and (b) directional graph of command and control network.

The directed graph—Figure 1.3(b)—illustrates the basic flow of information within an integrated air defense system consisting of a single battalion-level search radar and two fire control units, each with three missile batteries and a single fire control radar. The graph in the figure is derived from intercepted data links between the battalion C2 control center (BTN C2), its search radar (SRCH), two supporting fire control units (FIRE CTL), their associated fire control radars (RDR), and missile batteries (BTY). The graph also identifies the types of information that flow from node to node: The BN C2 center receives track cues (TRK CUE) from the search radar and passes radar cues to the appropriate fire control unit, which forward track data (TRK) to the fire control radar. When the fire control radar acquires and locks onto the target, it reports to the fire control center (LOCK), which issues a request to fire (REQ) to the battalion center and awaits approval to fire (APV). Upon receipt of approval, the fire control unit issues a fire command (FIRE) to the appropriate battery, based on the reported firing status (STATUS). This simple example illustrates a directed graph that follows a typical sequence of transactions, representing hidden states that may be discovered by observing the flows to infer the operating sequence. In Chapter 2, an approach is developed to match activity sequence templates with the observed sequence of activities of such a network to identify the nodes and operating patterns.

Later in Chapter 4, we describe the translation of these static SNA structures to dynamic models to explore the potential effects of influence operations to manipulate targeted organization. These dynamic behavior simulations allow planners to study the effects of planned operations on the decision dynamics within an organization [26].

SNA characterizes these networks, first, by computing basic structural properties, or quantitative indices, of each individual actor (the most common are summarized in Table 1.4), and then computing indices for subnets (groups) or an entire network. The indices in the table are measures to provide insight into the relative influence of actors based on their relationship links (*degree*, the number of links to

Table 1.4 Common Social Network Structural Actor Index Measures

Graph Type	Index Category	Actor Indices and Descriptions	Example Applications in Influence Analysis
Nondirected graphs	Centrality measures	Degree centrality (C_D)—A measure of the number of links an actor has to other actors	Actors with high C_D and/or C_C have access to many other actors and may be candidates for recruitment or exploitation
		Closeness centrality (C_C)—An aggregate distance measure (e.g., number of links) of how close the actor is to all other actors within the entire network; inverse of the sum of the shortest distances between each actor and every other actor in the network	
		Betweenness centrality (C_B)—A measure of the extent to which an actor is located between other actors; the extent to which the actor is directly connected only to the other actors that are *not directly connected* to each other	Actors with high C_B may be critical liaisons or bridges; they may be effective targets to reduce flows of information across the net
Directed graph (digraphs)		Centrality degree (C'_D)—Measure of the outward degree; the number of actors who are recipients of other actor outflows	Actors with high C'_D and C'_C may be critical decision makers; they may issue commands and polices
		Centrality closeness (C'_C)—An aggregate distance measure (e.g., number of *directed* inflow or outflow links) of how close the actor is to all other actors within the entire network	
	Prestige measures	Degree prestige (P_D)—A measure of the number of links that transmit resources to the actor; a measure of how many actors provide to the prestigious actor	Actors with high prestige may be the focus of critical reporting, finances, or other resources; they may be key reservoirs or decision makers
		Proximity prestige (P'_D)—A measure of the number of actors adjacent *to* or *from* an actor	

other actors in the network; or *indegree,* the number of inbound links, and *outdegree,* the number of outbound links in directional graphs). These indices also allow the analyst to measure the *structural equivalence* of actors, the extent to which the actors have a common set of links to other actors in the network. On the basis of structural equivalence, the network analyst may identify actors by their roles within an organization or even compare structurally equivalent actors within different organizations (e.g., compare an insurgent financer with a venture capitalist).

The structural indices in the figure may also be extended (with appropriate modification) to characterize groups of actors (e.g., a unit or department within an organization), and these groups can be described by additional properties, such as the following:

- *Clustering coefficient* measures the density of connections by quantifying the ratio of the total number of links connecting nearest neighbor actors to the total possible number of links in the group.
- *Degree distribution* is distribution of probabilities that any actor, a, in a group has k nearest neighbor actors.

These indices and overall graph topology can also be used to infer additional behavioral characteristics of the organization:

- *Heterogeneity*—the degree to which actors' roles are similar to one another;
- *Redundancy*—the degree of overlap in functionality such that the removal of a single actor or link does not prevent the organization from functioning;
- *Latency*—the degree of potential delay in organization response to stimuli and ability to adapt to unforeseen events.

Using actor, group, and entire network indices, the analyst seeks to answer key questions and characterize the following aspects of the organization:

1. *Organization structure*—How large is the organization? Is it linked to other organizations? Are subgroups with the organization identifiable? Is the overall network a random or highly organized network? If the organization is highly organized (*scale free*) with numerous high centrality actors (*mediators* or *hubs*), for example, it is relatively immune to random attacks but susceptible to failures at those central actors. More random organizations do not suffer from such vulnerabilities but also may not afford the efficiency of scale-free structures.
2. *Organization function*—What is the function of the organization? How does information flow (orders, reports, finances, other resources)? Does it have daily, weekly, or other cycles of behavior (e.g., reporting, traffic, delivery patterns)? What are the major subjects of information interaction?
3. *Groups within the organization*—What are the major subnets or groups? Are their functions distinguishable? What are their relationships? The analyst computes group indices and views the structure using network visualization to identify structural patterns across identified groups. The structure of the organization, correlated with supporting intelligence (e.g.,

content of messages or rate of transactions relative to known external events), aid in the discrimination of groups and their functions.
4. *Structural characteristics*—Which actors are the sources? The sinks? The mediators with high betweenness? The decision-makers with high prestige? The leaders with high centrality? The actors whose transactions deal with the most important information [27]?
5. Strengths and vulnerabilities—What are the potential strengths and vulnerabilities of the organization? Which actors or relationships are most susceptible (critical nodes or links)? What are the deviations from normal behavior, the pathologies of the organization that may be exploited?

While we have applied the methods here to quantify the properties of human networks at the social level of the organization, the structural analysis method can be applied to networks at all three abstract levels of description of the organization described earlier. At the information level, network analysis can identify information flows within a computer network; these analysis network principles are also applied in the supporting area of computer network exploitation (CNE) to understand the information structure within computer and communication networks. At the physical level, network analysis can also aid the analyst to identify the structure of transportation, electrical power distribution, and other physical networks.

Representative Analytic Process and Tools

A range of supporting automated tools may be employed by collectors, analysts, planners, and operations officers to collect data, organize, analyze, model, and then plan operations against adversary organizations. In this section, we introduce a structured analytic process to perform organization analysis, representative of the approaches that may be adopted by an intelligence organization with extensive sources and target sets. We discuss the analytic tools used to characterize and develop organization models in this chapter; we reserve the discussion on of the use of these models in planning influence operations for Chapter 4.

Consider the categories of analytic tools in the context of a representative workflow used by an intelligence cell of a high-technology nation such as the United States seeking to characterize an adversary organization and develop a target model for planning operations against it. The workflow (Figure 1.4) proceeds from the collection of data toward the production and delivery of intelligence products. (This *workflow* includes a narrative and the supporting graphical description of the "general" or "typical" process steps of the analytic business cycle; the description articulates the actions of people and their associated automated processes, or tools.)

The analytic workflow proceeds from collection to organization intelligence production in the following general steps:

1. Collection support is performed by the aggregation of data from many sources into a data warehouse (or data mart of heterogeneous data databases) with federated access. Sources include classified messaging or cable feeds, extracted technical data (e.g., network intelligence data), and

Know Thy Enemy

Workflow steps

1. Data are collected from multiple sources, translated to common formats, and transferred to data warehouse
2. Users may directly query the warehouse for known entities, sources, relationships

---

3. Unsupervised data mining is conducted to identify correlated entities; discover entity, group, organization candidates
4. Refine search, extraction criteria based on manual discoveries

---

5. Entities and relationships are extracted based on named entities and relationships, as well as organization ontologies, templates, and models
6. Link and network fragments are discovered, qualified, and assembled
7. Groups (subnets) are detected based on group criteria
8. Alternative network hypotheses are generated

---

9. Networks data are visualized in alternative forms: network structures, event-relationship-transaction timelines, causal traces, and so on
10. Social network analysis is conducted to identify net structural properties for identification of key actors, vulnerabilities, and so on

---

11. Target development creates organization target description data:
    - Mission, intent, membership, resources, locations
    - Activities, watch list entries
    - Network structure, properties, vulnerabilities

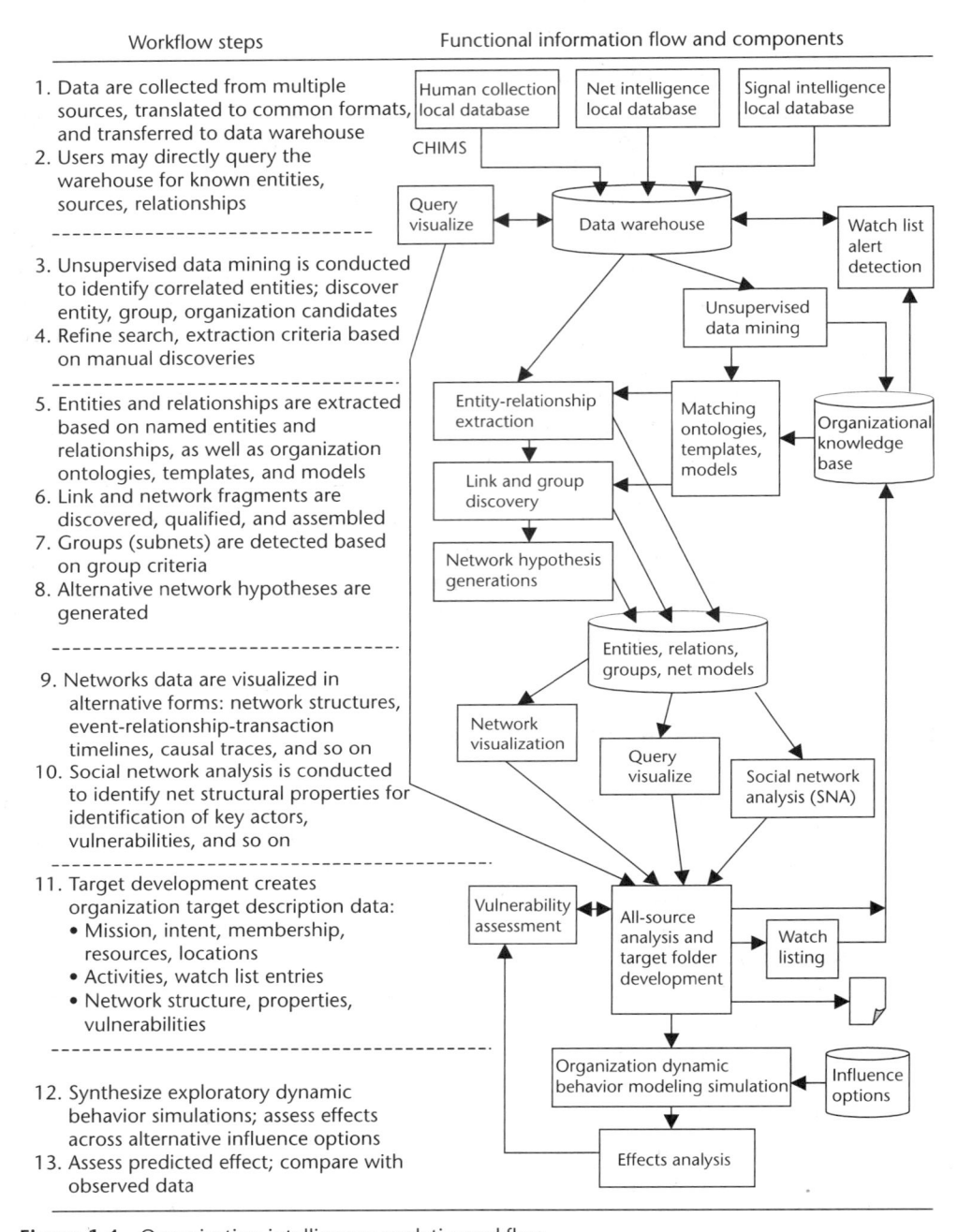

12. Synthesize exploratory dynamic behavior simulations; assess effects across alternative influence options
13. Assess predicted effect; compare with observed data

Figure 1.4 Organization intelligence analytic workflow.

commercial/public data sources. Organization data is also collected by field counterintelligence/HUMINT teams. For example, U.S. teams use the counterintelligence/HUMINT information management system (CHIMS) in a rugged laptop containing the following integrated functions [28]:

- Digital camera, scanner, fingerprint, and other biometric data input;
- Text-based language translation support;

- Analytic capability to create time-event charts, link analyses, telephone charts, and association matrices;
- Report creation via templates and a free-form word processor and ability to transmit results to post on HUMINT Web sites.

These capabilities allow field teams to rapidly gather, analyze, document, and forward entity-relationship data from surveillance observations, interviews, interrogations, and captured documentation.

2. *Query and visualize*—Users must be able to directly query the data warehouse for data elements by organizations, named entities, events, relationships, sources, or other relevant terms.

3. *Unsupervised data mining*—Users may perform automated induction, identifying correlated patterns within the data that may provide a discovery of previously unidentified patterns (e.g., clusters of events in time, similarly structured groups, or clusters of similarly correlated entity-relationships).

4. *Refinement*—Accumulated knowledge about targeted organizations is structured in user-created dictionaries and a computational ontology (a formal description of the concepts and relationships that can exist for organizations) to guide subsequent processes. The ontology describes the classes of entities of interest, specific named entities (e.g., the names of watchlist members and coreferences to their aliases), and the classes of relationships that exist among entities within human organizations.

5. *Entity-relationship extraction*—The first step in automated discovery of networks is preprocessing to perform machine language translation (if required), format recognition, and normalization. Next, entity and relationship word patterns are detected, identified, resolved (e.g., name resolution recognizes the alternative forms of Bill, William, Will, and resolves different forms of the same named entity), extracted, and tagged to accumulate structured data in an extracted database ready for link discovery.

6. *Link discovery*—The process of comparing extracted data searches through link fragments (e.g., entity-relationship-entity; Lt. James-called-HQ) and (a) applies pattern matching methods to match fragments to known patterns of interest, and (b) connects fragments to form larger linked subnets. The discovery process must cope with inexact (approximate) matches and abstract (A-attacked-B ~ A-engaged-B) matches to grow network links for evaluation by analysts. In the process of constructing networks, a pattern-learning process identifies the structure of successfully assembled nets and leverages this knowledge (e.g., net structures, key named entities, and relationship types) in subsequent searches and link attempts.

7. *Group (subnet) detection*—Group detection tools identify functional groups on the basis of structural cohesion properties (e.g., centrality indices) and joint activities. Joint activities include behavior such as access to common accounts, calls to common phone numbers, colocation in the same cities at the same times, correlated communication or travel activities even if not to common channels or places, and so forth [29].

8. *Hypothesis generation*—Results of the prior three stages are combined into net and group hypotheses for presentation and evaluation by analysts. This automated process evaluates candidate links in the context of structural or behavioral properties with regard to the organization being targeted. Known patterns of actors, groups, and events are defined in templates to produce alerts when matched [30]. The hypothesis-generation process is illustrated in Chapter 2 by matching C2 activity sequence templates with observed activities and in Chapter 3 by comparing communication patterns in large population data to locate command structure.

9. *Network visualization*—Visualization tools enable analysts to rapidly browse, manipulate, and view large-scale network data in alternative views that allow analysts to perform a variety of types of analyses, such as the following:
   - Link analysis to study the alternative types of relationships between entities and the timing of the relationship events;
   - Contextual analysis by comparing to external situation data (e.g., world events, media tone, threat levels, and resources);
   - Change detection by comparing network patterns over time;
   - Temporal analysis of event sequences to determine dependencies and causality;
   - Causal analysis to identify a chain of events (and associated entities) from initial causes to direct and secondary effects;
   - Geospatial analysis by plotting entity and event locations to discover spatial patterns of behavior;
   - Author connectivity analysis by identifying linkages between sources of information in reports, messages, or documents to locate clusters of common interest.

10. *Social network analysis*—SNA tools compute the indices for all actors in a network or designated group (a selected subnet) and allow the analyst to hypothesize links and explore structure beyond that available in the current evidence. Note that some tools integrate SNA functions and the preceding entity-link discovery processes within a common capability, as in the Microsoft/University of Calgary Social Network and Relationship Finder (SNARF) that derives social networks by computing social measures for every e-mail author (actor) found in a collection of e-mail messages. In Chapter 3 an alternative approach is described to identify a command structure by communication patterns within a large data set.

11. *All-source analysis and organization target development*—An analytic team performs all-source analysis, considering the evidence from all sources, qualitative and quantities analyses, and context to synthesize the current estimates of the organization composition, resources, structure, and behavior to produce current reporting to plans and operations. The team considers alternatives (e.g., organization bigger than current available evidence, weaker than estimated, or different intentions than currently observable).

12. *Exploratory simulation development*—Based on the prior analysis, a dynamic simulation of the organization and its doctrine and supporting infrastructure (e.g., computer network and communications channels) is performed for exploratory analysis of its dynamic behavior, limitations, and constraints. The simulation can be evaluated against known cases of behavior or may be tested by probing the organization to obtain responses for comparison and refinement (e.g., jamming communication channels or blocking financing to observe which actors respond and the time to recover).
13. *Effect analysis*—Planners use the refined behavioral simulation to assess the range of effects anticipated from alternative influence operations plans. This process is described more completely in Chapter 4.

Representative tools available in the United States are listed in Table 1.5, organized by the enumeration, relation, and simulation categories introduced earlier, and the general categories of processes in the workflow. (The trademark names of the tools belong to their respective holders; detailed description of the tools can be found by an Internet search.) Many of the tools in the workflow have been integrated into enterprise systems, such as the United States' National Ground Intelligence Center (NGIC) Pathfinder suite. Pathfinder provides a degree of automation and data exchange to manage a wide variety of intelligence data (text, graphics, audio, video, and images), allowing the analyst to normalize, index, search, visualize, sort, arrange, compare, group, match, and model large quantities of data for applications such as organizational analysis.

Application Considerations

While this chapter has focused on the basic concepts to obtain access and exploit information about organizations, there exist numerous practical implementation issues that confront the operational implementation of these concepts. Before we proceed, it is important to recognize the most significant barriers that challenge analysis:

- *Access to organization data*—Securing general information about an organization's existence and political intentions can often be readily obtained from open or gray (unpublished or limited distribution) sources, but gaining access to deeper information (participants, plans, resources, communications, and so forth) and penetration of the organization (physical, information, or social access) poses a much greater challenge, requiring greater operational commitment. The size, level of operational security, maturity, and commitment of members all influence the degree of difficulty in moving from a general knowledge to operationally useful knowledge to permit influence of the organization. The process of gaining reliable contact with the organization may also require significant time to develop covert channels of access.
- *Access versus influence*—There is a tension between the intelligence activities to covertly access and exploit channels of organizational information and the operational activities to deny, disrupt, or destroy those channels to influence the organization. Operational calculations must consider the potential for

Table 1.5 Analytic Tools in Support of Organizational Analysis

Tool Category	Tool and Functions	Example Commercial and Government Tools
Enumeration	*Field collection*—Collect narrative and structured data on individuals, relationships, and events for rapid codification and transfer to databases	CHIMS
	Organization Intel databases—Local data stores to ingest and index multiple sources of data	Heterogeneous Relational Databases; SQL; Oracle
	Processing translation—Language translation	Apptek Machine Translation
	Processing extraction—Detect and extract named entities and relations from unstructured data	Net Owl; Clear Forest; Convera Retrievalware
	Data warehouse—Hold unstructured data and extracted index data	NCR Terradata
Relation	*Link discovery*—Detect relevant network fragments and nets; screen and assemble groups and networks using knowledge in ontologies of the problem domain to determine relevance	Pathfinder; LGAT; Semio Taxonomy; Saffron Net; InXight ThingFinder and SmartDiscovery
	SNA—Computation of network metrics for individual actors, groups, and entire networks; perform comparison of groups and nets by structural equivalence properties to detect groups that meet structural criteria	UCINET; InFlow; DyNet; VisuaLyzer; MetSight
	Network visualization—Present entities and relationships in multiple viewing perspectives: network node-link, timeline and causal views	I2 Analyst Notebook's Case and Link Notebook; Visual Analytics, StarLight, InXight StarTree; Orion Magic
	Geographic visualization—Overlay network data on geospatial locations of entities and events	ESRI's ArcView, ArcIMS; Mapvision
	Relational database—Store organization model hypotheses	Oracle 9i
	Organization metrics—Assess organization performance, effectiveness, and structural vulnerabilities against standard fitness criteria	OrgCon
Simulation	*Organization state models*—Hidden Markov models track behavior changes of organizations that exhibit distinct states with observable transitions (e.g., C2 op sequences)	See models described in Chapter 2
	Organization behavioral models—Dynamic models of actor behavior via representing transactions of social resources or influence (e.g., dynamic Bayes networks or agent-based simulations)	SAIP (SAIC Influence net); DyNet
	Organization constraint models—Dynamic models of the processes and hard infrastructure controlled by organizations (e.g., manufacturing, security, mobility, weapons, or C2)	iThink (system dynamics); Extend (time-discrete), AlphaSim (Colored Petri Nets)

intelligence channels to be closed when influence operations begin, shutting off valuable feedback information on the effects of influence activities.

- Orchestrated denial and deception—In addition to operational security and counterintelligence to deny information about an organization, the organization may employ active deception measures to misdirect attention from its true activities and interests while simulating behaviors that are intended to mislead analysis of the organization's true character, structure, resources, or capabilities. An active counterdeception program may be required to deal with sophisticated and sustained denial and deception.

- *Organizational dysfunction*—In addition to denial, the organization may be operationally dysfunctional to a degree that distorts collected information and defies an approach to analysis that presumes the organization to be rational, logical, and effective. Consider conclusions of the *Commission on the Intelligence Capabilities of the United States Regarding Weapons of Mass Destruction* that describe the difficulty in understanding Saddam Hussein's organizational culture built on internal secrecy, lies, and fear (functionally efficient in security, but operationally dysfunctional) [31]:

The failure to conclude that Saddam had abandoned his weapons programs was therefore an understandable one. And even a human source in Saddam's inner circle, or intercepts of conversations between senior Iraqi leaders, may not have been sufficient for analysts to have concluded that Saddam ordered the destruction of his WMD stockpiles in 1991—and this kind of intelligence is extremely difficult to get. According to Charles Duelfer, the Special Advisor to the Director of Central Intelligence for Iraq's Weapons of Mass Destruction and head of the Iraq Survey Group, only six or seven senior officials were likely privy to Saddam's decision to halt his WMD programs. Moreover, because of Saddam's secretive and highly centralized decisionmaking process, as well as the "culture of lies" within the Iraqi bureaucracy, even after Saddam informed his senior military leaders in December 2002 that Iraq had no WMD, there was uncertainty among these officers as to the truth, and many senior commanders evidently believed that there were chemical weapons retained for use if conventional defenses failed . . . Moreover, in addition to dominating the regime's decision-making, Saddam also maintained secrecy and compartmentalization in his decisions, relying on a few close advisors and family members. And Saddam's penchant for using violence to ensure loyalty and suppress dissent encouraged a "culture of lying" and discouraged administrative transparency. As a result, the ISG concluded that instructions to subordinates were rarely documented and often shrouded in uncertainty.

- *Dynamics, situation dependencies and other factors*—Organizations, by their very nature, are not stationary in their structure or behavior due to the social and environmental dynamics in which they operate. This poses an additional challenge to the organizational analyst—the necessity to understand situational-dependent factors, to continually track organizational changes, and to be aware that phase shifts may occur (e.g., reorganization) that change everything.

Endnotes

[1] March, J., and H. Simon, *Organizations*, New York: Wiley, 1958.
[2] Allison, G., *Essence of Decision*, Boston, MA: Little, Brown, 1971. (See also Allison, G. T., and P. Zelikow, *Essence of Decision: Explaining the Cuban Missile Crisis*, 2nd ed., Boston, MA: Addison-Wesley-Longman, 1999.)
[3] Jervis, R., *Perception and Misperception in International Politics*, Princeton, NJ: Princeton University Press, 1977.

[4] Janis, I. L., and L. Mann, *Decision Making: A Psychological Analysis of Conflict, Choice, and Commitment,* New York: Free Press, 1977. (See also Janis, Irving L., *Crucial Decisions.* New York: Free Press, 1989).

[5] Weick, K. E., *The Social Psychology of Organizing,* 2nd ed., Reading, MA: Addison-Wesley, 1979, and Weick, K. E., *Sensemaking in Organizations,* Thousand Oaks, CA: Sage, 1995.

[6] Giles, L., *Sun Tzu on the Art of War: The Oldest Military Treatise in the World,* translated from the Chinese with Introduction and Critical Notes, London, U.K.: Luzac and Co., 1910, Chapter 3, "Planning Attacks."

[7] Giles, L., *Sun Tzu on the Art of War: The Oldest Military Treatise in the World,* translated from the Chinese with Introduction and Critical Notes, Chapter 13, "Using Spies."

[8] DOD Information Operations Roadmap, October 30, 2003, p. 11.

[9] This taxonomy identifies IO as the integrated employment of three operational elements—EW operations, network-warfare (NW) operations, and influence operations—to affect or defend decision-makers and their decision-making process. Air Force Doctrine Document (AFDD) 1, *Air Force Basic Doctrine,* November 17, 2003, p.6. The definition is reaffirmed in The Air Force's *Concept of Operations for Information Operations,* February 6, 2004.

[10] Department of Defense, October 30, 2003, p. 39, redacted and published under FOIA to George Washington University National Security Archive, January 26, 2006.

[11] These three fundamental levels are introduced in the author's text: Waltz, E., *Information Warfare Principles and Operations,* Norwood, MA: Artech House, 1998; the three levels are also adopted in Joint Publication 3-13 Joint Information Operations, February 6, 2006.

[12] Kehm, H. D., "Some Aspects of Intelligence Estimates," *CIA Studies in Intelligence,* Vol. 1, Winter 1956; CIA Historical Review Release, 1994, p. 37.

[13] This table is based, in part, on "Intelligence Support to IO" in Joint Publication 3-13 Joint *Information Operations,* February 6, 2006, pp. III-1–III-3.

[14] *The Commission on the Intelligence Capabilities of the United States Regarding Weapons of Mass Destruction,* Washington, D.C., March 31, 2005, Footnote 629, Interview with NIO/NESA November 8, 2004, p. 239.

[15] Intelligence and Analysis on Iraq: Issues for the Intelligence Community ("Kerr Report"), July 29, 2004; approved for release August 2005, p. 7.

[16] For an introduction to the analysis-synthesis process, see Waltz, E., *Knowledge Management in the Intelligence Enterprise,* Norwood, MA: Artech House, 2003, Chapters 5–7.

[17] This process may be described in terms of the U.S. DoD Joint Directors of Laboratories (JDL) *Data Fusion Model* that distinguishes these three levels in similar terms. Enumeration corresponds to the Level 1 process, relation to the level 2 process and simulation to the level 3 process.

[18] Board on Army Science and Technology, *Network Science,* Washington, D.C.: National Academy of Science, 2006, p. 7.

[19] Remarks of the Director of Central Intelligence, George J. Tenet: "Opening Remarks," *The Conference on CIA's Analysis of the Soviet Union, 1947–1991,* Princeton University, March 8, 2001.

[20] For an example psychological study of a wide population, see Leidesdorf, T., "The Vietnamese as Operational Target," *Studies in Intelligence,* Fall 1968, pp. 57–71.

[21] Janis, I. L., and L. Mann, *Decision Making: A Psychological Analysis of Conflict, Choice, and Commitment,* New York: Free Press, 1977; Janis, I. L., *Crucial Decisions,* New York: Free Press, 1989; J., Robert, *Perception and Misperception in International Politics,* Princeton, NJ: Princeton University Press, 1977; Jervis, R., *System Effects: Complexity in Political and Social Life,* Princeton, NJ: Princeton University Press, 1977.

[22] Vroom, V. H., and P. W.Yetton, *Leadership and Decision-Making,* Pittsburgh, PA: University of Pittsburgh Press, 1973.

[23] For a survey of leadership taxonomies, see Cameron, J. A., and D. A. Urzi, "Taxonomies of Organizational Behaviors: A Synthesis," CSERIAC-TR-99-002, March 31, 1999.

[24] Gardener, T., J. Moffat, and C. Pernin, "Modelling a Network of Decision Makers," *DoD CCRP Conf.,* June 2004.

[25] In this section, we adopt the terminology and metrics from the standard reference text on the subject: Wasserman, S., and K. Faust, *Social Network Analysis: Methods and Applications,* Cambridge, U.K.: Cambridge University Press, 1994.

[26] Breiger, R., K. Carley, and P. Pattison, *Dynamic Social Network Modeling and Analysis: Workshop Summary and Papers,* National Research Council Committee on Human Factors, Board on Behavioral, Cognitive, and Sensory Sciences, Division of Behavioral and Social Sciences and Education, Washington, D.C.: The National Academies Press, 2003; Saunders-Newton, D., and T. A. Brown, "'Have Computer, Will Model:' A Taxonomy of Credible Computational Model Uses for Social Science Analysis," *Proc. of 69th Military Operations Symp.,* June 2001; Wasserman, S., and K. Faust, *Social Network Analysis: Methods and Applications,* Cambridge, U.K.: Cambridge University Press, 1994; Prietula, M. J., K. M. Carley, and L. Gasser, (eds.), *Simulating Organizations: Computational Models of Institutions and Groups,* Menlo Park, CA: AAAI Press/The MIT Press, 1998.

[27] For an example SNA of a terrorist network, see Krebs, V. E., "Uncloaking Terrorist Networks," *First Monday,* http://www.firstmonday.org/issues/issue7_4/krebs/; see also Sparrow, M. K., "The Application of Network Analysis to Criminal Intelligence: An Assessment of the Prospects," *Social Networks,* Vol. 13, 1991, pp. 251–274, and Klerks, P., "The Network Paradigm Applied to Criminal Organizations," *Connections,* Vol. 24, No. 3, 2001, pp. 53–65.

[28] Strack, Michael, M., "Operational Test of the CHIMS Software and Hardware: Report from Fort Hood," *Military Intelligence Professional Bulletin,* July–September 2002.

[29] For an overview of representative tools developed on the DARPA Evidence Extraction and Link Discovery (EELD) Program, see *Link Analysis Workbench, SRI International,* AFRL-IF-RS-2004-247, Rome, NY: AFRL, September 2004.

[30] Pioch, N., et al., "A Link and Group Analysis Toolkit (LGAT) for Intelligence Analysis," *Proc. of the 2005 International Conference on Intelligence Analysis Methods and Tools,* Office of the Assistant Director of Central Intelligence for Analysis and Production, McLean VA, May 2005.

[31] *The Commission on the Intelligence Capabilities of the United States Regarding Weapons of Mass Destruction,* Washington, D.C., March 31, 2005, p. 155.

CHAPTER 2
Learning from the Enemy: Approaches to Identifying and Modeling the Hidden Enemy Organization

Kari Kelton, Georgiy M. Levchuk, Yuri N. Levchuk, Candra Meirina, Krishna R. Pattipati, Satnam Singh, Peter Willett, and Feili Yu

As the previous chapter makes clear, in order to impact an opponent's organization, one needs a significant amount of information about that organization, particularly about its structure and processes. Yet, no organization is eager to reveal its inner workings to an opponent. The challenge is to find ways to infer such hidden information from limited and imperfect observations of the adversarial organization.

In this chapter, we discuss how one can learn the structure (i.e., decision hierarchy, allocation of resources, and roles to decision-makers, communication and information structures, and expertise of decision-makers) and processes (allocation of tasks to decision-makers, task execution schedule, decision policies, and so forth) of an enemy organization based on observed sequences of communication transactions between the enemy actors and activity events, which may be noisy and incomplete. Once the structure and processes are learned, they can be used to classify new communication transactions and events, detect and diagnose abnormalities in an enemy organization, predict future actions of adversary organization, and develop successful counteractions against the enemy or the probes and data-collection strategy to gather more information about the adversary actors and organization.

An Elusive Enemy

Analysis of the behavior of organizations, ranging from the structured command systems of a conventional military to the decentralized and elusive insurgent and terrorist groups, suggest that a strong relationship exists between the structure, resources, and objectives of those organizations and the resulting actions. The organizations conduct their missions by accomplishing tasks that may leave detectable events in the information space. The dynamic evolution of these events creates patterns of the potential realization of organizational activities and may be related, linked, and tracked over time [1, 2]. The observed data, however, is very sparse, creating a challenge to connect a relatively few enabling events embedded within massive amounts of data flowing into the government's intelligence and counterterrorism agencies [3].

To counteract the enemy organization, knowledge of the principles and goals under which this organization operates is required. This knowledge provides the ability to detect and predict the activities of the enemy and to select the appropriate countermeasures. However, certain countermeasures require additional knowledge of the specifics of organizational structure and distribution of responsibility within the organization to be successfully directed at the most important enemy nodes.

Ultimately, the best source of information about the enemy is the enemy himself. While getting into the mind of an enemy is impossible, thinking similarly is desirable to truly predict its next moves. When dealing with an enemy organization, we are not merely interested in learning about individuals, but in how they are organized as a team and what they can do together. The structure of an enemy organization defines its capabilities, while its goals define the mission(s). Just like the brain structure can be discovered using MRI scans, the structure of an enemy organization can be discovered from observations of interactions and activities of its members; these may be observed as part of normal activity monitoring, or they may be induced with intentional probes. In the context of discovering a covert organization, the scope of the probes is very limited. Therefore, one needs to use nonintrusive observations obtained as part of normal monitoring; these are tightly coupled with the intent of the adversary.

The problem of structural discovery is very complex: the observed data does not relate to the structure directly; instead, it relates to its manifestation in the form of activities and processes that are enabled by the organizational structure(s) and that are performed by the organization's members. Therefore, the algorithms to reconstruct the organization from observations alone would need to search through a very large space of possible structures. Given historic data and the availability of subject-matter experts, we can instead pose the problem as one of hypothesis testing, where a set of predefined hypotheses about the enemy organization and its subelements, albeit very large, is given. The problem then becomes one of rank ordering these predefined hypotheses on the basis of how best they match (or explain) the observed data.

The basic premise is that organizations leave detectable clues or enabling events in the observation space, which can be related, linked, and tracked over time. We label the enabling events associated with enemy organizational processes and activities—attacks by physical and information means, financing, acquisition of weapons and explosives, travel, and communications among suspicious people—as *transactions*. By transactions, we mean *who is doing what to whom, when, where, and how*. This type of model does not rely solely on the content of the information gathered, but more on the significant links between data (people, places, things) that appear to be suspicious. A pattern of these transactions is a potential realization of an organizational activity, such as planning a force-on-force engagement, preparing to hijack or kidnap, suicide bombing, or an attack on an infrastructure target. The "signature" of this pattern is shaped by the underlying organizational structure and the organization's goals.

Domain of Organization Identification

Generally speaking, an organization is a group of people intentionally brought together to accomplish an overall, common goal or a set of goals. Organizations can range in size from two people to tens of thousands. One of the most common ways to look at organizations is as social systems [4]. This view is becoming common among professionals who study, model, design, teach, and write about organizations. Simply put, a system is an organized collection of parts that are highly integrated in order to accomplish an overall goal. Each organization has numerous subsystems, as well. Each subsystem has its own boundaries of sorts and includes a variety of inputs, processes, outputs, and outcomes geared to accomplish an overall goal for the subsystem. Common examples of subsystems are departments, programs, projects, teams, processes to produce products or services, technologies, and resources.

Organizations exist in many domains—military, business, civic, political, religious, as well as virtual. These organizations have different decision-making principles, levels of decentralization, formalization, and adherence to strict organizational rules and doctrines. There are many difficulties associated with identifying the organizations that have many informal relationships among their members and change dynamically over time. To make our discussion more concrete, in this chapter, we focus on the command and control (C2) organization, which is designed to manage personnel and resources to accomplish the mission requiring their collective employment. Such organizations are distinguished by relatively formal structures and limited variability over time, and they are common to both friendly and adversary military forces. Given specific functions and principles of individuals together with the structural form in which they are organized, myriads of the different potential organizations can be constructed. All of them are based on the underlying C2 principles defining how individuals interact in the organization and what actions they perform [5]. These interactions can be utilized to detect and understand organizational relationships.

Several assumptions frame the adversary identification problem space. The models to identify the C2 organization are based on using the observed interactions among the decision-makers of the organization and the task executions and engagements by decision-makers, units, and resources of the organization. All decision makers and units of the adversary, as well as neutral actors in the environment, are assumed detected and tracked over time. However, specific roles of the actors are not known, and the solution approaches need to be able to map the actors to the specific roles that have been developed and stored in an a priori library. Also, we assume that categories of the actions and interactions among actors have been defined and stored. The observed interactions and activities of actors are assumed to have been analyzed and coded. The text exploration and data mining approaches can be used to obtain this data, and they are not the focus of this chapter. Instead, the problem solution needs to account for inherent uncertainty of the data collection and specifically uncertainty of text analyses and coding methods.

Identifying an enemy C2 organization—command hierarchy, communication networks (formal and informal), control structure (amount, distribution, and access to resources), and roles of individuals—is at the core of information operations. But while commercial tools for data collection and visualization are available, currently only manual procedures exist for identifying an enemy C2 organization.

Consider, for example, the information operations and the problem of identifying an adversary organization faced by the U.S. Army. The U.S. Army conducts operations using doctrinal military decision-making process (MDMP) [6]. One of the important steps in the MDMP process, intelligence preparation of the battlefield (IPB), requires the assessment of the enemy's command and control structure to improve the military's understanding of the enemy decision-making processes. Currently, the intelligence operations officer provides input to help the planning officer develop the IPB templates, databases, and other products that portray information about the adversary and other key groups (Figure 2.1) in the area of operations and area of interest. These products contain information about each group's leaders and decision-makers. Information relevant to enemy C2 includes:

- Religion, language, and culture of key groups and decision-makers;
- Agendas of nongovernmental organizations;
- Size and location of adversary and other forces and assets;
- Military and civilian communication infrastructures and connectivity;
- Population demographics, linkages, and related information;
- C2 vulnerabilities of friendly, adversary, and other forces and groups.

The linkage information is also produced manually from the data on activities (using an activity matrix template) and intelligence on the relationships between individuals. Using this information, a link diagram is developed to show the interrelationships of individuals, organizations, and activities.

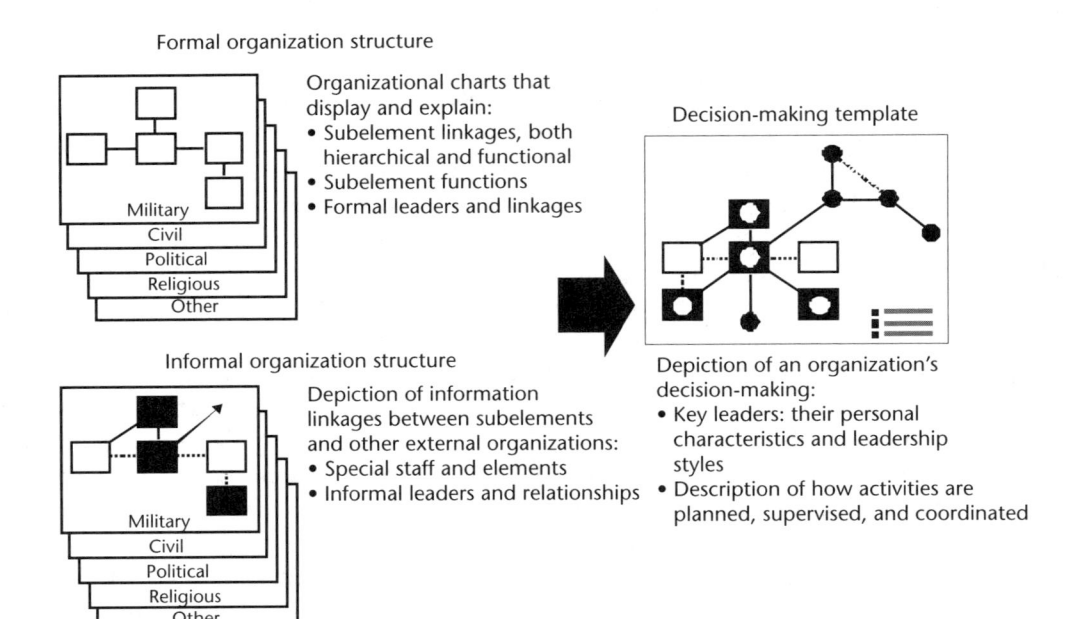

Figure 2.1 Sample information operations doctrinal template. (*After:* [7].)

Knowledge of connections (e.g., communication, command) between individuals and specific roles of individuals in the covert organization is needed because of the following effects. First, connections provide a means to share information and resources, and coordinate task execution. Second, captured individuals can share information about those to whom they are connected. Since it is a given that members of a cell share information and can compromise one another, the relevant question might be how interconnected are the cells that make up the organization? And third, capturing individuals, destroying bottleneck resources, or disabling organizational connections would allow disruption of enemy's operations and decision-making processes for preemptive actions.

The relationships among adversaries and the patterns of their activities change over time. This requires continuous or periodic updates of the knowledge about organizational structure to execute effective countermeasures. The rapidly accelerating technologies of communications and computers are overflowing the intelligence analysts with information at all levels of decision-making. Current labor-intensive manual processes to discover enemy organizations fail to keep up with dynamic environments [8]. Therefore, analytical tools are needed to reduce the complexity of organizational discovery to allow analysts to focus on information most essential for decision-making and search through only a limited number of most likely hypotheses.

Adversary Identification Research

Organization identification and analyses research has been applied to understanding the terrorist teams. Terrorist groups have evolved from hierarchical, vertical organizational structures to more horizontal, less command-driven groups. However, research suggests that terrorist organizations have limited forms, because they also operate under a set of principles that can set boundaries and provide guidelines for decision-making [9, 10]. Recent findings have also shown that terrorist organizations are analogous in structure, motivation, and operating environment to criminal street gangs [11], thus focusing the analyses on the set of known and well-documented forms. These types of organizations are different from traditional command and control organizations, because they are not "managed" in a strict sense, they often do not adhere to traditional doctrinal processes, and they are changing significantly over time in terms of size, personnel, and structure. However, many underlying organizing principles, such as sharing information, managing resources, coordinating activities, and synchronizing engagement, are analogous to C2 teams. While academic and applied studies of terrorist network structures have generated little actionable results to date [12], this research is new, and it is important to understand different approaches and their generalizations to the case of C2 organization identification. Specifically, two classes of techniques (Table 2.1) have been widely used to detect and analyze adversarial organizations: data mining (DM) and social network analysis (SNA).

A major challenge facing all antiterrorism intelligence agencies is to develop accurate and efficient analysis of the huge volumes of data. *Data mining* is a powerful approach that enables intelligence operators to explore large databases quickly and efficiently, extracting implicit, previously unknown, and potentially useful

Table 2.1 Summary of Current Research in Adversary Identification

Research Approach	Applications	Approach	References
DM	Criminal clustering	Identify closely related groups of criminals	[13]
	Text mining	Automatically identify from text documents the names of entities, relationships, and events of terrorist groups	[14]
	Spatial and temporal mining	Gather intelligence about the geographic areas of the enemy's concentration and detect behavior trends	[15]
SNA	Relational analysis, positional analysis, clustering	Identify the key members in a terrorist network	[22]
	MTML model	Examine the structural tendencies of various relations among actors	[19]
	Network formation	Explore the dependency between the organizational behavior and the formation of network linkages on the existing structural relationships inside and outside the adversarial network	[24–27]
	Network inference	Construct the network shape from incomplete and noisy observations	[25, 30]
Decision aids	Network exploration	Network visualization	[20, 28, 29]

knowledge from data. Unfortunately, very little research available to the unclassified community is directly related to the adversarial network structure mining. However, applications of data mining in crime investigation provide very rich information sources. The application categories that can be potentially utilized in adversarial network detection are criminal clustering, intelligence text mining, and crime spatial and temporal mining. Criminal clustering is used to identify groups of criminals who are closely related. This type of clustering relies on relational strength that measures the intensity and frequency of relationships among criminals [13]. Intelligence text mining helps to automatically identify from text documents the names of entities, relationships, and events of terrorist groups [14]. Some artificial intelligence technologies are utilized to improve the accuracy and efficiency of text mining, such as statistical modeling technology and machine learning technology. Crime spatial and temporal mining is the means to analyze patterns of adversaries' spatial and temporal characteristics. It aims to gather intelligence about geographic areas of the enemy's concentration and detect behavior trends. Statistical models and clustering techniques are commonly used in this area [15].

Social network analysis has been applied to study both terrorist and criminal organizations in the context of adversarial social networks [16–20]. Adversarial social networks are defined as the networks of multiple organizations within a population or a community competing for the same or similar resources and seeking to drive out their competitors [21].

The SNA usually involves relational analysis, positional analysis, clustering, and visualization, and it can also be employed to identify the key members in a terrorist network by computing centrality measures [22]. To expand the scope of adversarial network structure research, recently a multitheoretical multilevel (MTML) model was proposed in [23], which decomposed the framework into theories of self-interest, mutual interest and collective action, contagion, cognitive, exchange and dependency, and homophily. The MTML model focused largely on the social mechanisms that explain the creation, maintenance, and dissolution of network linkages within single networks by examining the structural tendencies of various relations (such as communication linkages, knowledge linkages, and trust relations) among the actors within that network and the attributes (such as gender, level in the hierarchy, and level of expertise) of the actors within the same network. Applying SNA methods to analyze the structure of al Qaeda and related terrorist organizations, Skillicorn showed how matrix decompositions can be used to extend the standard repertoire of social network and link analysis tools. For example, the authors studied the inclusion of other information about individuals and higher-order information about the relationships among them [19].

While some of the SNA researchers explored the dependency between the organizational behavior and formation of network linkages on the existing structural relationships inside and outside the adversarial network [24–27], most SNA analysis tools are focused on data *visualization* rather than *inference* from data, which is needed for structural analysis and identification. Moreover, some analyses have to be done manually (e.g., Krebs mapped a terrorist network into an association matrix and drew a terrorist network manually [17]). Challenged by huge amounts of data about terrorism, computer-aided analytic tools are emerging (e.g., CrimeNet Explorer [20], the Atypical Signal Analysis and Processing (ASAP) Tool [28], and Analyst Notebook [29]). Still, these tools mostly rely on visualizing and exploring single dyadic relationships (relationships between two actors) and are unable to distinguish between the hypotheses developed by analysts given the observed data. The baseline methods do not account for the constraints that shape the organizational structures, which are due to cultural, political, economic, social, historic, and most importantly doctrinal settings.

Since standard social network analysis is insufficient to address the issue of the dynamic and evolutionary nature of adversarial networks, recently some researchers integrated technologies from mathematics and artificial intelligence with SNA to enhance the capability of SNA. In [25], the authors combined multiagent technology with hierarchical Bayesian inference models and biased net models to produce accurate posterior representations of a terrorist's network structure. While utilizing probabilistic inference to construct the network shape from incomplete and noisy observations, Dombroski and colleagues [30] explored only single dyadic relationships with some triad closure updates and thus were not able to explore more complex topological configurations of the adversarial network realization.

Still, we find little information on computational modeling related to the detection of adversarial network structures in the presence of missing and noisy observations. The quantitative work by terrorism scholars forms only a tiny percentage of the total amount of work in organizational theory research. Although social network analysis is extensively used to explore adversarial networks, current work in this area often makes assumptions that ignore complex and dynamic aspects of adversarial behavior. New technologies including DM, probabilistic models, and artificial intelligence should be more actively used in this area.

Merely presenting and visualizing the network formed by observations does not solve the network identification problem due to the complex dependencies between individual elements of the networks and existence of missing, irrelevant, deceiving, and mislabeled attributes and links. Current research does not take full advantage of topological constraints on the organizational structures of the adversary and its relationship to enemy behavior. In this chapter, we aim at connecting the structural constraints of the organizations with the patterns of their behavior, thus integrating the structure and behavior identification in one common framework.

To identify the adversary, we first need to develop the quantitative representation of the adversary's organization, the relationships between its members, the behavior rules, and how this behavior can be structured, modeled, and related to the observations. That is, we need to understand what is it that we need to discover.

Identification Focus: Definition of Command and Control Organization

While all C2 organizations are designed to manage personnel and resources to accomplish the mission requiring their collective skills, the term is not limited to only one type of the organization and is common to both friendly and adversary domain. Given specific functions and principles of individuals together with the structural form in which they are organized, a myriad of different potential organizations can be constructed. All of them are based on the underlying C2 principles [5]. Since one of the most important findings from the research on organization theories is that there is no single "best" approach to (or philosophy of) command and control, many organizational constructs are possible.

Command and control refers to procedures used in effectively organizing and directing armed forces to accomplish a mission. The *command* function is oftentimes referred to as the art of an individual to set the initial conditions and provide the overall intent for mission execution. The *control* is referred to as those structures and processes devised by command to enable it and to manage risk and other entities in the organization. The commander in a C2 organization issues instructions to subordinates, suggestions to commanders of adjacent units, and requests and reports to supporting units and superiors. He develops and maintains a situational awareness of the area of his operations through reports presented by other people or by electronic systems [31]. The basic premise of command and control organization is the ability to distribute the responsibilities among its elements and to coordinate these seemingly independent entities for joint operations to achieve the objectives. The fundamental need for communications significantly constrains the options for both command and control, making communications infrastructure a critical feature of a C2 system. However, describing the communications links and nodes of a fighting force does not suffice to explain, understand, or predict successes and failures in command and control. We need to be able to represent, model, and identify the functions and objectives of the individual elements of the C2 organization.

Let us describe the command and control organization as a collection of C2 nodes and resources connected via command, control, communication, and task structures (Figure 2.2). The roles, responsibilities, and relationships among C2 nodes and resources constrain how the organization is able to operate. *C2 nodes* are

Identification Focus: Definition of Command and Control Organization

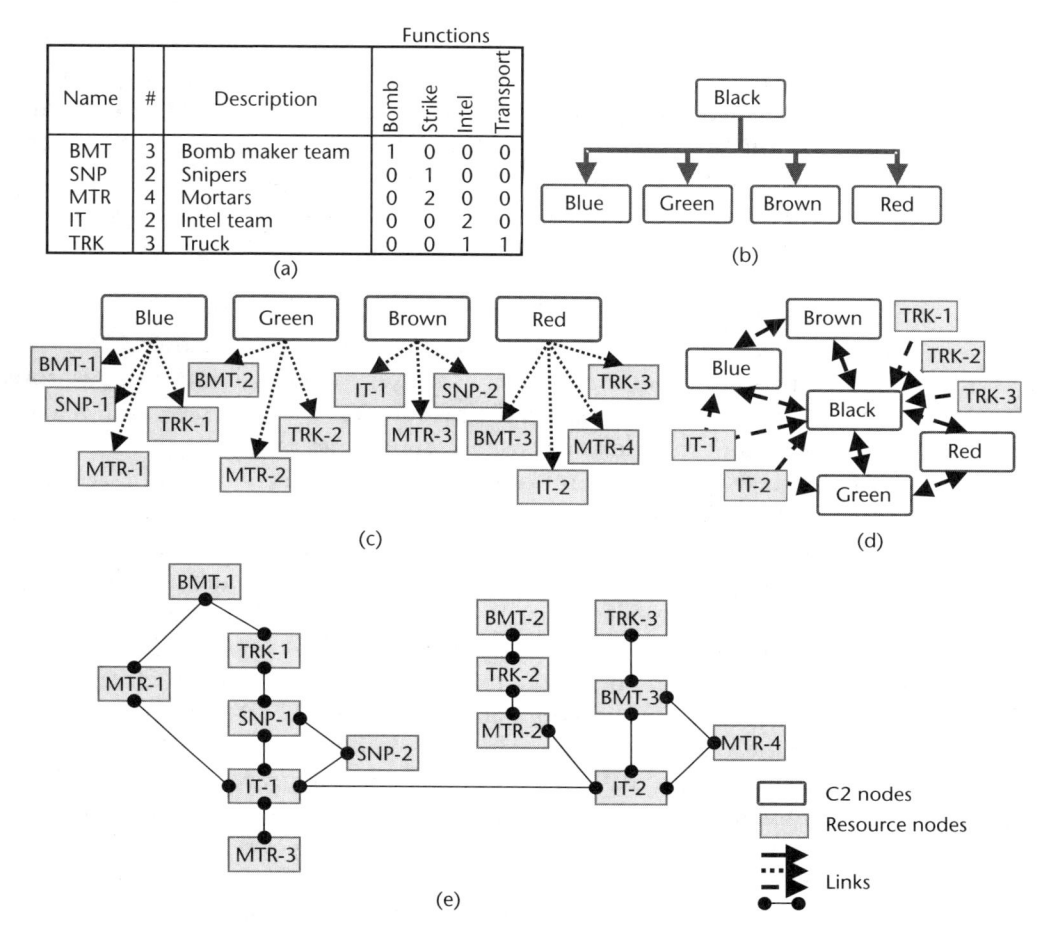

Figure 2.2 Example of a C2 organization: (a) resources of C2 organization; (b) C2 nodes and command structure; (c) control structure; (d) communication structure; and (e) task structure.

entities with information-processing, decision-making, and operational capabilities that can control the necessary units and resources to execute mission tasks, provided that such an execution does not violate the concomitant capability thresholds. A C2 node can represent a single commander, liaison officer, system operator, or a command cell with its staff. A set of physical platforms and assets, C2 nodes, and/or personnel can be aggregated to a *resource* (e.g., squad, platoon, or weapons system). A resource is considered a physical asset of an organization that provides resource capabilities and is used to execute tasks. The level of aggregation depends on the problem at hand. For example, in cordon and search missions executed by the company-size forces (62–190 soldiers), we can consider resources being the single squads (9–10 soldiers). The roles and responsibilities of the C2 nodes and resources identify possible operational and tactical policies: decisions they can make and actions they can perform.

Command structure, represented as a network with directed links, defines superior-subordinate relationships among C2 nodes of the organization, thus specifying who can send commands to whom. *Communication structure* is a network between the decision-makers of the organization, that defines "who can talk to whom," the information flow in the C2 organization, the communication resources that decision-makers can use (communication channels), as well as the security of the communication channels. A *control structure* is an assignment of resources to C2 nodes, and it specifies which commanders can send tasking orders to what assets. A *task structure* is a network among resources, where each link corresponds to operations jointly executed by these resources.

In Figure 2.2 we present an example of the enemy C2 military team consisting of 5 command elements and 14 units or resources. The commanders of this organization make decisions to manage assigned resources in a cooperative manner to achieve team objectives. Commanders are executing mission tasks and prosecuting the desired targets via allocation of their resources (military assets and weapons) and synchronization of their mission task execution and target engagements.

Figure 2.2(a) describes a set of resources—military units and assets controlled by commanders. The assets include bomb maker teams, sniper teams, mortars, intelligence and reconnaissance teams, and trucks. This chart also shows the functional capabilities of the units and resources in terms of bomb making, strike and small-arms attack, intelligence and monitoring, and transportation. The authority structure among five commanders is a flat hierarchy—Figure 2.2(b)—with a single commander ("BLACK") being the main commander of enemy forces. The assignment of assets and units to commanders—Figure 2.2(c)—determines the control structure of the C2 organization. Note that in hypothetical example of Figure 2.2, the main commander ("BLACK") does not control any resources directly. A communication structure (who can talk to whom) of the organization is depicted in Figure 2.2(d), along with the direction of unit-reporting observed events (information flow) beyond the control structure (we assume that units controlled by commanders also report their observations to these commanders). A partial task structure—a network between resources—is shown in Figure 2.2(e). The task structure is due to the joint task execution by resources; therefore, it evolves throughout mission execution and depends on how the commanders manage their resources to assign and execute tasks.

The purpose of the organization discovery is to recognize the command, control, communication, and task structures of the organization. However, the challenge is that most of the time we cannot observe the elements of the structures of the organization. Instead, we can obtain the intelligence due to the actions and activities of the organization. The specific actions depend on the structure of enemy command and control organization and are derived from the goals of the team. The mission goals and their relationships to the actions and tasks can be specified in the *behavior model* of the organization. Given a specific model, we can determine whether it could have generated the obtained intelligence.

From Structure to Strategy: Characterizing Behavior of a C2 Organization

We represent the behavior model as consisting of the mission model, which is independent of the organization, and the activity pattern derived from the mission and

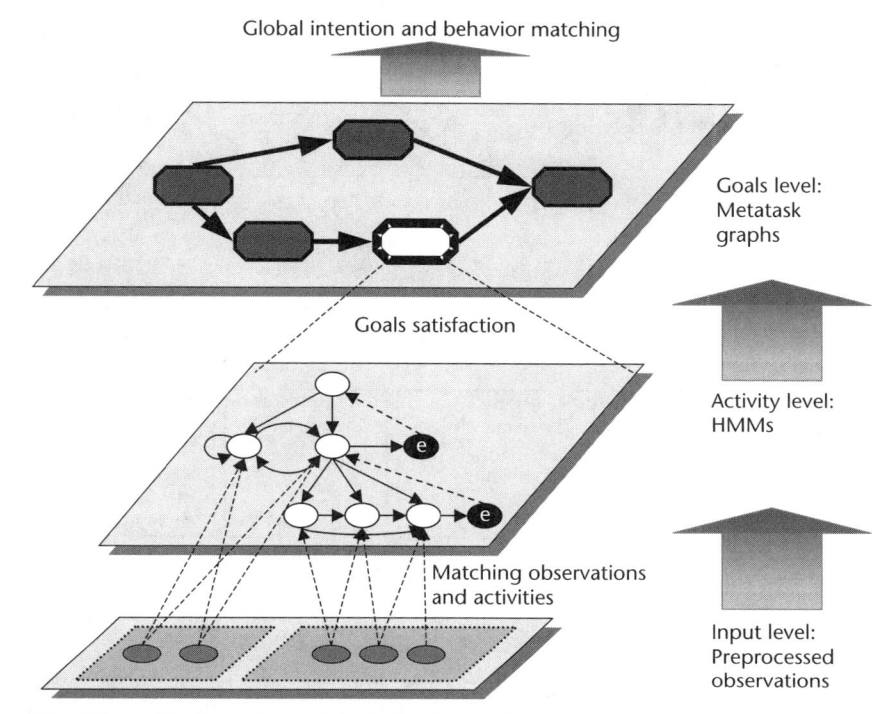

Figure 2.3 Hierarchical structure of the behavior model.

organization. Our model employs a hierarchical representation, with the observations at the lowest ("input") level, process and action pattern representations at the middle ("activity") level, and goal graph delineating the goal structure of the organization at the top ("goals") level (Figure 2.3).

We specify the *mission model* of the adversary as a precedence graph of specific goals or objectives that an enemy organization seeks to achieve. An example of the mission model is shown in Figure 2.4(a). Each goal (or meta-task) in a mission represents its intermediate phase. The precedence constraints in a meta-task graph define the mission plan and relative order among the goals. A goal defines *what* the enemy seeks to achieve but does not specify *how* this will be accomplished. To do this, we need to hypothesize a pattern of alternative processes and activities that the enemy might employ to achieve the concomitant goal(s). Several methodologies may be utilized to model these patterns, including applied cognitive task analysis [32], Bayesian networks, transition graphs, and Markov decision processes [33]. In our modeling, we first construct the decomposition of the mission goals into the physical tasks (or actions) that the enemy can perform—Figure 2.4(b). Using the knowledge of organizational theories [5, 34–38], task decision-making cycle (e.g., observe, orient, decide, and act (OODA) loop [39]), and team tasks for command and control teams (see Figure 2.4 for an example of team tasks and task decision cycle), we then define the probabilistic Markov transition diagram representing the *activity pattern* to achieve each goal (see Figure 2.5 for an example of activity pattern). The Markov transition model allows a representation of interdependencies among actions and processes, including parallelism, and temporal and spatial dependencies. To obtain the activity pattern, we first need to define the conditional

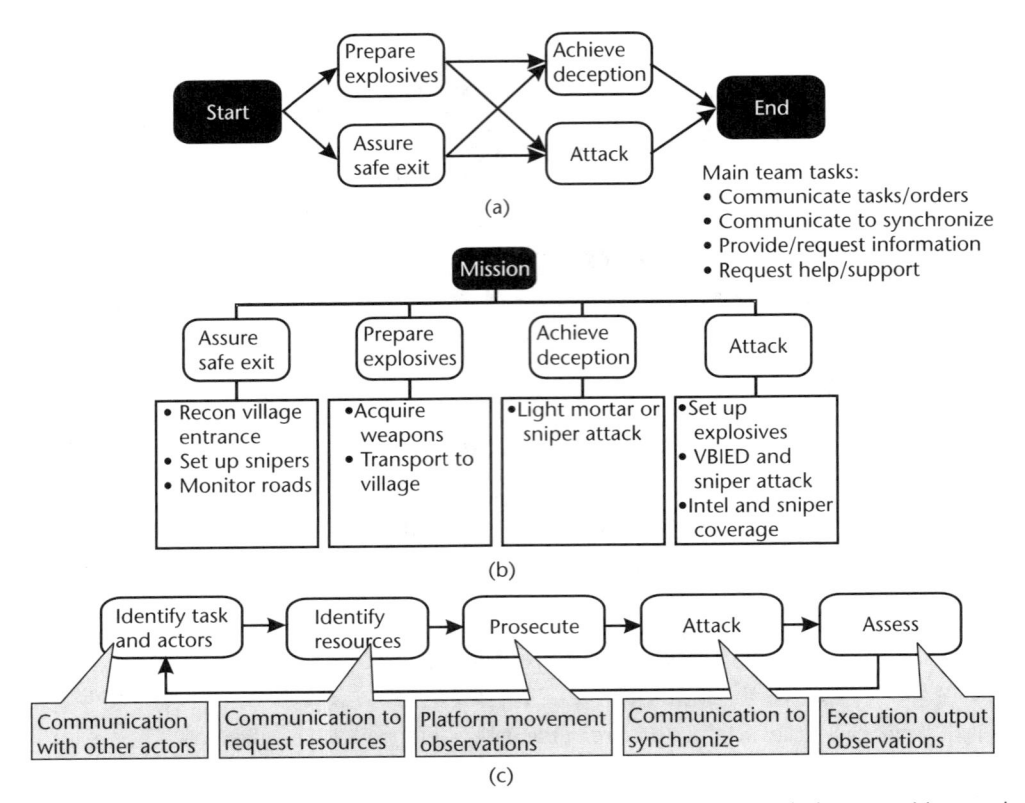

Figure 2.4 Example of mission parameters: (a) mission model, (b) goal-task decomposition, and (c) task decision-making cycle.

transition diagram, with state nodes corresponding to adversarial actions and communications, and conditional nodes corresponding to alternative behaviors and outcomes. The probabilities in the activity pattern relate to the likelihoods of alternatives and the duration of the underlying activities [40].

Goals can be performed in parallel or sequentially—so that the precedence constraints are satisfied. When the organization starts executing a goal, it invokes the operations of the underlying activity pattern. Thus, multiple goals could be *active* at the same time, so that their underlying activities emit the potentially observable transactions in a joint way. When the organization completes the underlying processes or actions, the corresponding goal is considered completed (objective is satisfied or not) and is terminated.

The states of activity pattern can represent the state of enemy resources, the geo-spatial location of enemy forces or enemy attacks, the action that an enemy team or an individual member performs, and so forth. Activity patterns can also be defined separately for the nodes of the organization, their resources, communication channels, elements of the environment, and so on. For example, the map of the environment can be divided into regions, and every region can be associated with an independent activity model or pattern (to track the state of the region) or become a state in a single activity model. In the latter case, the pattern of movement of the enemy unit in the environment, guided by a unit's maneuver capability, objectives that the unit needs to achieve, friendly forces, and physical structures in the environ-

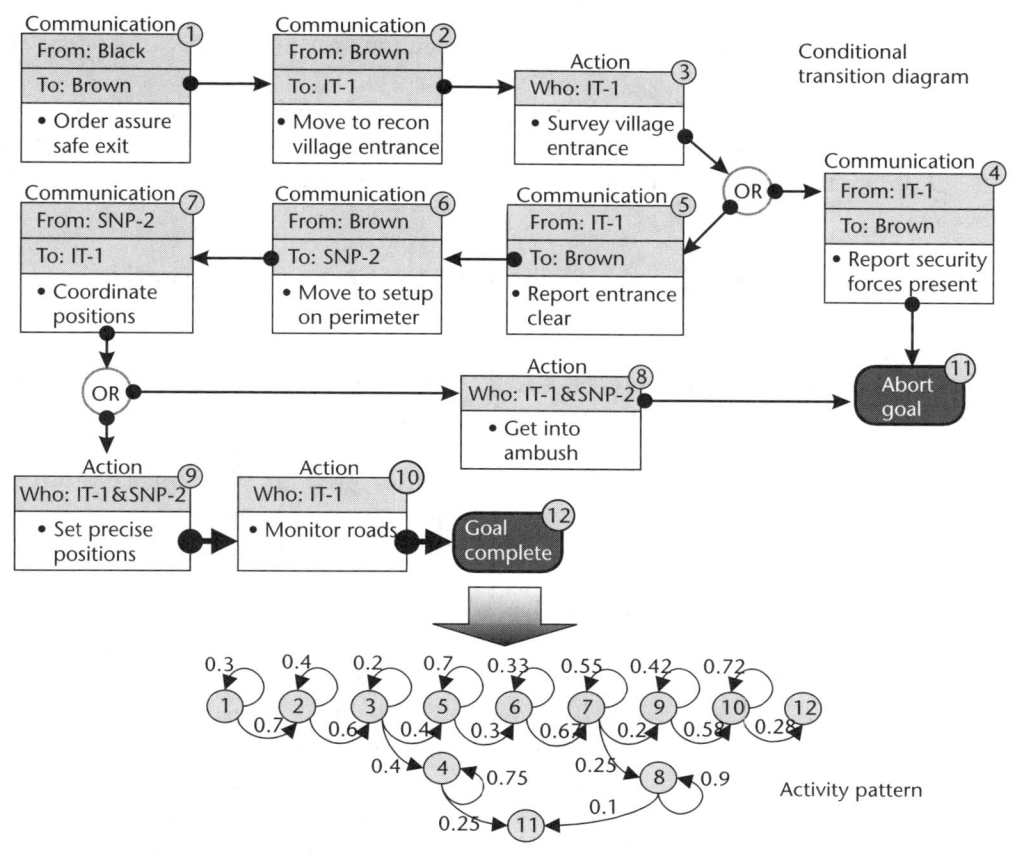

From Structure to Strategy: Characterizing Behavior of a C2 Organization

Figure 2.5 Example of activity pattern for *assure safe exit* goal.

ment (roads, rivers, houses) can be translated into an activity model and would represent the model of the unit maneuver over time. When multiple units operate in the environment, multiple activity models are active at the same time (called *parallel models*). In another example, the activity model states could represent the level of the insurgency in a region and hence devoid of geographical information. The activity model states can represent any level of action and decision granularity—from the strategic, operational, and tactical levels to elementary activities.

Since the observations are uncertain, the activity-observation model can be represented using a hidden Markov model (HMM) [40]. Hidden Markov models and their generalizations (hierarchical HMMs, factorial HMMs, and coupled HMMs) are probabilistic models that combine Markov chains with partially observable and noisy observation processes. The premise behind an HMM is that its true underlying process is not directly observable (hidden), but it can be probabilistically inferred through another set of stochastic processes (observed transactions, for example). In our problem, the process that is "hidden" is what occurs behind the curtain of an enemy organization; the observation process is an intelligence database containing any information that can be represented as observed activities. HMMs are perhaps a natural choice for this problem because we can evaluate the probability of a sequence of events given a specific hypothesized behavior model, determine the evolution of an enemy activity represented by the HMM, and estimate new HMM parameters that produce the best representations of the observed activities.

To understand how the organization and its behavior can be recognized, we need to understand the intelligence-gathering process and explore what types of observations can be obtained. We consider these issues next.

Looking Beyond the Smoke Screen: Definition of Observations

What can be gleaned about the adversary organization? What data is most useful to discover an individual's roles as well as important relationships within the organization, and which data is meaningless? What types of observations are more important to discovery and thus should be the focus of data collection efforts? These questions are at the heart of organizational identification problem. While it is desirable to have full access to certain types of information about the enemy to improve detection capabilities, in practice, we are constrained to focus on the types of information that *can* be collected.

In the normative representation of organizational structures and quantitative theories of organization identification, we restrict our models to several types of intelligence information that are feasible to collect. We assume that the observations can include the set of tracked (monitored) *individuals* whose positions in the organization we need to determine, information about these individuals, and identified adversary's resources. Tractable information regarding the individuals encompasses their attributes and resources (e.g., expertise of individuals, training, background, affiliation, and family ties). Information about an adversary's resources may include detection of the enemy's military assets and their capabilities, communication means, political connections, and financial capabilities. In addition, the observed information may also include *transactions* that involve these entities; these are comprised of *communications* among the individuals, including some classification of communication content (e.g., a request for or transfer of information, resource, action; acknowledgment; or direction), and the individuals' *actions*—the involvements in observable activities (functions or tasks performed, such as individuals committing the same crime, performing financial or business transactions, or using the resources in covert or open operations).

There are several challenges associated with the collection and utilization of intelligence that make the observations uncertain. First, in many cases, the individual enemy's operation triggers only a limited number of relationships in its organization; therefore, complete information about the enemy's interactions is not readily available. Second, the real world is full of uncertainty, with many activities overlapping in their objectives, some events being indistinguishable, and many observations not related to the activities that an adversary may perform; this results in noisy observations. Third, the technological limitations and the covert nature of the adversary's activities cause much of the data to be hidden and not readily available to the observers. In addition, realizing that they are being watched, the enemy might use deceptive tactics. Finally, to associate the observed events with the true processes, we oftentimes need additional information not contained in the observation of the event, such as association of monitored individuals with their functions or positions in the organization.

As a result, we must account for these types of uncertainties when modeling and associating the observations with real adversarial activities. This is enabled by a probabilistic model, where every observation is associated with a vector of probabilities that could have been generated by a hidden transaction.

The observations can be related to true activities and communications of the adversary organization. They could also be related to the structural elements of the organization. For example, if we intercept a command message from one individual to another, it must be that the first individual is a commander of the second individual. If we intercept the information exchange between two individuals, it means that they are connected through an existing communication channel. However, we encounter problems when the uncertainty of the observations increases, and we cannot make such conclusions with confidence.

Adopting the hypothesis-testing approach, where we need to test which of the organizations from a library of hypothesized C2 structures best explains the observations, addresses some of these problems. In this model, we need to relate a given structural link or an activity to the one that was observed. However, too many structural links and activities seem to be similarly related to the observation. This is especially true when the content of the observation is highly uncertain and sparse. What we can and need to use to improve the discovery is the information about participants of the observed action and/or communication. In this case, we encounter a new challenge: while the individuals of the adversary can be tracked over time and distinguished between one another, we do not immediately know how they are related to the individual positions in the hypothesized organization. This problem is illustrated in Figure 2.6. For example, the intelligence can be obtained from intercepted communication that the "individual X ordered individual Y to conduct operation," but what operation is not known. This observation can be associated with the command relationship between the nodes of an adversary organization (on the left of Figure 2.6) as well as with some hypothesized actions in the behavior model (on the right). First, we need to understand what nodes in a hypothesized C2 organization do the tracked individuals "X" and "Y" correspond to. If "X" is commander "BLACK," and "Y" is commander "BROWN," then this observation can be related to the structural relationship "BLACK is commander of BROWN." In this case, we are also able to associate this observation with hypothesized activity "BLACK ordered BROWN to assure safe exit in the village." If another association is made (e.g., "X" is commander "BROWN" and "Y" is unit leader of "SNP-2"), then other relations of this observation with structural links and with hypothesized activities are possible. Note that such relations are not necessarily unique. For example, if we had received intelligence that "unit Z has been observed maneuvering in the village," then if we assume that "Z" is unit "IT-1," then this could mean any of the actions "IT-1 surveys village entrance," "IT-1 monitors roads," or "IT-1 sets positions for sniper fire with SNP-2."

The associations of the tracked individuals "X" and "Y" with commanders and units and resources of a hypothesized C2 organization is called *node mapping*: assigning observed individuals to positions in a hypothetical organization. It might seem that being able to place a tracked individual at a specific position in the

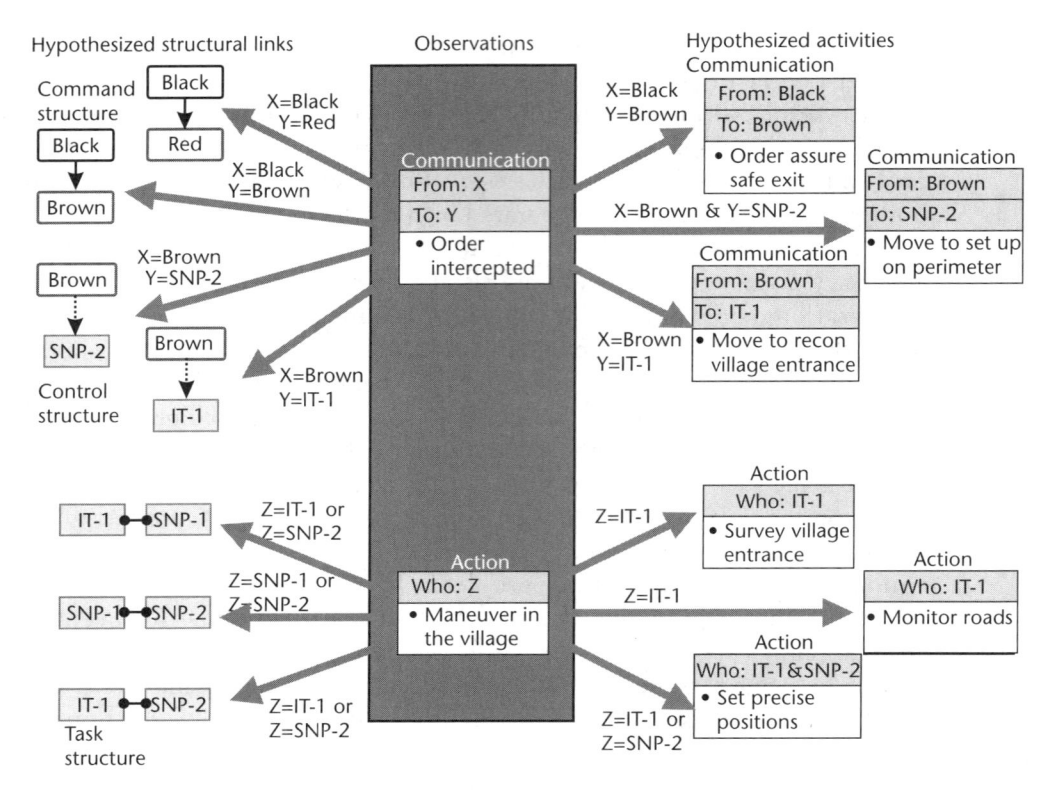

Figure 2.6 Example of relating observations to structural links and activities.

organization is adequate. But discovering positions of individuals in the organization in this fashion lacks the notion of organizational activity evolution and therefore cannot distinguish between the organizations and missions. Without the node mapping, we would not be able to relate the observations with hypothesized activities. Even if such associations were possible, they would be too broad and would influence future relations. This influence would be due to implicit association between tracked individuals (such as "X" and "Y") and the corresponding participants in the hypothesized activities. This association would prevent us from using HMMs to identify activity patterns, since HMMs assume conditional independence between observations.

This example illustrates the need to approach this problem in two steps. In the first step, we will obtain the mapping of tracked individuals to actual C2 nodes and resources of hypothesized organization. This will be achieved using all collected observations. We perceive this step as discovering the dots. In the second step, we need to link the dots together. In this step, we use the HMMs to match the time-evolving patterns of observed activities with behavior models in the hypothetical organization. This is referred to as activity pattern discovery, and it allows us to recognize the mission of the adversary. This is equivalent to actively tracking the behavior patterns of the enemy and reassessing which behavior is currently active with each incoming observation. This is a final step in discovering the acting organization. Next, we describe these two models and underlying data that supports their functioning.

Discovering the Dots: Finding Identities of Actors

How can we know the true identify of a tracked individual? This is a main question of the *node mapping* problem. The challenge is to relate the observed individual with a place in the organization—mapping him against a C2 node and its concomitant command role, expertise, dedicated responsibilities, access to people, control over resources, and so forth. Just as law enforcement agencies discover this knowledge by collecting pieces of data about an individual, that person's relationships to others, and participation in transpired events, we base our decisions on observed attributes of individuals, transactions between them, and attributes of those transactions.

A quantitative representation of the node mapping problem is posed as relating the nodes of an observed network (formed from tracked actors and transactions among them) to nodes of a hypothesized adversary organizational network. The time component is temporarily disregarded, with all observations aggregated and used for a single estimation evaluation. All collected observations are linked together to form the *data (observed) network*. The nodes of this network are tracked individuals, units of enemy fight force, and other resources. The links of this network are the structural relationships perceived to be realized in observed communications between tracked individuals, commands sent from one individual to another or to the unit, information requests and information transfers, joint task executions (discovered from action observations), and so on. Here, we utilize the classification of communication into 12 classes (see Table 2.2) of communications [41]. These classes were assumed to be fixed, and no other information about the intercepted communications was used. The classes allowed the determination of which organizational relationship the intercepted communication corresponded to, thus creating the concomitant link in the data network.

We need to find how to map the nodes of this data network to the *model network*—the hypothesized C2 organization with command, communication, control, and task substructures. Figure 2.7 illustrates the problem, where 10 nodes and units

Table 2.2 Communication Categories

Communication Category	Abbreviation
Info request	Inforeq
Info on task request	Infotkr
Info on asset request	Infoassr
Info on action request	Infoactr
Resource request	Resur
Coordination request	Coordr
Info transfer	Infotran
Info transfer about task	Infotkt
Info transfer about asset	Infoasst
Have/will perform action	Perfactt
Resource used to perform action	Userest
Have/will coordinate	Coordt
Acknowledgment	Ackn

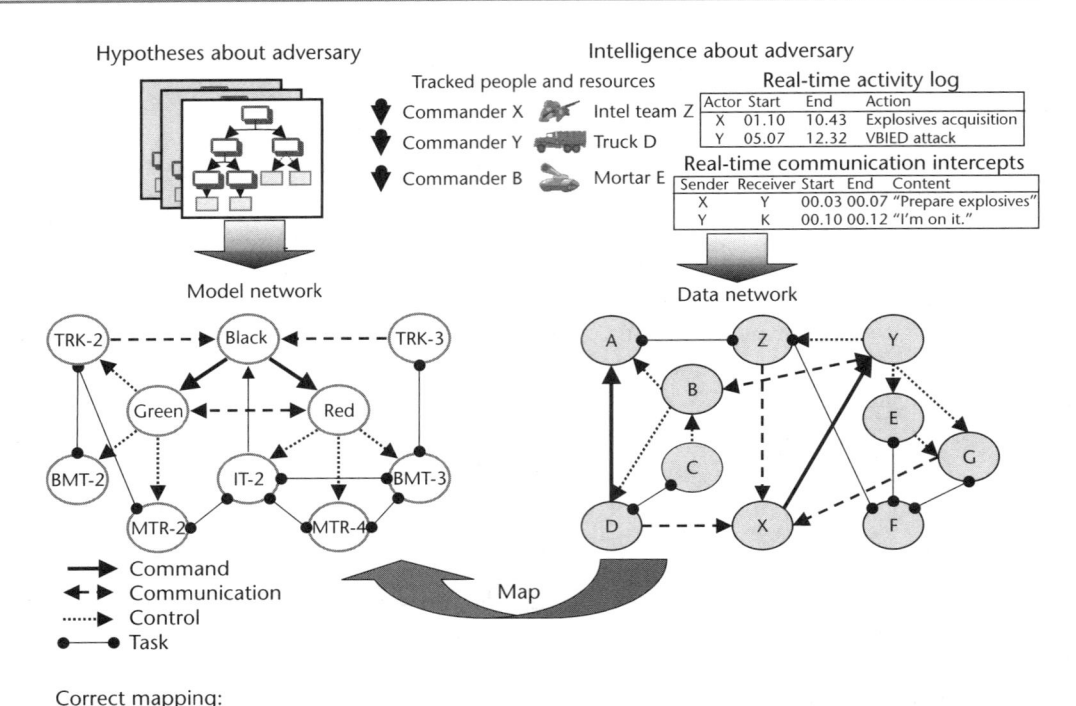

Figure 2.7 Example of node mapping problem.

Correct mapping:
A=MTR-2, B=Green, C=BMT-2, D=TRK-2, E=MTR-4, F=BMT-3, G=TRK-3, X=Black, Y=Red, Z=IT-2

of the adversary have been detected (A, B, C, D, E, F, X, Y, Z), and the communication intercepts and action observations of the adversary are aggregated to a data network. Matching the topology of this network to a hypothesized C2 model network produces the correct mapping: A = MTR-2, B = GREEN, C = BMT-2, D = TRK-2, E = MTR-4, F = BMT-3, G = TRK-3, X = BLACK, Y = RED, Z = IT-2. That is, we say that tracked agent "X" is commander "BLACK," while tracked resource "A" is a mortar resource (MTR-2), and agent "Y" is commander "RED," and so on.

To relate the nodes of the data network to nodes of a model network, we employ probabilistic node mapping [42, 43] to maximize the likelihood that the data network has been generated by the model network over all possible mappings from nodes of the data network to nodes of the model network. The mapping does not have to be one to one. It could be one to many, which is equivalent to discovering the individual of the enemy organization who performs multiple roles, or many to one, which is equivalent to finding the enemy command cell. To calculate the maximum likelihood solution, we employ the structural consistency approach [42]. The mapping is obtained to maximize the topological closeness between data and model networks. In simplified terms, it is equivalent to finding the consistency between the elementary substructures of the data and model networks. An example of elementary substructure is a clique—a node and all its neighbors (see Figure 2.8). The consistency is based on (1) correctly matching the nodes of the data network to the nodes of the model network (i.e., relating observed individuals to the C2 nodes in the hypothesized C2 organization), (2) maximizing the number of correctly identified relationships between nodes, and (3) minimizing the number of incorrectly identified relationships and missed (unobserved) relationships.

Discovering the Dots: Finding Identities of Actors

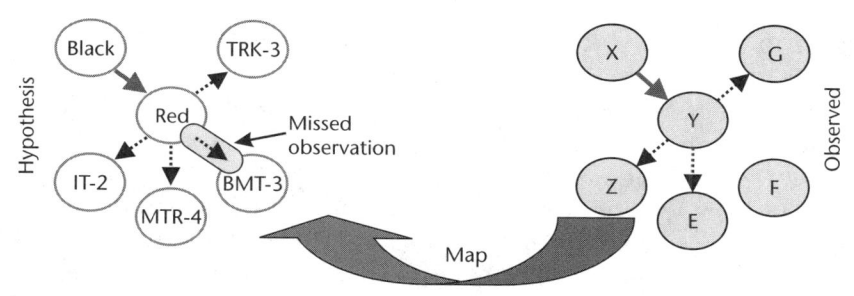

Figure 2.8 Example of node mapping solution for a clique.

Formally speaking, we represent a *hypothesized organization* as a graph $G_M = (V_M, E_M)$, a *model network* where V_M is a set of C2 and resource nodes, and E_M is a set of edges among them. Without loss of generality, we assume that we deal with a single network structure of the enemy organization. The edges can also be expressed in the form of an adjacency matrix: $M = \|M_{\alpha,\beta}\|$, where $M_{\alpha,\beta} = 1$ if and only if $(\alpha, \beta) \in E_M$. Observed data is aggregated to a *data network*—a graph $G_D = (V_D, E_D)$ with adjacency matrix, $D = \|D_{a,b}\|$. Here, V_D is a set of observed individuals and resources, and E_D is a set of observed relationships among them. We need to discover the mapping from actors to their roles in the organization—that is, from the nodes of the data graph to the nodes of the model graph. This is accomplished by finding an assignment matrix $S = \|s_{a,\alpha}\|_{a \in V_D, \alpha \in V_M}$, where $s_{a\alpha} = 1$ if data node a is mapped to model node α. We find an assignment matrix S that maximizes the likelihood function, $P(G_D \mid G_M, S)$, which is equal to the probability that the observation (data network) has been generated by the hypothesized organization (model network), given the roles of tracked individuals (mapping between nodes of data and model graphs). In this model, the uncertainty of observing relationships between network nodes is modeled using false alarm probability for observed but deceptive activities and the probability of a miss for unobserved secure or covert activities. While direct optimization of the likelihood is infeasible, an approximate solution can be obtained by relaxing the structural consistency measure to consider a subgroup match and then employing an iterative expectation maximization algorithm to find its solution [42]. Not only do we obtain the correspondence of tracked individuals to specific nodes in each hypothesized organization, we can also rank-order these associations for each organization using values of likelihood function, $P(G_D \mid G_M, S)$.

It might seem that being able to relate a tracked individual to a specific position in the organization is adequate. But discovering positions of individuals in the organization in this fashion lacks the notion of organizational activity *evolution*, and therefore cannot distinguish between the organizations and missions. The mapping algorithm uses all the observed data together. Its complexity does not allow the dynamic remapping to be performed with every incoming observation, which is needed for organizational monitoring to detect when the organization changes its structure or changes its mission. The mapping does, however, allow the association of observed transactions with activities in the behavior model of the adversary. This will be used to monitor and detect patterns of adversary activities and will serve as the final organization detection step. We discuss solving this problem next.

Connecting the Dots: Discovering Patterns of Interactions and Activities

In the previous section, we discussed how the identities of tracked individuals can be inferred from aggregated network observations. This allows the association of observed transactions, such as task execution by resource(s), involvement of C2 nodes in an activity or transaction, communications among individual C2 nodes, and joint operations of resources and/or C2 nodes with true activities in the organization. The mapping is obtained for each organizational hypothesis, and we need to distinguish which hypothesis is most likely. While the values of the likelihood function for the observations given the organizational structure can be used to rank-order the organizational hypotheses, this method would be weak due to its inability to account for evolution of activities in time. Only this evolution can describe the real missions of the acting organization. When in isolation, the activities are of limited significance; however, when activities are combined and patterns emerge, they become indicative of potential threats. Here, we discuss the second problem in the organization identification (i.e., how to discover the patterns of organizational activities from incomplete and noisy observations).

Real-world processes and activities that the adversary performs emit signals that are captured as a series of observations. Due to intelligence-gathering limitations, possible deceptions performed by the enemy, and overlapping activities of various nonadversary actors and groups, we need to model the activities of adversarial organization as only partially observable.

It is our purpose here to develop a *signal model* that can be used to distinguish between suspicious patterns of activity to detect real instances of adversarial activities. In doing so, our model must be able to (1) detect potential threats in a highly cluttered environment, (2) efficiently analyze large amounts of data, and (3) generate hypotheses with only partial and imperfect information. We are representing signal model as a transaction-based model, which identifies relationships between nodes in a network to describe their structures and functionalities. The point here is that if we can identify the types of activities, or transactions, that adversarial group may be involved in, then we can construct a model based solely on these.

We have chosen to apply HMMs because they constitute a principal method for modeling partially observed stochastic processes and behaviors—processes that have structure in time. Each HMM can be viewed as a detailed stochastic time evolution of a particular behavior. An HMM can sequentially process new information (a window of data) each time an observed transaction or event occurs. The window of observations could contain a single or a batch of transactions to improve the efficiency. HMMs provide a systematic way to make inferences about the evolution of adversary's activities. The premise behind an HMM is that the true underlying process (represented as a Markov chain representing the evolution of the transactions as a function of time) is not directly observable (hidden), but it can be probabilistically inferred through another set of stochastic processes (observed transactions, for example). Figure 2.9 shows a typical HMM. The gray colored circles represent states of a true process. This true pattern or process is a "hidden" process with series of true transactions describing the behavior of a particular adversary group. This true "hidden process" is observed through a noisy process represented by a series of observed transactions (white circles in Figure 2.9). Our objective here is to detect a

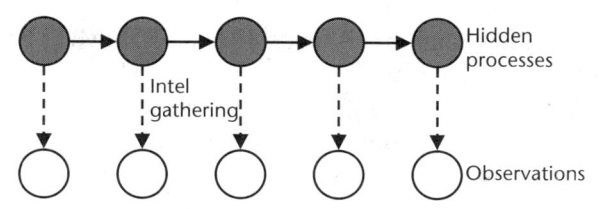

Figure 2.9 HMM representation.

hidden "true" pattern, which is a sequence of transactions (shown inside gray circles) via an observed process (white circles). We can infer the existence of a true pattern based upon a set of observations, as shown in Figure 2.9, because HMM states are statistically related to a noisy observation process.

Each state of HMM is a single activity or a set of activities of the enemy (communications, synchronizations, performing the action, and so forth). But we are not limited to the activity-based states. The states of HMMs can also represent the availability of enemy resources, the geo-spatial location of enemy forces or enemy attacks, the state of the enemy's belief about a situation or desire to perform an action, and so on. For an example of geography-based state representation, the map of the environment can be divided into regions, and every region can be associated with an independent HMM (to track the state of the region) or become a state in a single HMM. In the latter case, the pattern of movement of the enemy unit in the environment, guided by a unit's maneuver capability, objectives that the unit needs to achieve, friendly forces, and physical structures in the environment (roads, rivers, houses) can be translated into an HMM that would represent the model of the unit maneuver over time. When multiple units operate in the environment, multiple HMMs are active at the same time (parallel HMMs). In another example, HMM states could model the activity level of the insurgency in a region, hence devoid of geographical information. Thus, HMM states can model any level of the action and decision granularity—from the strategic, operational, and tactical levels to elementary activities.

HMMs have been very successful in modeling and classifying dynamic behaviors because they have clear Bayesian semantics and efficient algorithms for state and parameter estimation. HMMs have been extensively researched and used for the past several decades, especially in speech recognition [40] and computational molecular biology [44]. Several extended HMMs have been successfully employed to solve coupled sequence data analysis problems, such as complex human action recognition [45], traffic modeling [46], biosignal analysis [47], and financial data modeling. These new models usually aim to enrich the capabilities of the standard HMM by using more complex structures, while still being able to utilize the established methodologies—for example, the expectation-maximization (EM) algorithm—for standard HMMs. Recent advances in modeling complex HMMs and other hybrid stochastic pattern learning, together with improvements in speed of computing power, make it possible to recognize intricate behavioral patterns present in multimember teams.

Formally, a discrete HMM is a quantization of a system's configuration space into small number (N) of discrete states, $S_1, S_2, ..., S_N$, together with probabilities for transitions between states (many processes, including continuous ones, can be cast as continuous-state and/or observation-space HMMs, but generally the most interesting applications occur with discrete processes). A single finite discrete variable $q_t \in \{S_1, S_2, ..., S_N\}$ denotes the state of the system at current time t. The state changes (an approximation of the dynamics of the system) are described by a table of transition probabilities, $A = \{a_{ij}\}$, with $a_{ij} = P(q_t = S_i | q_{t-1} = S_j)$. The initial state distribution of the system is denoted by a vector, $\pi = \{\pi_i\}$, where $\pi_i = P(q_1 = S_i)$. The states of an HMM have a Markov assumption (i.e., given the present, the future does not need to know the past). The observations of the system state could be continuous or discrete. In the discrete case, a single finite discrete variable $o_t \in \{V_1, V_2, ..., V_M\}$ indexes the observation (output) at the current time t. The observation at time t depends only on the hidden state at that time; this dependency is described by observation probability matrix, $B = \{b_{ij}\}$, where $b_j(k) = P(o_t = V_k | q_t = S_j)$. Thus, a standard HMM is quantitatively described by a triplet, $\lambda = (\pi, A, B)$.

Using HMMs, we can address three main challenges of our organizational activity discovery problem. In challenge 1, also called the observation evaluation, we need to assess the likelihood that a given observation sequence was generated by the activity pattern in the library. Addressing this challenge is equivalent to scoring how well a given model matches a given observation sequence. This can be achieved by computing the probability $P(O|\lambda)$ of an observation sequence $O = \{o_1, o_2, ..., o_T\}$ given a model $\lambda = (\pi, A, B)$ using a forward/backward algorithm. Challenge 2 is how to find the sequence of (hidden) activities of a specific activity model that most likely generated (best explains) the obtained sequence of observations. There are several reasonable optimality criteria that could be used to address this problem. The choice of the criterion is a strong function of the intended use of the uncovered state sequence. For example, to find a sequence of hidden states $S = \{q_1, q_2, ..., q_T\}$ that maximizes the probability $P(S|O,\lambda)$ uses the Viterbi algorithm [40]. Challenge 3, termed learning problem, is about finding an "optimal" parameter set $\lambda = (\pi, A, B)$ for an HMM given example observation sequences (i.e., we infer the probabilistic structure of a HMM). This problem can be perceived as training HMM against a set of inputs. The solution is obtained by applying the Baum-Welch algorithm [40]—a special case of the EM algorithm.

While maximum likelihood criterion is an efficient discriminator when a single HMM is active, it is not suitable for detecting the presence of multiple HMMs. The latter is the case when several actors or subgroups of the enemy are acting at the same time, and the observations are the aggregated realizations of such activities. This problem is called a superimposition because the active HMMs seem to superimpose on one another to produce observations (Figure 2.10). To assess the existence of multiple superimposed HMMs and detect a change in their activities for large-scale models, we use modified Page's test [48]. The original Page's test was proposed in [49] as a series of sequential probability ratio tests as an efficient change detection schema for large hypotheses library testing.

A modified Page's test is a means to attack the change detection problem, where the main goal is to determine the time n_0 when the observation process switches from one distribution (set of active HMMs) to another distribution. A modified Page's test is optimal for detecting conditionally independent and identically distributed HMMs. The detection of an HMM in the presence of noise is shown in Figure 2.11 in the form of a CUSUM test statistic. The starting point of the CUSUM plot is asso-

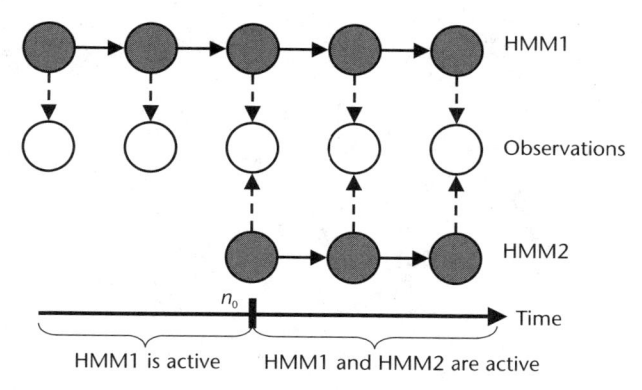

Figure 2.10 Illustration of superimposed HMMs.

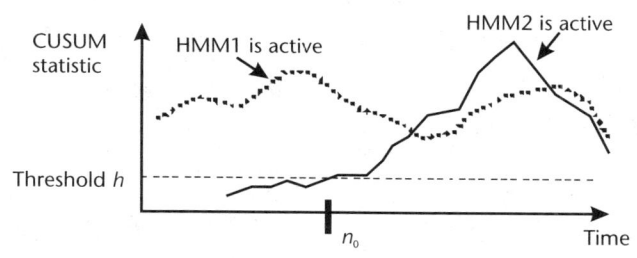

Figure 2.11 HMM detection via page test.

ciated with the first time this HMM is detected; thus, we believe (with certain probability) that the modeled adversary's activity is in progress. A peak probability usually results when this pattern evolves into the absorbing state of the HMM, and we obtain a maximum number of signal transactions for this HMM. Once the peak is attained, the numerous unrelated transactions will reduce the confidence in the detection. Thus, two reasons can decrease the probability in Figure 2.11. The decrease of the CUSUM statistic is caused by either noise transactions or simply that the adversary's activities have already reached their goal and thus do not warrant any further transactions.

To capture the natural and/or enforced hierarchy present in the organization activities, we can use *hierarchical* hidden Markov model (HHMM) representations (Figure 2.12). The hierarchical nature of the HHMM facilitates the creation of complex models that incorporate patterns over the short, medium, and long terms at varying levels of resolution and abstraction. This facilitates sensing, learning, and inferencing at multiple levels of temporal granularity—from point-wise observations (lowest level) to explanations (higher level). In this way, we can achieve an efficient segmentation of the observations. Rather than recognizing individual activities or atomic events, we seek to learn the overall activity patterns of a command organization that behaves according to its internal rules. The HHMMs allow for easy interpretation of queries at varying levels of abstraction, model reusability, and faster learning through independent submodel generation. The latter is due to the retraining needed when the problem space (e.g., environment) changes:

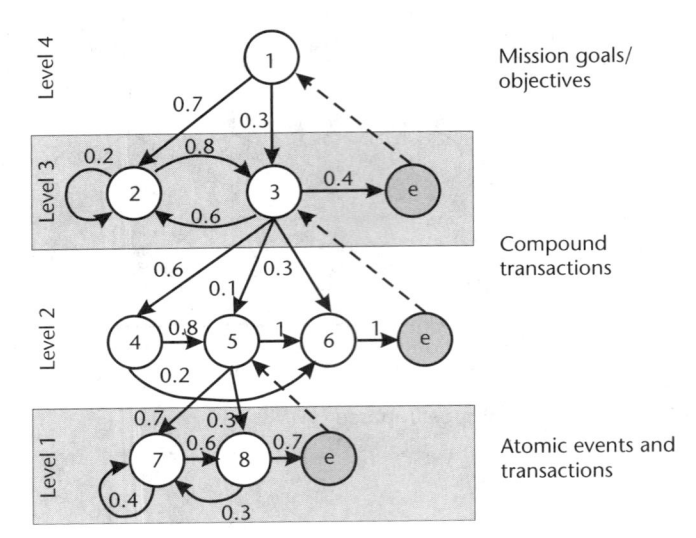

Figure 2.12 Example of hierarchical HMM.

HHMMs allow retraining some parts of the models independently while leaving others intact.

In our context, an HHMM trained on a single enemy operation is able to learn the hierarchical nature of the training data, populating the following:

- *Atomic events and transactions* (e.g., financial transactions, individual actions, and communication messages) denoting observations corresponding to the lowest states in the model topology;
- *Compound transactions* representing individual actions, team actions, and combinations of actions at the middle level;
- *Mission goals* at the top level.

Modeling minor dependencies in probabilistic structures as *in*dependent greatly reduces the number of parameters that must be inferred—learning is more accurate, and both estimation and inference are faster. Further, the segmentation of the structures facilitates atomic-HMM reuse.

Many interesting systems are composed of multiple interacting and interdependent processes, and thus merit a compositional representation of two or more variables. This is typically the case for systems that have structure both in time and space. With a single Markov state variable, Markov models are ill-suited to these problems. The three main types of complex behavior needing to be addressed in modeling true organizational processes include (1) interacting processes, (2) processes occurring in parallel, and (3) processes with interdependent initiation and/or termination. The multinomial assumption employed by HMMs introduces complexity issues. For example, to represent 30 bits of information about the history of a time sequence, a standard HMM would need 2^{30} distinct states. However, using distributed HMM state representation would allow us achieve the same task with 30 binary units. The distributed state representation can be obtained when the state can be naturally decoupled into features that together interactively describe the dynamics of the time series. Here, we utilize our multiple hypotheses tracking algorithm and employ recently developed HMM architectures that model specific dependencies of the processes on each other based on coupling principles, including factorial and event-coupled HMMs.

It might happen that we receive multiple observations about the same activity or event that took place (e.g., when there are multiple sources of intelligence). The problem can be addressed in one of two ways. First, if each observation source provides an independent assessment of the features of the observation (e.g., one source reports the participants of the communication, another source reports the content of the observation, and the third source pinpoints the geo-location of the communicating nodes), then we can consider conditional probabilities separately for each observation, and then use the "cross-product" of the different observations as a single observation data point. This approach would not work when the observations are correlated—for example, when multiple sources can report information on the same feature of the observed transaction. This will also be the case when features of the observed transaction or event are correlated. However, considering the "joint observation" and storing conditional probabilities is prohibitive due to the large number of possible observations (cross-product states). In this case, data fusion algorithms must be developed to extract the joint conditional probability for each feature vector. A simple approach is used in least squares estimation methods, when the features of observation are compared to the attributes of the true events. A more general method is to consider the neural network that will estimate the relationships between the observed features and all events [50].

All parameters for HMMs can be either specified by subject-matter experts or learned given *example observation sequences*. These sequences can be obtained by simulating the organizational activities and simulating data collection and intelligence gathering. We discuss these issues next.

Behavior Learning and Simulating Enemy Activities

The model described earlier requires internal knowledge representations to populate the library of behavior models (HMMs). The HMMs of the activities could be specified in a variety of ways:

1. Manual processes using subject-matter experts to define model parameters;
2. Automated processes using historical data;
3. Automated processes using input from activity simulators or data sources;
4. A hybrid approach combining any of these.

Due to the lack of availability of sample enemy activities, learning from historic data would appear to be difficult. Instead, the HMM learning can use the inputs from the organization simulator in the form of simulated activity to process sequences (including communication, individual and team tasking, and resource employment) and define the evolution of these activities using Markov transition diagrams. A major problem with data-driven statistical approaches, especially when modeling rare or anomalous behaviors, is the limited number of exemplar

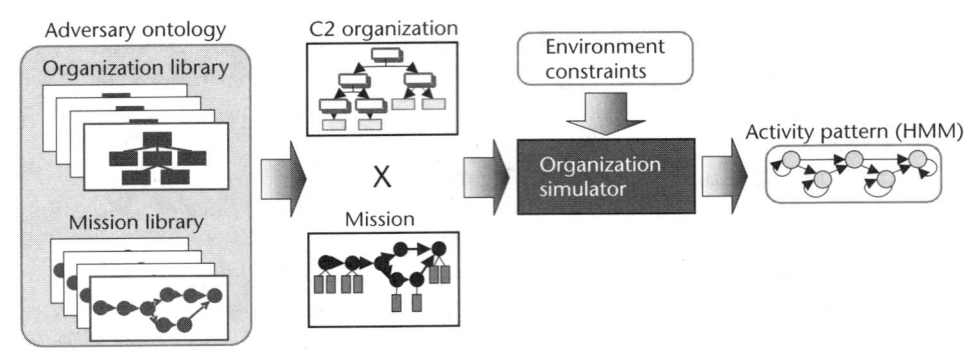

Figure 2.13 Automated HMM construction.

behaviors for training the models. A major emphasis of our work therefore is on efficient Bayesian integration of prior knowledge obtained by using synthetic prior models with evidence from data and on constructing HMMs using synthetic prior models (Figure 2.13). This methodology [51] is an extension of the approach proposed by Oliver in [52]. The organizational activity patterns, represented via HMMs, will be initially trained using activity sequences produced by synthetic agent-based simulations [53] that are aimed at mimicking the true enemy command organization's performance. Then, the models will be refined using data collected from the real world under surveillance. This interaction between organizational simulation tools and statistical hidden Markov model learning is a key feature of our approach.

Simulation is often used for understanding and predicting system behaviors [54]. The use of computational simulations to depict complex, dynamic, nonlinear, adaptive, and evolving systems has been accepted as a promising approach where conventional analytic tools fail. Although computer-based simulation is no more valuable than its underlying assumptions, significant benefits can be realized from its use. Simulation allows the discovery of hidden implications of the underlying assumptions. It also offers less abstraction from the details of a problem of interest due to the availability of computing power. Even in poorly understood environments, simulation yields beneficial insights by allowing the designers to concentrate only on selected details of interest; accordingly, the designers can focus on those crucial internal properties and not be burdened by unnecessary aspects of the problem. In this context, simulation can often reveal emerging aspects from the interactions of individual simulation components that are independent of the native properties of those components. That is, simulation allows emerging aggregate outcome that may not be easily explained or predicted from the individual components. These benefits put forth simulation as a strategy to analyze complex, dynamic, or emergent structures.

Recent research in organizational simulations has focused on computational *agent networks* as models of the coordinated team. In computational organization theory and artificial intelligence research literature, an agent is understood as a representation of a decision unit and is distinguished by the notion of autonomy [55]. Properties of an agent are defined in terms of a particular task and its environment, which influences an agent and is in turn affected by agent behavior. Depending upon the combination of tasks and the environment, the literature distinguishes between several categories of agents. An overview of agent models is provided by [56]. A recent overview of the computational modeling of organizations is provided in [57].

The use of computational organization theory and agent-based simulation spans a broad area ranging from commercial enterprises to military establishments. The applications range from analyzing mergers in differentiated product industries, modeling and simulating a deregulated electricity market [58], and discovering latent association in terrorist networks [25] to modeling and assessing novel military organizational concepts [53, 59–61]. There has been a tremendous growth in the availability of agent architectures. We capitalize on this work to obtain viable training data for learning activity patterns of hypothesized organizations and missions.

In choosing simulation as a research instrument, we need to consider suitability, adequacy, and validity of the corresponding models to our problem domain. Many specific agent architectures have been developed for particular problem contexts. These include the distributed decision-making agent (DDA) architecture [53, 59–61] and agents in model driven experimentation [62], which are intended for modeling and simulating the human command and control organizations. In the following, we present examples of general agent architectures that promise to be suitable and adequate for our problem context. One is *Swarm* [63] architecture, which was originally developed specifically for multiagent simulation of complex adaptive systems. The fundamental component that organizes the agents of a Swarm model is a "swarm," a collection of agents with a schedule of events over those agents that is also a representation an entire model. Swarm supports hierarchical modeling, whereby an agent can be represented as swarms of other agents in nested structures. Another architecture is *Brahms* [64]. The architecture allows the development and simulation multiagent models of human and machine behavior. Brahms is a full-fledged multiagent, rule-based, activity programming language. It has similarities to belief-desire-and-intention (BDI) architectures and other agent-oriented languages, but it is based on a theory of work practice and situated cognition. The third example is the Cognitive Agent Architecture (Cougaar) [65]. The Cougaar architecture is intended for the construction of large-scale distributed agent-based applications. It has demonstrated its feasibility to conduct rapid, large-scale, distributed logistics planning and replanning in extremely chaotic environments.

The Overall Process of Identifying Enemy Command and Control

In the previous sections, we discussed how to map individuals to the positions in a hypothesized organization, how to determine the currently active behavior model, and how to learn the parameters of a behavior model from simulated organizational behavior. Now, we need to summarize the enemy organization detection process.

Our quantitative organization identification process represents a hybrid model–based structure and process identification; it consists of the following four steps (Figure 2.14).

- Step 1, *node mapping and observation preprocessing*, employs a probabilistic node mapping to determine positions of tracked individuals in the enemy

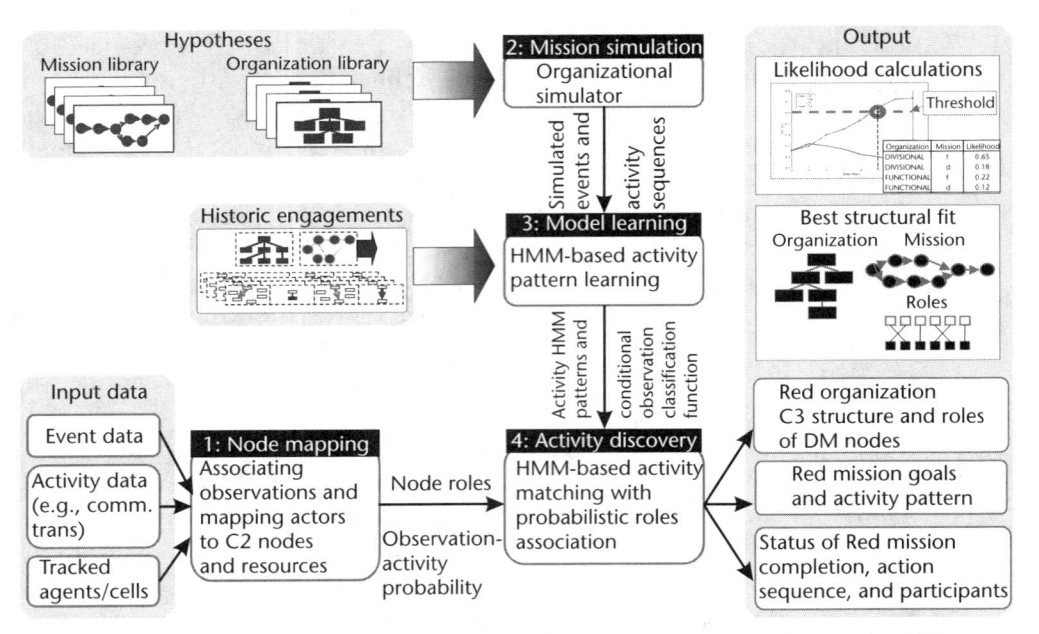

Figure 2.14 Hybrid model–based process to identify enemy structure, mission, and activities.

organization. In this step, we also associate observations with hidden behaviors by computing the conditional probability that the observation has been generated by the activity of the enemy organization.

- Step 2, *mission simulation*, simulates the behavior of an enemy organization versus a friendly organization executing a given mission, producing the sequences of possible activities and interactions, as well as the sequences of observations of enemy activities and mission events, given specific structure and behavior logic.
- Step 3, *model learning*, finds, using historic data or inputs from the organizational simulator, the evolution of enemy organizational processes and activities (including communication, individual and team tasking, information dissemination, and resource employment) via Markov transition diagram representation. Model learning also generates parameters to calculate conditional probabilities of observations given true events, actions, and communications.
- Finally, Step 4, *activity discovery*, identifies currently active organizations and missions via tracking events, activities, and processes using a hidden Markov model.

Step 4 finds the most likely HMM pattern that could have generated the observations or a set of these patterns and their rank-ordering using their likelihoods. Each active pattern corresponds to the hypothesized organization and mission from the predefined library. Together with the mapping of tracked actors to the nodes in concomitant organizations completes the identification of the adversary command and control organization.

Since we use the hypothesis-testing process, the model described here requires internal knowledge representation of organization and mission hypotheses library to feed several steps. These libraries can be specified by intelligence analysts familiar with a specific domain, gathered from previous histories of discovered adversary organizations and populated using synthetic generative methods utilizing organizational theories and interaction principles.

Several components of organization identification processes have been researched and successfully applied in our previous work. The quantitative node mapping model and solution algorithms have been developed in [66]. Model learning and activity discovery components have been implemented and tested in the adaptive safety analysis and monitoring (ASAM) system to detect and track the activity patterns of hostile organizations [1, 2]. The ASAM system was successfully applied to model enemy activities for various real-world events, including Indian airlines hijacking, Greek Olympics threats, boat attack scenarios, and various nuclear, biological, and chemical scenarios. Our mission simulation work was originally based on the optimized mission execution models [59, 60, 67] developed under the A2C2 program and later has been modified to mimic the behaviors of a human C2 team. Using cordon and search mission scenarios for company-size ground forces, we have calibrated the synthetic models to the human performance in [53, 61] and shown that the simulation can achieve the same trends and similar results across a range of performance and process measures.

Experimental Validation of the Organization Identification Process

The C2 identification process presented in this chapter is aimed at reducing the complexity of organizational discovery. This will allow analysts to focus on information most essential for decision-making and explore in detail only a limited number of most likely hypotheses. Before the decision aid based on this work is built, we need to evaluate whether this solution can significantly increase capabilities to make inferences regarding enemy command structures and explore how discovered information can be used by friendly forces to disrupt adversarial activities. In this section, we describe our current work focused on human-in-the-loop (HIL) experimentation that compares the accuracy of adversarial organization discovery obtained by a team of human analysts versus the automated C2 identification process.

Our evaluation method (Figure 2.15) leverages many years of similar model-based experimentation cycles executed for the A2C2 research program [67–71]. This work studied the ability to use models to develop optimized military organizational structures for different missions and to encourage organizational adaptation. The A2C2 program included iterative cycles of experimentation to evaluate and validate the modeling approaches. These experiments have been conducted using distributed dynamic decision-making (DDD) virtual environment [72]. DDD is a distributed real-time simulation platform implementing a complex synthetic team task that includes many of the behaviors at the core of almost any C2 team: assessing the situation, planning response actions, gathering information, sharing and transferring information, allocating resources to accomplish tasks, coordinating actions, and sharing or transferring resources. Successive DDD generations have demonstrated the paradigm's flexibility in reflecting different domains

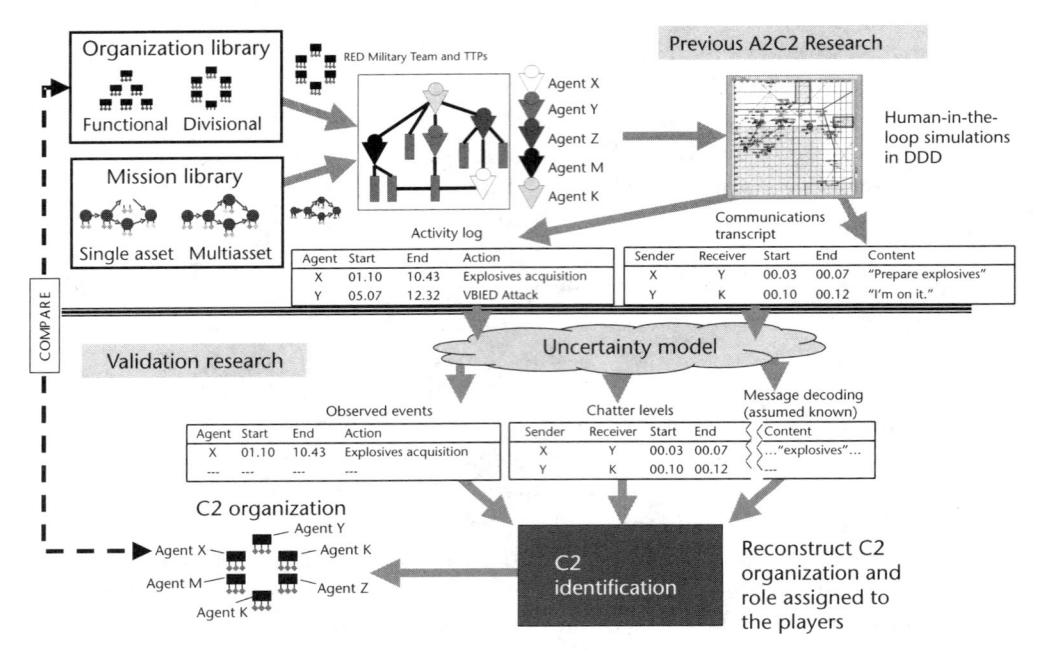

Figure 2.15 NetSTAR validation process.

and scenarios to study realistic and complex team decision-making. An outcome of A2C2 program that directly feeds our validation work has been the creation of DDD-based scenarios and organizational structures. The A2C2 experiments have catalogued a diverse set of outcomes from HIL runs for various teams, organizations, and mission conditions.

An HIL DDD run includes a team of participants playing roles of commanders in a predefined command and control team and performing the mission tasks in the DDD virtual environment using kinetic and nonkinetic assets/resources. Of particular interest to our validation work are A2C2 experiments with joint task force (JTF) organizations, which explored the range of possibilities to assign the command and control relationships, resource ownership, and individual responsibilities among commanders. Under the A2C2 program, we have tested both traditional and nontraditional C2 structures, thus providing rich data for the validation experimentation. For each HIL run from an A2C2 experiment, the data logs have been captured; they include task execution logs (who does what, where, and when) and the communication interactions among team players. The latter information has been coded into distinct categories corresponding to several types of formal and informal interactions in a C2 organization. This data can be directly used by our validation process (Figure 2.15) with the addition of the uncertainty model component that can take the task execution and communication logs from real experiment runs and make the data noisy—that is, introduce deceptive events (false alarms), create missing data (misdetection), and add noise and errors to other data elements. In the validation experiment, this noisy observation data is presented to both a human analyst team and an automated C2 identification model that must reconstruct the acting enemy C2 organization. The outcomes of human analyst team and automated identification model are then compared to judge the benefits of the proposed automated process in both identification accuracy and time required to identify (or manpower needs).

In order to properly evaluate the proposed process, we need to answer the following two questions:

1. Is it possible to judge the impact of *uncertainty* on the quality of the organization identification solution?
2. Is it possible to judge the impact of *problem domain* and *complexity* on the quality of the organization identification solution?

To address the first question, our study includes exploring various levels of uncertainty and the corresponding parameters (probability of false alarm, misdetection, and errors). To address the second question, we conduct comparisons according to the type of organization that needs to be recognized. Different information is needed to recognize different types of organizations. In our pilot studies, we found that when the low-noise commander-to-subordinate intercepts can be obtained, a functional organization, where a single commander controls resources of the same type distinct from other commanders, is easier to recognize than a divisional organization, where each commander controls a variety of resources but thus has similar capabilities to other commanders. The divisional organization is more complex than the functional in terms of resource control, but it can be easily recognized given the low-noise data of commanders' activity locations, since commanders' geographic responsibilities in divisional organization are distinct. Both functional and divisional organizations have elements that are encountered in today's C2 teams, and thus a study of such "hybrid" teams is essential to explore how difficult it is for human analysts to use multiple types of information for C2 discovery.

Another objective of our research is to study a tradeoff between the complexity of the identification problem and the uncertainty level that the analysis can allow to yield high recognition accuracy. This is important to understand where current observation data is incomplete and improvements in data collection capabilities are needed to recognize the acting organization.

References

[1] Singh, S., et al., "Stochastic Modeling of a Terrorist Event Via the ASAM System," *Proc. of IEEE Conference on Systems, Man and Cybernetics*, The Hague, the Netherlands, October 2004.

[2] Tu, H., et al., "Information Integration via Hierarchical and Hybrid Bayesian Networks," *IEEE Trans, on System, Man and Cybernetics, Part A: Systems and Humans*, Vol. 36, No. 1, 2006, pp. 19–33.

[3] Popp, R., et al., "Countering Terrorism Through Information Technology," *Communications of the ACM*, Vol. 47, No. 3, 2004, pp. 36–43.

[4] McNamara, C., *Field Guide to Consulting and Organizational Development: A Collaborative and Systems Approach to Performance, Change and Learning*, Authenticity Consulting, LLC, February 2005.

[5] Alberts, D. S., and R. E. Hayes, *Understanding Command and Control*, Washington, D.C.: CCRP Publications, 2006.

[6] Wade, N. M., *The Battle Staff SMARTbook: Doctrinal Guide to Military Decision Making and Tactical Operations*, 2nd ed., Lakeland, FL: The Lightning Press, 2005.

[7] FM 3-13, "Information Operations: Doctrine, Tactics, Techniques, and Procedures," *Headquarters, Department of the Army*, Washington, D.C., November 28, 2003.

[8] Popp, R., and J. Yen, (eds.), *Emergent Information Technologies and Enabling Policies for Counter-Terrorism*, New York: Wiley-IEEE Press, 2006.

[9] Arquilla, J., D. Ronfeldt, and M. Zanini, "Networks, Netwar, and Information-Age Terrorism," in I.O. Lesser, (ed.), *Countering the New Terrorism*, Santa Monica, CA: RAND, 1999.

[10] Morgan, M., "The Origins of the New Terrorism," *Parameters*, Spring 2004, pp. 29–43.

[11] Turnley, J. G., and J. Smrcka, *Terrorist Organizations and Criminal Street Gangs*, Report, Advanced Concept Group, Sandia National Laboratories, November 2002.

[12] Qin, J., et al., "Analyzing Terrorist Networks: A Case Study of the Global Salafi Jihad Network," *Proc. of IEEE International Conference on Intelligence and Security Informatics*, Atlanta, GA, May 2005, pp. 287–304.

[13] Stolfo, S. J., et al., "Behavior Profiling of Email," *Proc. of NSF/NIJ Symp. on Intelligence & Security Informatics*, 2003.

[14] Grishman, R., "Information Extraction," in R. Mitov, (ed.), *The Oxford Handbook of Computational Linguistics*, New York: Oxford University Press, 2003, pp. 545–759.

[15] Brown, D. E., J. Dalton, and H. Hoyle, "Spatial Forecast Methods for Terrorism Events in Urban Environments," *Proc. of the 2nd Symp. on ISI*, 2004, pp. 426–435.

[16] Klerks, P., "The Network Paradigm Applied to Criminal Organizations: Theoretical Nitpicking or a Relevant Doctrine for Investigators? Recent Developments in the Netherlands," *Connections*, Vol. 24, 2001, pp. 53–56.

[17] Krebs, V. E., "Mapping Networks of Terrorist Cells," *Connections*, Vol. 24, 2001, pp. 43–52.

[18] Sageman, M., *Understanding Terror Networks*, Philadelphia, PA: University of Pennsylvania Press, 2004.

[19] Skillicorn, D., *Social Network Analyses Via Matrix Decompositions: al Qaeda*, Report, http://www.cs.queensu.ca/home/skill/alqaeda.pdf, August 2004.

[20] Xu, J., and H. Chen, "CrimeNet Explorer: A Framework for Criminal Network Knowledge Discovery," *ACM Transactions on Information Systems*, Vol. 23, No. 2, 2005, pp. 201–226.

[21] Van Meeter, K. M., "Terrorists/Liberators: Researching and Dealing with Adversary Social Networks," *Connections*, Vol. 24, 2001, pp. 66–78.

[22] Freeman, L. C., "Centrality in Social Networks: Conceptual Clarification," *Social Networks*, No. 1, 1979, pp. 215–240.

[23] Monge, P. R., and N. S. Contractor, *Theories of Communication Networks*, New York: Oxford University Press, 2003.

[24] Baker, W. E., and R. R. Faulkner, "The Social Organization of Conspiracy: Illegal Networks in the Heavy Electrical Equipment Industry," *American Sociological Review*, No. 58, 1993, pp. 837–860.

[25] Dombroski, M. J., and K. M. Carley, "NETEST: Estimating a Terrorist Network's Structure," *CASOS 2002 Conference*, No. 8, 2002, pp. 235–241.

[26] Rothenberg, R., "From Whole Cloth: Making Up the Terrorist Network," *Connections*, Vol. 24, 2001, pp. 36–42.

[27] Weiser, B., "Captured Terrorist Manual Suggests Hijackers Did a Lot by the Book," *New York Times*, October 28, 2001.

[28] Hollywood, J., et al., "Connecting the Dots in Intelligence: Detecting Terrorist Threats in the Out-of-the-Ordinary," RAND brief, http://192.5.14.110/pubs/research_briefs/RB9079, 2005.

[29] http://www.i2inc.com/Products/Analysts_Notebook/default.asp.

[30] Dombroski, M. J., P. Fischbeck, and K. M. Carley, "Estimating the Shape of Covert Networks," *Proc. of the 8th International Command and Control Research and Technology Symposium*, National Defense War College, Washington, D.C., 2003.

[31] Coakley, T. P., *C3I: Issues of Command and Control*, Washington, D.C.: National Defense University, 1991, pp. 43–52.

[32] Militello, L. G., and Hutton, R. J. B. "Applied Cognitive Task Analysis (ACTA): A Practitioner's Toolkit for Understanding Cognitive Task Demands," *Ergonomics*, Vol. 41.11, 1998, pp. 1618–1641.
[33] Oliver, R. M., and J. Q. Smith, *Influence Diagrams, Belief Nets and Decision Analysis*, New York: John Wiley & Sons, 1990.
[34] Atkinson, S. R., and J. Moffat, *The Agile Organization: From Informal Networks to Complex Effects and Agility*, Washington, D.C.: CCRP Publications, 2005.
[35] Carley, K. M., and M. Prietula, (eds.), *Computational Organization Theory*, Hillsdale, NJ: Lawrence Erlbaum Associates, 1994.
[36] Keeney, R., and H. Raiffa, *Decisions with Multiple Objectives*, New York: Cambridge University Press, 1993.
[37] Swezey, R., and E. Salas, (eds.), *Teams: Their Training and Performance*, Norwood, NJ: Ablex Publishing Corporation, 1992.
[38] Wasserman, S., and K. Faust, *Social Network Analysis: Methods and Applications*, Cambridge, U.K.: Cambridge University Press, 1994.
[39] Boyd, J. R., "The Essence of Winning and Losing," *Presentation to the AF 2025 Study Group, Maxwell Air Force Base*, Montgomery, AL, Oct. 1995.
[40] Rabiner, L., "A Tutorial on Hidden Markov Models and Selected Applications in Speech Recognition," *Proc. of the IEEE*, Vol. 77, No. 2, 1989, pp. 257–286.
[41] Entin, E. E., F. J. Diedrich, and B. Rubineau, "Adaptive Communication Patterns in Different Organizational Structures," *Proc. of the Human Factors and Ergonomics Society 47th Annual Meeting*, Denver, CO, 2003.
[42] Luo, B., and E. R. Hanckock, "Structural Graph Matching Using the EM Algorithm and Singular Value Decomposition," *IEEE Trans. on Pattern Analyses and Machine Intelligence*, Vol. 23, No. 10, 2001, pp. 1120–1136.
[43] Wilson, R., and E. R. Hanckock, "Structural Matching by Discrete Relaxation," *IEEE Trans. on Pattern Analyses and Machine Intelligence*, Vol. 19, 1997, pp. 634–648.
[44] Krogh, A., et al., "Hidden Markov Models in Computational Biology: Applications to Protein Modeling," *Journal of Molecular Biology*, Vol. 235, 1994, pp. 1501–1531.
[45] Lühr, S., et al., "Recognition of Human Activity Through Hierarchical Stochastic Learning," *1st IEEE International Conference on Pervasive Computing and Communications*, Dallas, Fort Worth, TX, 2003.
[46] Kwon, J., and K. Murphy, *Modeling Freeway Traffic with Coupled HMMs*, Technical Report, University of California at Berkeley, May 2000.
[47] Rezek, I., and S. J. Roberts, "Estimation of Coupled Hidden Markov Models with Application to Biosignal Interaction Modeling," *Proc. of IEEE Int. Conf. on Neural Network for Signal Processing*, Vol. 2, 2000, pp. 804–813.
[48] Chen, B., and P. Willett, "Superimposed HMM Transient Detection Via Target Tracking Ideas," *IEEE Trans. on Aerospace and Electronic Systems*, Vol. 37, No. 1, 2001, pp. 946–956.
[49] Page, E., "Continuous Inspection Schemes," *Biometrika*, Vol. 41, 1954, pp. 100–115.
[50] Abrash, V., et al., "Incorporating Linguistic Features in a Hybrid HMM/MLP Speech Recognizer," *Proc. of International Conference on Acoustics, Speech, and Signal Processing*, Adelaide, Australia, 1994.
[51] Levchuk, G. M., and K. Chopra, "NetSTAR: Identification of Network Structure, Tasks, Activities, and Roles from Communications," *Proc. of the 10th International Command and Control Research and Technology Symposium*, McLean, VA, June 2005.
[52] Oliver, N. M., B. Rosario, and A. P. Pentland, "A Bayesian Computer Vision System for Modeling Human Interactions," *IEEE Trans. on Pattern Analysis and Machine Intelligence*, Vol. 22, No. 8, 2000, pp. 831–843.
[53] Meirina, C., et al., "Normative Framework and Computational Models for Simulating and Assessing Command and Control Processes," *Simulation Modeling Practice and Theory: Special Issue on Simulating Organizational Processes*, 2006.
[54] Simon, H. A., *The Sciences of the Artificial*, 3rd ed., Cambridge, MA: MIT Press, 1996.
[55] Maes, P., "Artificial Life Meets Entertainment: Life-Like Autonomous Agents," *Communications of the ACM*, Vol. 38, No.11, 1995, pp. 108–114.

[56] Wooldridge, M., and N. R. Jennings, "Intelligent Agents: Theory and Practice," *The Knowledge Engineering Review*, Vol. 10, No. 2, 1995, pp. 115–152.

[57] Prietula, M. J., K. M. Carley, and L. Gasser, "Computational Approach to Organizations and Organizing," in Prietula, M. J., K. M. Carley, and L. Gasser, (eds.), *Simulating Organizations: Computational Models of Institutions and Groups*, Cambridge, MA: MIT Press, 1998.

[58] Chaturvedi, A. R., and S. R. Mehta, "Simulation in Economics and Management," *Communications of the ACM*, Vol. 42, No. 3, 1999, pp. 60–61.

[59] Levchuk, G.M., et al., "Normative Design of Organizations—Part I: Mission Planning," *IEEE Trans. on Systems, Man and Cybernetics Part A*, Vol. 32, No. 3, 2002, pp. 346–359.

[60] Levchuk, G. M., et al., "Normative Design of Organizations—Part II: Organizational Structures," *IEEE Trans. on Systems, Man and Cybernetics Part A*, Vol. 37, No. 3, 2002, pp. 360–375.

[61] Popp, R., et al., "SPEYES: Sensing and Patrolling Enablers Yielding Effective SASO," *Proc. of the 2005 IEEE Aerospace Conference*, Big Sky, MT, 2005.

[62] Handley, H., Z. R. Zaidi, and A. H. Levis, "The Use of Simulation Models in Model Experimentation," *Proc. of 1999 Command and Control Research and Technology Symposium*, Naval War College, Newport, RI, 1999.

[63] http://www.swarm.org.

[64] http://www.agentisolutions.com.

[65] http://www.cougaar.org.

[66] Levchuk, G. M., and Y. Levchuk, "Identifying Command, Control and Communication Networks from Interactions and Activities Observations," *Proc. of the 2006 Command and Control Research and Technology Symp.*, San Diego, CA, June 2006.

[67] Levchuk, G. M., et al., "Congruence of Human Organizations and Missions: Theory Versus Data," *Proc. of the 2003 International Command and Control Research and Technology Symposium*, Washington, D.C., June 2003.

[68] Diedrich, F. J., et al., "When Do Organizations Need to Change (Part I): Coping with Incongruence," *Proc. of the 2003 International Command and Control Research and Technology Symposium*, Washington, D.C., 2003.

[69] Entin, E. E., et al., "When Do Organizations Need to Change (Part II): Incongruence in Action," *Proc. of the 2003 International Command and Control Research and Technology Symposium*, Washington, D.C., 2003.

[70] Entin, E. E., et al., "Inducing Adaptation in Organizations: Concept and Experiment Design," *Proc. of the 2004 Command and Control Research and Technology Symposium*, San Diego, CA, 2004.

[71] Kleinman, D. L., et al., "Scenario Design for the Empirical Testing of Organizational Congruence," *Proc. of the 2003 International Command and Control Research and Technology Symposium*, Washington, D.C., 2003.

[72] Kleinman, D. L., P. Young, and G. S. Higgins, "The DDD-III: A Tool for Empirical Research in Adaptive Organizations," *Proc. of the 1996 Command and Control Research and Technology Symposium*, Monterey, CA, June 1996.

CHAPTER 3
Who's Calling? Deriving Organization Structure from Communication Records

D. Andrew "Disco" Gerdes, Clark Glymour, and Joseph Ramsey

In the previous chapter, we explored the approach that generates multiple hypotheses of the structure and processes of the adversary's organization, estimates the characteristics of the observable events and communications based on each hypothesis, and then compares the actual observables with the predictions. In order to generate the hypotheses, the approach requires a library of organizational structures and missions. Creating and validating such a library is a labor-intensive task. Besides, there is a danger that the library is biased toward known organizational and procedural patterns, and will produce misleading results when applied to an organization with significant novel features. Clearly, it would be desirable to find an approach that allows us to infer the structure of an opponent's organization without the burden of creating, validating, and trusting such a library. Such approaches do exist, particularly with respect to analysis of communications intercepts, and we discuss some in this chapter. Naturally, there has to be a price to pay for avoiding the effort to build the library: such approaches may yield less information and apply to narrower subclasses of problems.

Furthermore, data recording and use for search for hostile actors can potentially misidentify a significant number of innocents. Depending on the place and conditions under which such monitoring occurs, it can cause a broad range of legal, diplomatic, and ethical concerns. Accordingly, it seems to us of considerable importance to investigate how far, and by what means, potentially threatening groups and cells can be identified from time-stamped communications while minimizing both the number of persons whose communications are monitored for content and perhaps also minimizing, for those communications that are monitored, the specificity with which they are examined. This chapter addresses aspects of these questions.

After considering why conventional social network analysis does not help with the problem, we use two simulated, unpublished but unclassified, databases of terrorist activities created by the Department of Defense [1] to show how simple algorithms enable the statistical separation of such command, control, and reporting networks from among a much larger population of other communicators, using only communicator identities (e.g., phone numbers) and time of communication. Of course, in a large society, not all or even most such networks will be hostile; our aim is merely to investigate filters that eliminate most communicators and communications, leaving a residual in which command, control, and reporting networks, including hostile networks, form a high proportion. If a hostile agent is known, the procedures can be used to help separate those with whom the agent conspires from those who have innocent contact with him or have innocent contact with those who have contact with him. We emphasize that social statistics is not magic, and the procedures we describe still carry a significant cost in false positives.

The Tasks

This chapter concerns the following tasks:

1. There is available a large body of communication logs, in each of which the communicators are identified, the time of the communication is recorded, and text is recorded. The meanings of the texts are assumed to be unknown because of coding or language, although syntactic features can be identified, and it is unknown which of the communicators initiated a communication. Among these communicators there are one or more hostile groups (hereafter known as *perpetrators*) who operate with a quasi-military command, control, and reporting (CCR) structure connected with plans, while others are simply normal communications among people—including some or all of the people in CCR groups. The task is to provide an algorithm that, with this information and nothing more, will separate the CCR groups and their members from everybody else, as well as possible, and to estimate the command and reporting relationships among the individuals in a group thus identified.
2. The same criteria as with the first item are used, but using minimal information about the content of messages.
3. The same data is used, along with the correct identification of a set of target communicators—known perpetrators—to identify those in CCR relationships with this initial set.

The tasks are impossible to fulfill completely. We can only hope for a computerized procedure that identifies some of the relevant hostile groups and their members (perpetrators), while excluding most of the others (hereafter known as *innocents*). We describe such procedures and their results when applied to data from military and quasi-military scenarios synthesized under the auspices of the Department of Defense.

CCR Structures and Their Graphical Representation

Organizations take on a variety of forms, and each group has a unique dynamic. However, groups are frequently organized in a similar manner. Groups most often have their leaders and their followers. In business we can observe a president-vice president-manager structure (as in Figure 3.1), and in military organizations, a general-captain-sergeant structure. We refer to these as CCR structures.

CCR Structures and Their Graphical Representation 65

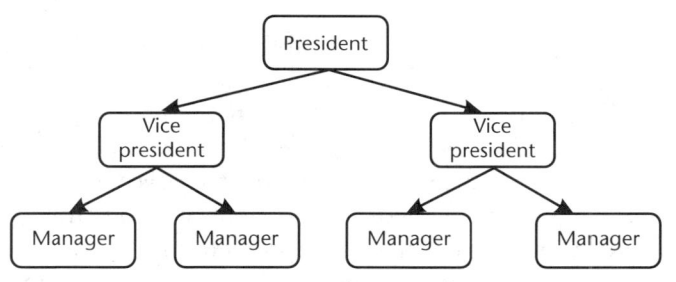

Figure 3.1 A common structure from business defined by top-down communication.

We do *not* assume that there must always be a unique top-most commander or that subordinates always report to a single commander. We of course allow that a population may have multiple such groups functioning independently, or in a flat communication hierarchy, or forming a super hierarchy and follower roles. These roles will in turn affect the communication within the group, which is ultimately what we are concerned with.

As in Figure 3.1, CCR structures are directed graphs whose vertices or nodes represent individuals or groups and in which a directed edge indicates either a relation in which the node at the tail passes orders to the node at the head of the edge or the tail node reports to the head node.

When discussing CCR structures, or social structures of any kind, graphs can act as an aid to the understanding and visualization. Visually, we can quickly identify patterns and structure that would be difficult to discover algorithmically, and that might be less obvious to us if the information were presented in another manner. In the graphic representation of social structures, the nodes of the graph represent actors in the network. Lines connecting nodes in a graph represent a relationship of some kind. In some cases, a line would indicate "knows," as in "A knows B" (again, see Figure 3.1). A line connecting two actors in CCR structures indicates communication between those two actors. Arrows represent the inferred direction of that communication between the actors of the network. In the case of command and control structures, such communication is interpreted as commands, requests, orders, or simply the transfer of knowledge (e.g., from professor to student). In the case of reporting structures, communication is interpreted as reports from subordinates to leaders. Whether a given network's structure is said to be a command or control structure or a reporting structure is based on the analysis of the resulting structure, as will emerge in the discussion.

Social networks have long been represented by undirected graphs, and such networks are easily produced from a body of communication logs. It is worth noting that the relations they specify are not transitive and specify no direction. If A communicates with B, and B communicates with C, it does not follow that A communicates with C; it may be that B was the source of both communications, or that B received a message from A and from B, or that B received a message from A but communicated with C about something else entirely. Undirected graphical communication networks provide very limited information about communication chains, whether of commands or of reports. It is further worth noting that such networks contain no temporal information; if A is linked to B, and B to C, the undirected graph provides no information about the time relations between A-to-B communications and B-to-C communications.

We will graphically represent CCR relations using directed, possibly cyclic, graphs in which a directed edge between A and B indicates the claim that A sends a command or report to B, and chains of directed edges sometimes indicate transitivity of command authority or reporting relations and sometimes do not. Special markings are required to provide a representation of the different transitive or nontransitive cases. Edges in both directions are allowed between two persons (e.g., if A reports information to B, and B likewise to A, or if A commands B, and B sends reports to A—see Figure 3.2). Edges can be marked C, for commands, or R for reports, or, as we will usually do, left unmarked if unknown.

In a social network, authority figures tend to have influence over a greater number of actors than do other actors in the network. Influence, in this case, need not be a direct relationship but rather may transfer through other actors in the network. As a result, the structural characteristics of such authority figures tend to make them central nodes in the network (i.e., we should expect that nodes representing authority figures tend to be the source or the sink of much communication).

The Ali Baba Scenarios

The first fictitious scenario consists of approximately 800 synthesized unclassified documents that replicate intelligence reporting on suspected terrorist activity in southern England. The documents tell the story of the planning activities of the main terrorist cell of the fictitious Ali Baba organization that took place over several months in 2003, as well as a good amount of "chatter." The documents consist of police reports, Foreign Broadcast Information System reports, tactical reports (local communication interceptions and activity observations), interrogation reports, and newspaper articles. Of the 800 documents, approximately 75 were those that reported directly on the terror-related activities. Several hundred of the messages are communications and activities of peripheral members, including friends and family of the terrorist cell. The remaining messages are innocent, but possibly interesting, chatter coming predominantly from university campuses in Cambridge and Oxford [1]. An important relevant feature is that the scenario involves a single active cell, with a leader but no other organizational structure that would be expected to be revealed in communication network structure.

The second scenario of the Ali Baba data set is similar in nature to the first but is larger and updated to more accurately depict a realistic (yet synthesized) government database on terrorist and suspected terrorist activity. This data set is composed of more than 2,000 documents: 2,000 tactical reports and 133 reports of shipments and activities at precisely identified geographic locations. Because it is a more extensive data set, its network of actors is also much larger than in the first scenario. The scenario describes the activities of the Ali Baba organization in the fictional country

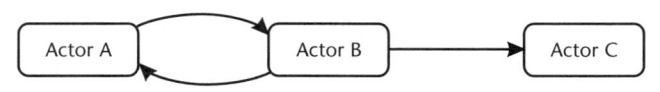

Figure 3.2 A CCR structure between actors A, B, and C.

Pavlakistan. The organization plans to blow up the Pavwan dam. The second scenario also includes plans of another organization to blow up a bridge. The communications between more than 1,000 actors are described in the data set. In both scenarios, communications of perpetrators use informal codes (e.g., the term *birthday cake* might refer to a bomb).

Given a day, or perhaps less, a competent analyst culling through the raw data, or spread sheets listing who communicates with whom on what dates on what topics, can form a reasonable conjecture about scenario 1. Scenario 2 is more difficult, but the possibility of such inferences makes it extremely important that, in using such data to test algorithms, the computerized procedure be entirely "hands off." That is, while an algorithm might have its parameters tuned or optimized using one data set, the data set used for testing should be entirely distinct, and no tuning—data deletion, prior knowledge, parameter adjustments, and so on—should be allowed. Data in, predictions out. With that caveat, the Ali Baba data sets form the best unclassified test case we know of.

Social Network Analysis

SNA focuses on the patterns of *relationships* between *actors* and examines the structure of these observed actors in the network [2, 3]. The field relates very closely to graph theory, although it does remain heavily grounded in its sociological roots. Relationships in a network are established by exchanges of both tangible items (e.g., goods, services, and money) and intangibles (e.g., information, support, orders, or influence). This focus on patterns of relationships separates SNA from other techniques of social analysis. Because the terms *relationship* and *interaction* are not rigorously defined in the field, the kinds of interactions that constitute a relationship are left up to the individual researcher. Working on social research from a networking approach gives the researcher formal definitions for aspects of the social structural environment. The formal definitions of impressions and terms also allows for increased understanding and communication within the field.

Much of the data collected in social network research is gathered through questionnaires, interviews, and direct observation. Questions such as "Whom do you work with?" and "Who are your friends?" as well as "To whom do you go for advice?" are posed to participants in a study. Documents and publications may also be reviewed to determine relationships. In more recent years, with the rising use of e-mail and more technically advanced telephone systems, companies can maintain better records of who is interacting with whom. Some recent research has gathered e-mail records to model the social networks within organizations. Such an analysis better reflects actual rather than perceived connections, as a relationship is established by documented interaction. These techniques of data collection are less available to the researcher when performing studies outside of the business world, as such thorough logs are not often kept. When studying organizations outside of business, other data collection techniques must be employed. The Ali Baba data set, for example, simulates intercepted communications and observations of meetings. This provides a good basis for building a social network to be evaluated. What such data sets often lack, however, is information about the direction of the relationships.

Network analysis arose as a method to describe systems of many elements that are related in some way. The elements of the system may or may not hold a relationship with the other elements. Social network analysis views groups of people as such a system. The individual has classically been the focus of attention for psychologists, traditional sociologists, human resource managers, and researchers, among others. While social network analysis does take the individual into account, its primary focus is the relations between the actors being studied. It should be noted that even though social network analysis is most often used for the evaluation of people, social networks are not limited to people. The techniques can be used to describe anything that can have a relationship with something else. For example, Johnson and Krempel used network analysis to measure the relationship between the various actors in the Bush administration and words uttered in the two months following 9/11 [4].

Many graph theoretic concepts and techniques can be applied to social network analysis and to network analysis in general. A subset of these concepts is frequently used to describe social networks. For the sake of clarity, the terms one will encounter in reading this chapter are defined as follows:

- *Adjacency*: Two nodes or actors connected by a line or relational tie are said to be adjacent.
- *Subgraph or subnetwork*: A subgraph or subnetwork of a graph or network, respectively, is a subset of the nodes or actors and all of the ties connecting the members.
- *Degree*: The number of ties incident at a given node is the degree of the node.
- *Density*: The number of ties that exist in a graph divided by the number of possible ties in the graph.
- *Path*: A path is a series of ties that connect two nodes in a network that does not cross the same node more than once.
- *Geodesic*: This is the shortest path between two nodes.
- *Clique*: A maximal set of actors in which every actor has a relational tie to every other actor. Cliques have the properties of maximum density (1.0), minimum distances between actors (1), and *can* be overlapping—that is, an actor can reside in two cliques simultaneously (e.g., an actor can be in his family-clique and his workgroup-clique). See Figure 3.3.

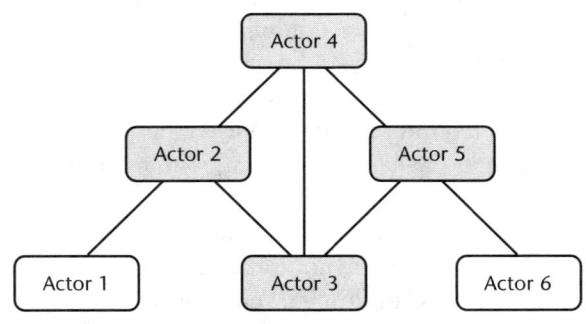

Figure 3.3 An example of two cliques in a social network. Actors 2, 3, and 4 form one clique and actors 3, 4, and 5 form another clique.

Social Network Analysis

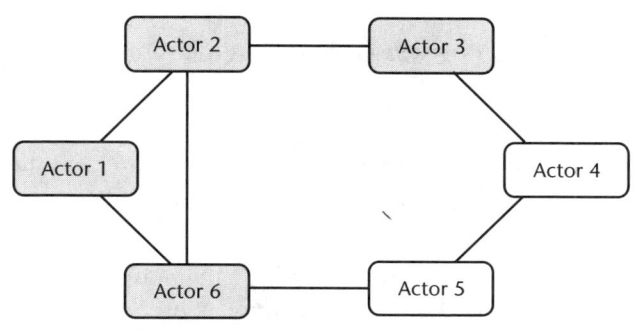

Figure 3.4 An example of a 2-clique in a social network. Actors 1, 2, 3, and 5 form a 2-clique.

- *n-clique*: An *n*-clique describes a set of actors that are within distance *n* of each other. This is a relaxation of the distance and density requirements of cliques. Note that a 1-clique is simply a clique. See Figure 3.4.

The field of graph theory has many more concepts, but these areused most frequently to describe social networks and to make simple evaluations of them. Because of how well they describe social networks, some terms from graph theory have crept into the vocabulary of the layperson. The term *clique* is often used to describe small groups of closely associated groups of people, for example.

Graphs are probably the clearest way to view smaller social groups (up to 50 or 100 actors), but it becomes increasingly complex to spot patterns without the aid of calculation as the number of actors goes up. Of course, the complete graph does not need to be drawn in order to perform calculations on the network; a representation, such as an adjacency or association matrix or edge list, is sufficient.

One of the primary uses of graph theory in network analysis is the attempt to identify the significant actors within a social network. The notion of significant actors can be taken to mean different things, and their significance is interpreted differently by each researcher. These significant actors are usually taken to have some level of importance or prominence within the network. The measure of importance is derived solely from an actor's location within the network. An actor who is located in a strategically advantageous position in a network is regarded to be the more important or prominent than an actor is a less strategically advantageous position. The concepts that allow us to locate such characters are reasonably simple to grasp, although the calculations themselves can be rather challenging. Centrality measures are the graph theoretic calculations that, with some amount of interpretation, allow an analyst to make conclusions about which actors can be regarded as important.

Centrality is a concept based on the ideas of degree, closeness, betweenness, and rank of the nodes in a network. A central actor is one involved in many ties. There are two major bases for measuring the centrality of a node: path based and walk based (see Figure 3.5):

1. Path-based methods include:
    - *Degree centrality* is based on the number of nodes adjacent to a given node. The node with the highest degree centrality is that which has the most neighbors.

- *Closeness centrality* is based on the sum of geodesic distances to all other nodes in the network. It can be seen as an inverse measure of centrality ($c_i = \Sigma_j d_{ij}$).
- *Betweenness centrality* can be approximately defined as the frequency with which a node lies along the path between two other nodes ($b_k = \Sigma_{ij} g_{ikj}/g_{ij}$).

2. Walk-based methods, decidedly much more challenging, include among others the *eigenvector centrality*, an iterative version of degree centrality (i.e., the centrality of a node is measured in proportion to the sum of the centralities to which it is tied).

As one can see in Figure 3.5, each centrality measure returns a different result for the question "Which node has the highest centrality?" In some networks, the different measures of centrality may return the same node, but this network was designed in order to demonstrate the possibility for a perceived discrepancy. Each measure returns a different result because the calculation takes different features into account. This allows the analyst to find the most appropriate measure for what she is trying to learn.

Which measure of centrality gives us the best predictor of prominence is dependent upon how the analyst defines prominence and what he or she is hoping to learn. Degree centrality helps the analyst locate the most active nodes. Without information such as edge weights, the best measure of activity is the degree of the actor. An actor of high degree centrality is certainly among the most visible in a network. This actor is likely to be recognized by others as a major medium for information and is likely to be considered a significant hub. Actors with relatively lower degree centrality can be viewed as peripheral and, to some degree, unimportant. A similar calculation can be done for subgroups within the network. For example, the degree centrality for a given clique can be measured [3].

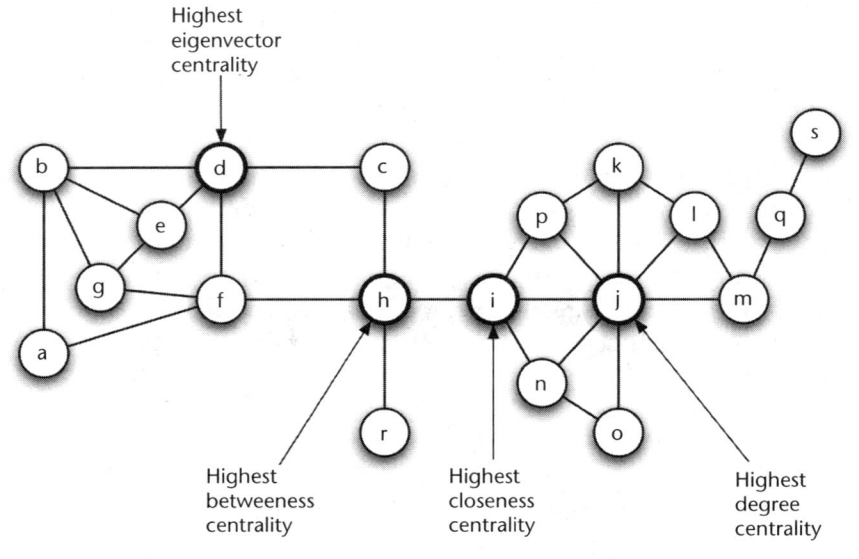

Figure 3.5 The result of different approaches to calculating centrality.

The closeness centrality measure allows for a slightly different interpretation of importance. With this measure, importance is gauged by the ability of an actor to readily contact others. An actor with the "highest" (or best) closeness centrality needs to rely on fewer actors to relay information. With the highest closeness centrality, an actor is theoretically capable of getting more done more quickly, which puts the actor in a position of some prominence.

The measure of betweenness centrality gives the analyst a look at who is involved in the most communication channels. An actor with a high betweenness centrality measure in a communication network is entrusted with more information than any other actor. In a basic social network, an actor of high betweenness centrality is able to control the interactions of more individuals and is capable of bringing other actors together or keeping them apart. As it relates to covert networks (terrorist or criminal organizations), the removal of an actor with high betweenness centrality could be quite damaging to plan formation, relationships, and activity [3].

The interpretation of the walk-based measure of eigenvector centrality is similar to that of degree centrality but takes more into account than just immediate neighbors. If lasting, cascading influence is more significant than immediate influence in the network the analyst is studying, then perhaps eigenvector centrality would be the best indication for prominence in the network.

The different methods of measuring centrality and, by interpretation, importance gives the analyst some powerful tools for learning some key information about the nature of a social network. However, when applied to large groups about which little is known, how useful is the social network analysis toolkit?

We initially pursued social network analysis to learn about the Ali Baba data set because it provided us an obvious starting point for sorting through the large number of documents and even larger number of actors (see Figure 3.6). Disregarding actors and dyads not connected to a larger network allowed us to eliminate those actors unlikely to be involved in any plots. The identification of cliques and actors central to the network and subnetworks allowed us to further narrow the scope of investigation. We found that classic social network analysis methods positively identified many of the actors named in the ground truth as known perpetrators, but also identified many innocents. Social network analysis lacked methods for determining more about the structure of a network, namely, the directionality. Undirected networks are quite simple to create from communication logs; it can be assumed that two actors engaged in communication have a relationship of some type—but not much more.

Previous Work Exploiting Time in Social Networks

Our procedures rest on a simple idea about the communications of people involved in a command and control network who also communicate with others who are not so involved (spouses, relatives, friends). Certain communication probabilities reflect the chain of command. The idea is that *if person C is subordinate to person B, who is subordinate to person A, then it should be more probable that B communicates with C shortly after A and C have communicated.* Reporting relationships work the same way. The hard work is to make this idea precise and measurable and of use in a computable search procedure.

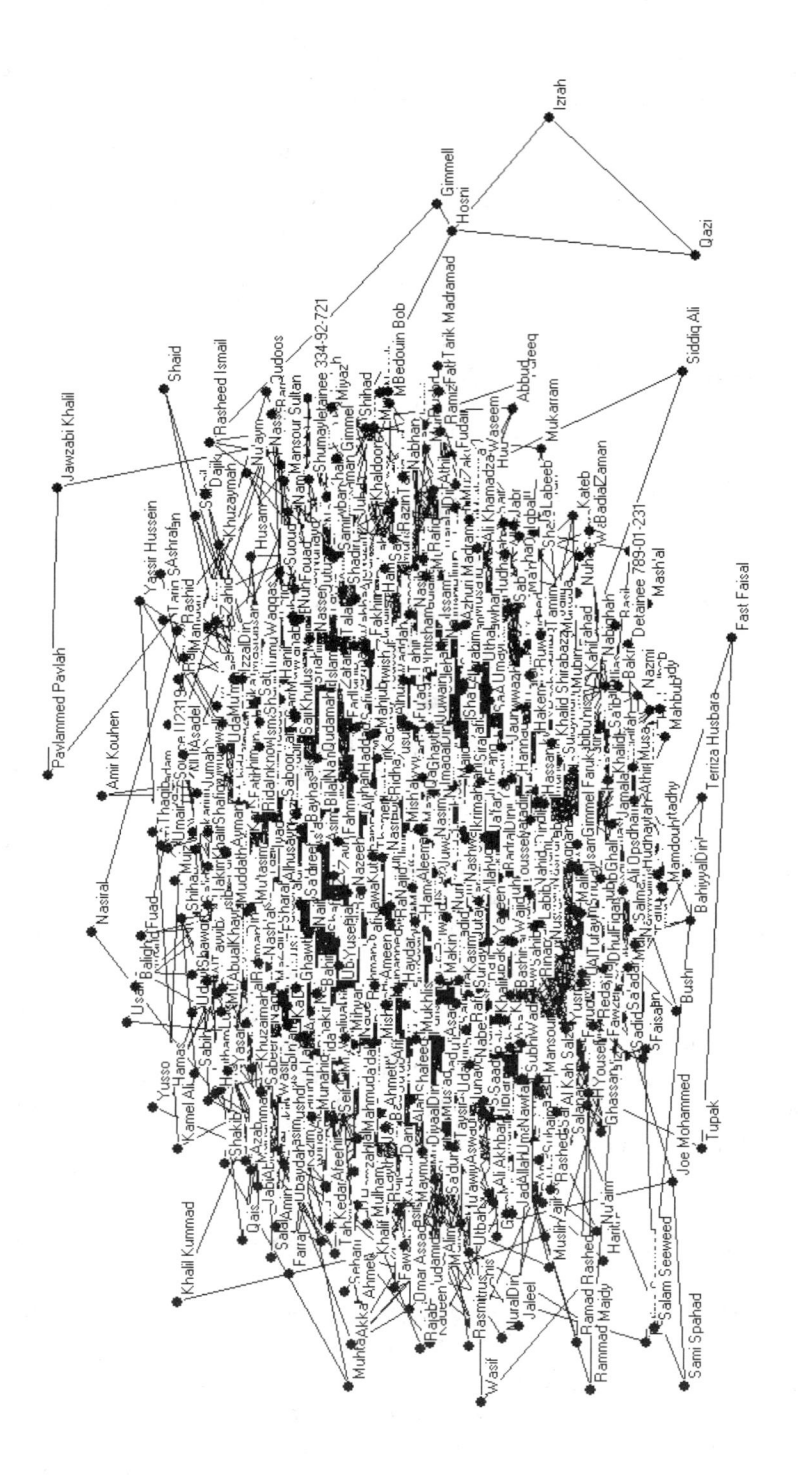

Figure 3.6 A total of 832 actors in the second Ali Baba dataset after a trim in which actors connected to zero or one other actor are removed.

Malik Magdon-Ismail, a researcher at Rensselaer Polytechnic Institute, proposed the idea of *streaming*. In streaming, the social network is grown as communication continues over time [5]. However, he provided no algorithms or attempts at solutions.

Other groups have thought to use the time-series nature of data to learn what they could about social networks. But most research taking time into account has focused on the evolution of social networks and their subgroups and cliques. Some analyses take a central actor from the network and then grow the network around the individual. Research based on surveys often takes a similar "snowballing" approach, though not from time-series data. Much like observing crystal growth, the analyst can observe how an actor's relationships and, in turn, realm of influence grow over time. Some research has gone into calculating the probability that a given network will arrive at a certain state at some time. By observing the dynamic formation of networks, the researcher can gain insight into the choices actors make in the creation of and severing of ties and the effect these decisions make on the structure of the network and therefore the opportunities presented to the actor [6].

Recent research on the observation of groups over time has focused on the persistence of subgroups and cliques within networks. In light of the focus on terror networks, researchers hope to uncover hidden subgroups within larger networks. Subgroups that maintain connection patterns tighter than can be expected to randomly occur and with greater persistence than the average within the group are considered to be part of a group that is "up to something" [7]. Some success has been noted in the unveiling of such hidden networks, at least with simulated data. Baumes et al. [7] first employed HMMs in order to discover hidden groups.

The focus of many social network researchers has been the Internet. In part, this focus is due to the availability of data. Because of the time-ordered nature of blogs, discussion forums, and e-mail, it is relatively easy for the researcher to collect data on link creation at a single point in time, rather than having to observe a series of websites over time to observe when new links are created. The algorithms developed by researchers sought to evaluate whether networks grew in a bursty way, as Kleinberg described [8]. The term *bursty* refers to periods of increased activity; in the case of network growth, it is periods of noticeably higher than average link creation. Some amount of directionality can be, and certainly is, tracked by this and similar research. However, because the data is hypertext link–based, directionality is already included.

While this research is certainly interesting and perhaps important to the field, it still lacks the discovery of directionality that we seek. The only research that seems to touch on algorithms similar to ours comes from motif discovery in genetics and from link analysis.

The Windowing Down the Lines (WDL): The Algorithm

We will describe the more complex of our algorithms in some detail. The fundamental idea is that of conditional frequencies, mentioned earlier, but implementation details are somewhat complicated, and the complications matter.

The Scenario

The algorithm for command structure identification is based on scenarios similar to the following:

> Actor Y calls many people. Actor Y only calls Actor X with instructions to "deliver a cake." Actor Y makes many such calls to Actor X over time. Actor X calls many people, but only calls Actor Z after receiving a call from Actor Y; Actor X gives Actor Z instructions to pick up the cake at place A and deliver it to Actor W at place B. Shortly after, Actor Z calls Actor W to report that the "cake will be delivered to place Y." Actor Z never calls Actor W otherwise. This sequence is repeated many times, amid many other irrelevant calls involving Actors W, X, Y, and Z separately. In some calls, part or all of the content topic is missing. The time of communication is always known. Figure 3.7 is a representation of the resulting command structure.

Within a data set, many such command structure chains may exist. The union of these smaller graphs is a tree or web defining the command structure of an entire social network.

The Data

The data we use for the algorithm for command structure identification is in the form of communication logs. Any number of actors can be involved in the network as a whole, but the algorithm is designed on the assumption that each conversation has two known participants. Each of n topics is recorded, and, finally, the date-time of communication is noted. An example communication log can be seen in Table 3.1.

Formatting the Data

In order to calculate command structures, the algorithm takes a list of vectors L of the type:

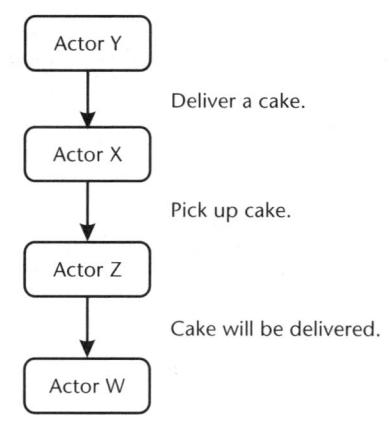

$$<\text{Actor } A, \text{Actor } B, \text{Topic } 1, \ldots, \text{Topic } n, \text{time } I>$$

Figure 3.7 The command structure of the example scenario in which Actors W, X, Y, and Z discuss the delivery of cakes.

Table 3.1 Sample Communication Log Data As Used by the Command Structure Identification Algorithm

First Associate	Second Associate	First Topic	Second Topic	Date-Time
Actor 1	Actor 3	Topic 4	Topic 3	2005-01-01T03:36:57
Actor 5	Actor 3	Topic 4	Topic 5	2005-01-01T09:56:54
Actor 3	Actor 6	Topic 0	Topic 6	2005-01-01T13:34:04
Actor 6	Actor 1	Topic 1	Topic 5	2005-01-01T23:53:06
Actor 2	Actor 9	Topic 5	Topic 3	2005-01-02T05:25:51

The vector represents the communication occurrence at time t between Actor A and Actor B during which the topics topic 1 through topic n were discussed.

In realistic scenarios, we know that the topics of communication are not always known or cannot be easily identified. Therefore, in the development of this algorithm, we took that into account and assume that the topic of communication is not always known.

The vectors of L are organized for calculation into sets of a different structure at different stages during the algorithm: $C(X)$ and $T(X)$:

- $C(X)$ is the set of all vectors in which Actor X is involved in the communication with any other actor.
- $T(X)$ is the set of all time-ordered pairs of vectors from $C(X)$ in which Actor X communicates with some Actor Y at time t and some Actor Z at time t', where Actor $Y \neq$ Actor Z and time t occurs before time t'. $T(X)$ is organized as pairs of vectors:

<{Actor Y, Actor X}, Topics, t >, < {Actor X, Actor Z }, Topics, t >

The time ordering of the algorithm is based on a time-windowing or time-binning system. The data is organized into time windows of a user-specified length, which is identified as t in the algorithm.

Time Windowing

Initially in designing the algorithm, we allowed for a dynamic calculation of the time spacing of communication. Each $Y \rightarrow X \rightarrow Z$ was calculated by looking back a certain amount of time, e, from each {X, Z} communication to see if a {Y, X} communication had taken place. While this seems like a good approach, consistent with the intuition of how the algorithm works, it causes the run time of the algorithm to increase drastically. Additionally, the time-windowing method provides the calculation with the denominator required to determine frequency. In a typical command structure, it is expected, and often is the case, that actor dyads lower on the chain of command communicate more frequently than dyads higher up. Much of this communication often occurs within a tight frame of time but, with the dynamic calculation of time differences, it appears as though more communication is taking place without command influence than really is. This problem is corrected to some degree with time windowing, an idea based on the binning often used in data mining (see Figure 3.8). More intuitively, the motivation for time-windowing communication comes from the observation that bursts of communication often occur between actors in a network. Bursts of communication come in many forms, but they come frequently in the form of an overzealous boss seeking results or an insecure subordinate seeking clarification of an assignment. When calculated in a purely dynamic way, such bursts invalidate the calculation of command chains.

Time windowing takes all communication between a given start and end time and presumes all communication is taking place concurrently. When more than one communication takes place between any pair {Actor X, Actor Y} within a given time window, the algorithm discounts such bursts. Viewing both multiple and single communications as the same equates the overzealous boss and the boss that is able to communicate all necessary information at once.

The start time for the time windowing is the first time that appears in the data set. All time windows are calculated from that time until the end time, defined as the last date appearing in the data. The first time window is bounded by the start time and the length of time e from the start time. The second time window is begins where the first ends, and so on. When running the algorithm, only time windows t and t' are considered for the calculation of a given command chain. That is, for each {X,Z} communication occurring in time window t', only {Y,X} communication from the immediately preceding time window t is considered.

In order to completely understand the motivation for time windowing, we review the model in more detail. Referring to Figure 3.9, it can be seen that the dyad {Actor 3, Actor 6} communicates as part of two directed triads: Actor 1 → Actor 3 → Actor 6 and Actor 3 → Actor 6 → Actor 7. This is a similar situation to the earlier one. The dyad {Actor 3, Actor 6} will exhibit more communication than either the dyads {Actor 1, Actor 3} or {Actor 6, Actor 7}. Without time windowing, both directed triads are found quite unreliably (i.e., only with specific data sets with the right time window and probability thresholds specified). With time windowing, the directed triads are found quite reliably, especially with the addition of time overlapping, as reviewed in the following section.

Referring to Figure 3.9 again, dyads like {Actor 1, Actor 2} present some trouble with the use of dynamic time calculation. According to the assumptions of the command structure identification algorithm, it should not be able to find this dyad nor the directionality of its tie. Because of placement in time, certain data sets calculated with certain parameters identified this dyad as part of a triad with incorrect

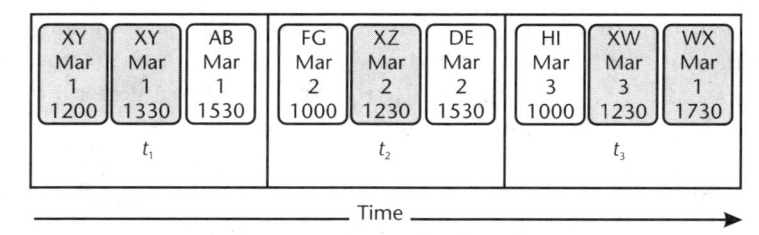

Figure 3.8 Logged communication separated into time windows. In this case, the communications are binned by days. Notice that more than one communication between a pair of actors can occur within the same time window.

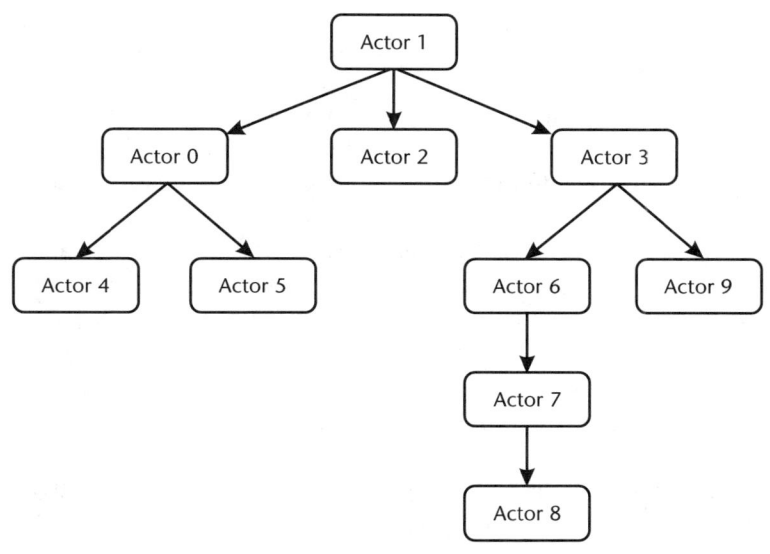

Figure 3.9 A command structure model.

directionality or as part of a triad that did not even exist. With the use of time windowing, it is quite rare for the algorithm to find dyads that it is not meant to be able to discover.

While reducing the appearance of being truly dynamic, making use of time windowing instead of calculating time differences between individual communication records has been shown in our testing to produce results closer to the known model or ground truth. Additionally, by using time windowing, a significant amount of computation time is saved as compared to the more dynamic approach.

A drawback to using time windowing is a certain amount of information loss. If communication between two dyads of a directed triad always occurs in a very tight window, the command chain might not be discovered. To compensate for this, we introduced a method of time window overlapping.

Time Overlapping

With the time windowing arrangement, however, there is a certain amount of information lost. For example, if all {Y,X} communication preceding {X,Z} communication occurred within a time frame much shorter than the average, it is possible that the relationship would not be detected, though it is possibly important. It would be prudent to test overlapping time windows as well. With the testing of overlapping time windows, less information would be lost.

To test overlapping time windows, a few definitions had to be established and modifications needed to be made. First, a well-defined time-window overlap counting method was created. Time-window overlappings are counted as the number of time windowings being tested. Therefore, testing one time window overlapping is the same as testing a single time windowing. Second, a definition of a *time shift* needed to be established. A time shift is the amount of time offset from the start time of the first time windowing. The length of a time shift is calculated as the duration of the time window divided by the number of overlaps being tested. For example, with a time window of one day specified and two overlaps being tested, the time shift would be 12 hours. Testing three overlaps, the resulting time shift would be 8 hours, and so on. When testing the first time windowing, the start time defines the beginning of the first time window, and the second time window begins after an amount of time, e, the user-defined time window, from the start time. In testing the next overlapping, the first time window again begins at the start time. The second time window, however, starts after the time shift is added to the start time. The first time windowing has n time windows, each subsequent windowing will have $n + 1$ windows, allowing for the shorter first and $n + 1$th window. A visualization of this can be seen in Figure 3.10.

This method of time shifting is based on the intuition that testing time shifts greater or less than the time window divided by the number of overlaps results in either testing the same set of time windows twice, not testing all possibilities, or both. This presents an inefficiency and imbalance in the calculation of the final graph, as will soon become clear. Testing with more randomized time shifts presents issues involved in random testing that we are not prepared to handle at this time, although this might be an area worth exploration. A visualization of the time windowing overlap can be seen in Figure 3.11.

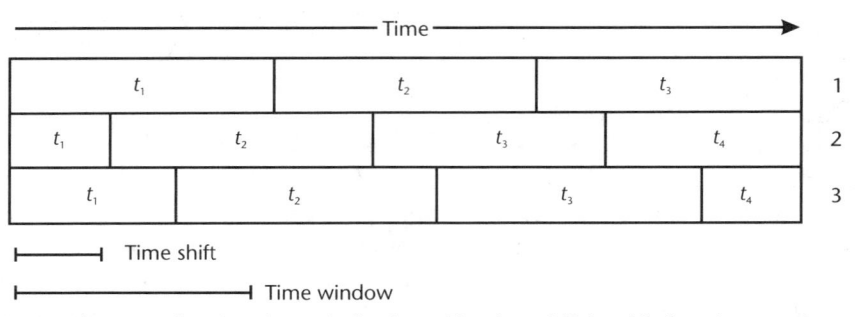

Figure 3.10 Three overlapping time windowings. The time shift is added to the start time to define the starting time for the second window. In subsequent time windowings, the time shift is added to the second window start time from the previous windowing.

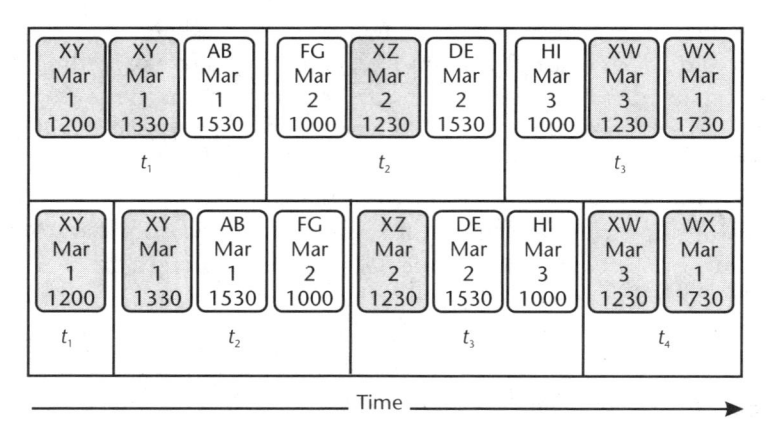

Figure 3.11 Data separated into two overlapping time window sets and the resulting groupings of events. In this example, a time window of one day and a time shift of 12 hours have been utilized.

There is a noticeable difference between the results of running the algorithm with one time windowing and with the results of a two overlap run, like that diagramed in Figure 3.11. However, with each added overlapping, the results did not become markedly different. In repeated tests on a variety of data sets, there seemed to be no advantage to calculating more than six overlaps; however, fewer than four often gave unreliable results.

In order to compute probabilities using overlapping time windows, the time shift is first calculated based on the time-window duration and number of overlaps specified by the user. Then the probabilities for the time windowing are calculated. The probability value for each possible triad $X - Y - Z$ is stored. For the next overlapping, the time shift is added to the start time. The time window from the original start time to this point is considered the first time window, and each subsequent time window is calculated using the original time window duration. Then the algorithm is run again and values of each triad added to the previous value. This is repeated for each time window overlapping that needs to be specified.

In order to calculate the triads with time overlaps, the algorithm takes the many different probability values into consideration. The probability values for each triad $X - Y - Z$ are summed and then divided by the number of overlaps performed. This resulting value is compared to d, the probability threshold value. If the value is indeed greater than *d*, the triad is added to the graph.

The idea of testing overlapping time windows improved performance of the algorithm and satisfied our desires to have our algorithm's results describe the network as accurately as possible. With time-window overlapping, the algorithm can catch relationships that it would otherwise not see with just straight time windowing.

With all of the foundations laid, we now cover the command structure identification algorithm.

The Algorithm

The basic algorithm for identifying command structures is as follows (e is the user specified time window and d is a user specified probability threshold):

1. Graph G = empty.
2. Actors = enumeration of all actors in L, a list of all vectors representing all logged communications.
3. For each Actor X in Actors:
   If the probability of <{X, Z}, *t'*> given <{Y,X}, *t* > minus the probability of {X,Z} in any time window is greater than the probability threshold *d*:
   Then add the directed triad Y ? X ? Z to G.
4. Return G.

Intuitively, in order to calculate the ordering $Y \rightarrow X \rightarrow Z$, we need to learn if X and Z communicating is at all related to the communication of Y and X. We can calculate this by determining whether it is more likely for X to communicate with Z following a communication between Y and X than for X and Z to communicate in general. We calculate this using probabilities that are defined in terms of frequencies. First, we need to calculate the probability that X and Z communicate in a time window, given that Y and X communicate in the previous time window. This is expressed in the algorithm as:

The probability of $<\{X, Z\}, t'>$ given $<\{Y,X\}, t>$

This probability calculation is made as follows: for each time window t', if X and Z communicate and X and Y communicate in the previous time window t, add 1. Then, divide by the number of time windows in which X and Y communicate.

To balance out the calculation, we must ensure that the probability that the relationship exists is greater than the chance that the relationship doesn't exist. This is expressed in the previous algorithm as:

The probability of $\{X, Z\}$ in any time window.

In order to do this, we calculate the likelihood that X and Z will communicate in a given time window. This value is simply the number of time windows in which X and Z communicate divided by the total number of time windows.

The difference between these two probability calculations is compared to the probability threshold d. If the difference is greater than the probability threshold, the directed triad is added to the graph.

With an understanding of the basic notion of the algorithm, the full algorithm can be presented:

1. Graph G = empty.
2. For each time window overlapping:
   - L, a list of all vectors representing all logged communications.
     - Compute time windows.
     - Assign time windows to each vector in L.
   - A, an enumeration of all actors in L.
   - For each Actor X in A:
     - Calculate $C(X)$, the set of all vectors in which Actor X communicated with any other actor.
     - Calculate $T(X)$, the set of pairs of vectors from $C(X)$ in which Actor X communicates with some Actor Y in time t and any other Actor Z at time t'.
   - For each possible triad $Y \rightarrow X \rightarrow Z$,
     - For each time window t', if X and Z communicate and X and Y communicate in t, add 1. Then, divide by the number of time windows in which X and Y communicate.
     - From the previous value, subtract the number of time windows in which X and Z communicate divided by the number of time windows.
     - Store value.
3. For each possible $Y \rightarrow X \rightarrow Z$:
   - If the average probability difference value is greater than the probability threshold d:

- Then add $Y \to X \to Z$ to G.
4. Return G.

Simple Simulation Tests

Generally speaking, when using test data that is created with the algorithm it is designed to test in mind, it is not rare to see the expected results. Such data are usually generated in order to test the thing or set of things that an algorithm or system is designed to deal with. Algorithms and systems are not often designed to try to deal with the whole world at once. Accordingly, data generated based on well-defined models can hardly be expected to accurately reflect complete real-world scenarios. What testing on generated data does provide, though, is a basis for ensuring that one's algorithm or system is performing calculations as expected and can at least deal with some amount of noise.

A model containing noise was created. Noise in the model takes on the form of dyads that communicate with one another but not a third actor. The algorithm should disregard or simply not be able to detect these links.

The data that we generated in order to test our algorithm was primarily designed to test the ability of the algorithm to discover a command structure despite a moderate amount of noise. Noise in the generated data provided a representation of the "chatter" that might occur in real-world scenarios. While the noise we generated was not quite as extensive as real-world chatter, we suspected our algorithm would deal with the noise fairly well. For example, dyads that only talk between themselves are disregarded. We found later with the tests on the Ali Baba data sets that when the algorithm was not able to completely disregard sets of actors, it did find several separate command structures.

With this in mind when testing the command structure identification algorithm with generated data, it was not very surprising to find that it was able to reliably discover the command structures we expected to see. After ensuring that the proper calculations were being performed, the task of learning what kinds of parameters worked best came next. As mentioned earlier in this chapter, we found after several tests that in most cases there was little difference between using six time-window overlaps and any subsequent higher number of time-window overlaps. Using fewer than four overlaps, especially with data reasonably evenly spaced as our model data, often provided unreliable results. This discovery allowed us to begin to adjust the other parameters with a fixed number of time-window overlaps.

We then set the probability threshold very low in order to learn what range of time windowings would discover the command chains sought. In the model data sets, communication between dyads in a command chain occurs in succession with a specified average separation time. Because of this fact, it did not take long to realize that the best time windowing for finding all of the command chain triads was a length of time slightly longer than the average separation time.

Having found both reasonable numbers of time-window overlaps and time-window lengths, we tested possibilities of reasonable probability thresholds. With this parameter, it was hard to establish what would constitute a believable probability threshold. In motif discovery algorithms, a probability threshold as low as 0.001 is often used. That is, there is a 1 in 1,000 more of a chance that a sequence is caused by another sequence than the sequence simply randomly occurring.

Not surprisingly, this seemed a little low for our purposes, so we tested with values in the 0.10–0.50 range. With probability threshold values greater than 0.30, edges were rarely found, depending on the data set used. With almost every data set, all of the directed triads that conformed to the model and that the algorithm was able to find had probability values of between 0.15 and 0.30. Therefore, at least when testing the models, we determined that 0.15 is the best probability threshold. With the probability threshold of 0.15, 90 percent of the known actors of a command structure were identified as part of that command structure.

After achieving satisfactory results and with a better understanding of what parameters to use when calculating a command structure graph, we determined that it would be appropriate to test the algorithm with more realistic data.

Evaluation of WDL with the Ali Baba Datasets

We evaluated the Ali Baba dataset using the algorithm in two ways. The first way disregarded the topics of discussion. The second way took the topics of conversation into consideration when calculating the graph.

Without Topics

The results of running the command structure identification algorithm on the Ali Baba data set were promising. Depending on the metric used to measure the level of success of a specific set of parameters, different directions were taken. If the numbers of positively identified, false-positively identified, and nonidentified perpetrating actors are taken as the sole measures of success, it is not entirely clear how well the algorithm is performing. In both Ali Baba scenarios, the numbers of known perpetrators found are always lower than the numbers of actors falsely identified. However, as noted earlier, the algorithm is not so much designed to find which suspects are more likely to be perpetrators as it is designed to determine how actors within a network are communicating. The best measure of results, therefore, is to inspect the resulting graphs to note the groupings. We found that the command structure identification algorithm grouped actors together fairly well, as can be seen in Figure 3.12.

The result of running the command structure identification algorithm on the second Ali Baba scenario data set with the parameters—12-hour time window, three time-window overlappings, and a 0.35 probability threshold—demonstrates the ability of the algorithm to discover command structures reliably. In evaluating the groupings discovered, each grouping that contained known perpetrators was composed of at least 50 percent known perpetrators. The groupings in the results also match the groupings provided in the ground truth (i.e., the cliques found were composed of actors known to be involved in the same cliques). Of all the known perpetrators that the algorithm discovered, five were identified as leaders in the ground truth. There are nine leaders named in the scenario for each of the nine subgroups. The algorithm placed the five known leaders higher in the command chains than any

Evaluation of WDL with the Ali Baba Datasets 83

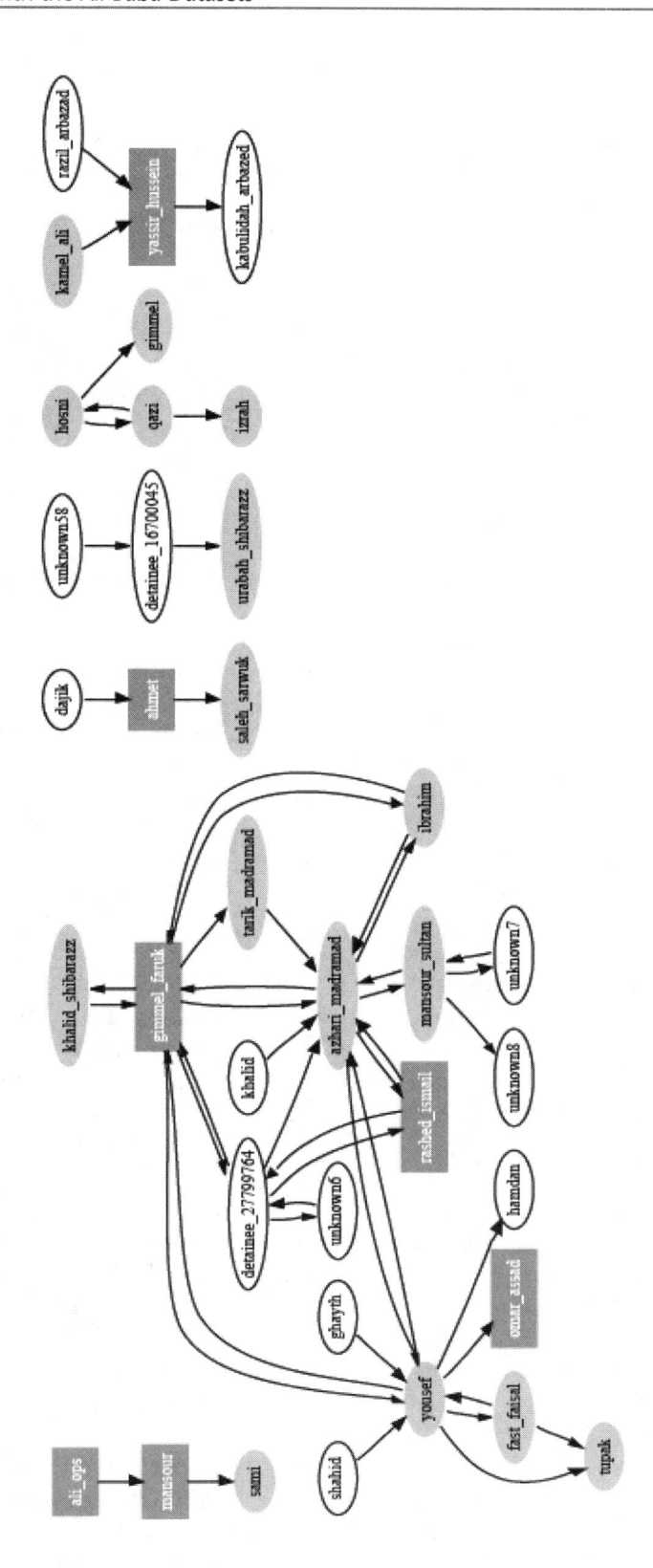

Figure 3.12 This figure focuses on the identified actors who are known to be perpetrators according to the ground truth. Gray, rectangular nodes are known to be leaders, and light gray, elliptical nodes are known to be general perpetrators.

other actor whose role is known. This alone is encouraging for the prospect of discovering command structures.

With results in hand, we proceed to evaluate the success of the command structure identification algorithm and the prospects of using it for realistic applications.

With Topics

In order to attain more accurate results, we felt it prudent to explore the inclusion of topics in our search. To do this, we added a few additional steps to the algorithm along with a few additional parameters. Our attempt to include the topic of discussion as a requirement proved to be simple to implement, does not require knowledge of the language of discussion, and produced useful results.

In addition to the steps laid out earlier, we include the creation of a list of list of vectors W that includes all words discussed in the corpus and the number of unique conversations in which each appears:

<word, number of conversations in which word appears>

Once the vector is constructed, we remove those words from W that appear in more than a user-specified percentage f of the conversations.

f = number of times the word appears in the corpus divided by the number of conversations in the corpus

Then, we remove the words from each vector L that no longer appear in W. When calculating triads, we store an additional probability value. In each $T(X)$ pair, the number of intersecting words is divided by the average number of nonremoved words in the $\{X,Y\}$ and $\{X,Z\}$ conversations. In calculating the graph, the average of the stored probabilities for each triad is compared to a user threshold h. If the average exceeds h in addition to the other criteria, then the triad is added to the graph.

The results of using topics were indeed promising, as can be observed in Figure 3.13. Within a few trials, it was quickly observed that the frequency of less interesting chains appearing was reduced. Fewer known perpetrators appear in the resulting graph. Some of this is attributable to the fact that the topics used in these trials were taken from the spreadsheets with manually extracted topics rather than from the complete documents.

The procedure with topics finds 33 percent of the perpetrators in the Ali Baba scenario, with a false positive rate of less than 4 percent of the initial search population; innocents were about two-thirds of the individuals named in the WDL output using topics. Put another way, the chance that a random member of the initial population is a bad guy is about 6 percent; the chance that a random member of the WDL selected population is a bad guy is 37 percent. In principle, the procedure is language independent, as long as strings and communicators can be reidentified across communications. However, some knowledge of the language helps in identifying topics, generalizations, and synonymies. The WDL output graphical structure gives some indication of the organizational structure within each cell or group, although we have not assessed that accuracy quantitatively. A directed path in the graph does not necessarily indicate a communication chain: A may call B, and B call C, while X calls

Evaluation of WDL with the Ali Baba Datasets

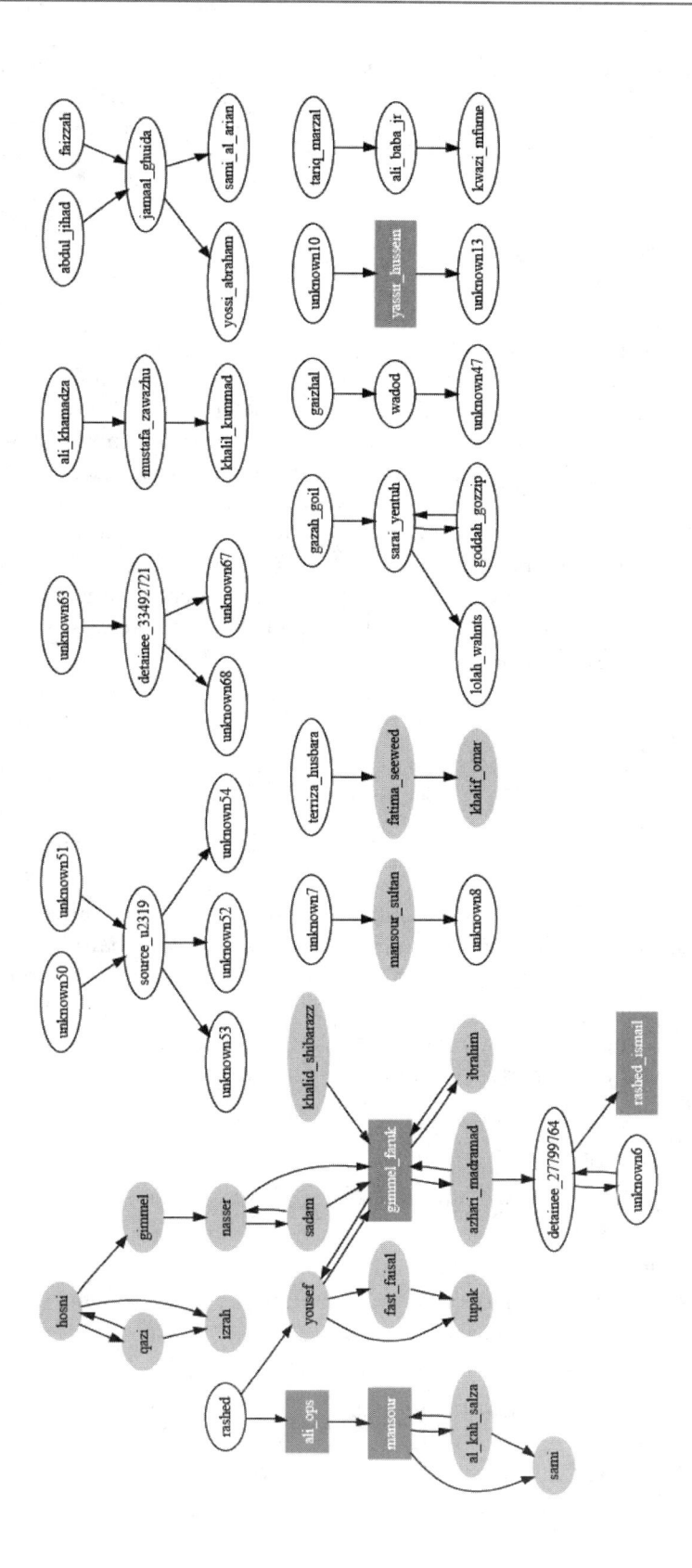

Figure 3.13 Results of using the command structure identification algorithm on the second Ali Baba scenario data with the inclusion of topics. Gray, rectangular nodes represent known leaders; light gray, elliptical nodes represent known perpetrators; and plain, elliptical nodes are actors whose status is unknown.

B, and B calls Y, so that there is an X to B to C directed path but not a communication chain. The information identifying communication chains is, however, in the output of the algorithm and can be visually presented interactively (e.g., by allowing the user to select any two nodes and highlighting the communication paths between them).

The RSF Algorithm

Rather than focusing on changes in probability of communications of a type, one might use time series focused on frequency of a communication series involving at least three actors. One such algorithm, the Ramsey Spy Finder (RSF), has been implemented and tested on the Ali Baba data by Joseph Ramsey [9] with interesting results. The essential idea is that one counts how often the sequence "A calls B, then B calls C" occurs, within any very long time bound (Ramsey used 150 days for Ali Baba). The most frequent such chains (Ramsey used less than the highest 1 percent) are then entered into a directed graph. Actors and links D such that D calls A, then A calls B, or B calls C, and then C calls D—that is, actors and links that extend the communication chains but occur with much lower frequency—are then added. The algorithm is described next.

Parameters:
Min width w_1
Max width w_2
seed threshold s
extension threshold e
topic threshold p_t

The algorithm is as follows, where C is an Ali Baba (or other) data set, r_i is the i'th record in C, T is a map from triads to lists of record pairs, S and R are sets of record pairs, and G is a graph. Also, t1 is the maximum time interval considered between r_1 and r_2 for any record pair $<r_1, r_2>$, t_2 is minimum number of record pairs needed to support adding a triad to the seed set, and t_3 is the minimum number of record pairs needed to support adding a triad to the extension set.

RSF(D, w_1, w_2, s, e, p_t)

1. L ← ∅
2. T <− count-word-frequencies(C, p_t)
3. **for** i = 1 to |C|
4.   **for** j = i + 1 to |C|
5.   **for** each triad <a, b, c> such that a, b are actors in r_i,
6.   b, c are actors in r_j, a ≠ c, and w_1 < time(r_j) − time(r_i) < w_2
7.   **if** <a, b, c> is not in domain(T)
8.   L(<a, b, c>) ← .
9.   L(<a, b, c>) ← L(<a, b, c>) + <r_i, r_j>
10. S ← ∅
11. **for** each <a, b, c> in domain(L)

12.   **if** |L(<a, b, c>)| > s
13.   S ← S ∪ {<a, b, c>}
14.   R ← domain(L) \ S
15.   Sort R in nonincreasing order of |L(<a, b, c>)| for <a, b, c> in R.
16.   **while** S changes
17.   **for** i = 1 to |R|
18.   **if** R[i] = <a, b, c> is of length at least e
19.   **if** <b, c, d> or <d, a, b> exists in S for some d
20.   S ← S ∪ {R[i]}
21.   R ← R \ {R[i]}
22.   G ← ∅
23.   **for** each <a, b, c> in S
24.   Add a→b→c to G
25.   **return** G.

Note that in RSF, one can vary the following parameters independently of one another:

- Maximum interval;
- Seed and extend thresholds.

RSF uses the procedure count-word-frequencies described next. C is a list of conversations; f is the maximum ratio of word occurrence to conversations allowed.

Count-word-frequencies(C, f)

1. u ← ∅
2. **for** each c ∈ C
3.   **for** each topic of c
4.   S ← words(c)
5.   **for** each word s
6.   u(s) ← u(s) + 1
7. **for** each word s in domain(u)
8.   **if** u(s) / |C| > f
9.   Delete s from u.
10. **return** n

With parameters set to optimize results from Ali Baba I, when run on Ali Baba II without topic information, RSF discovers 17 perpetrators with 25 false positives. Adding topic information actually gives worse performance, identifying only 11 perpetrators [9].

References

[1] Jaworowski, M., and S. Pavlak, "The Ali Baba Data Set Background Description and Ground Truth," unpublished, 2003.

[2] Scott, J., *Social Network Analysis: A Handbook*, Cambridge, U.K.: Sage Publications, 2000.

[3] Wasserman, S., and K. Faust, *Social Network Analysis: Methods and Applications*, Cambridge, U.K.: Cambridge University Press, 1994.

[4] Johnson, J., and L. Krempel, "Network Visualization: The 'Bush Team' in Reuters News Ticker 9/11–11/15/01," *Journal of Social Science*, 2004, pp. 221–235.

[5] Magdon-Ismail, M., "Modeling and Analysis of Dynamic Social Communication Networks," *Decision Sciences and Engineering Systems Seminar*, Rensselaer Polytechnic Institute, November 8, 2005, http://www.cs.rpi.edu/~magdon.

[6] Dutta, B., and M. O. Jackson, *Models of the Strategic Formation of Networks and Groups*, New York: Springer-Verlag, 2002.

[7] Baumes, J., et al., "Discovering Hidden Groups in Communication Networks," *2nd NSF/NIJ Symposium on Intelligence and Security Informatics (ISI 04)*, Tucson, AZ, June 11–12, 2004.

[8] Kleinberg, J., "Bursty and Hierarchical Structure in Streams," *Proc. of 8th ACM SIGKDD International Conference on Knowledge Discovery and Data Mining*, 2002.

[9] Ramsey, J., *Algorithms for Command and Control Graphs from Communication Logs*, Technical Report, Laboratory for Symbolic Computation, Department of Philosophy, Carnegie Mellon University, 2006.

CHAPTER 4

Means and Ways: Practical Approaches to Impact Adversary Decision-Making Processes

Ed Waltz

Having discovered the information about an adversary organization, such as we have seen in the previous chapter with respect to the Ali Baba organization, what would one do with the information? One possibility would be to use it for the purposes of further, more detailed collection of communications between the Ali Baba members. This could help determine their future planned actions and take defensive measures. Another would be to use the information to impact the performance and behavior of the organization, and that is in fact the focus of our book. But how would one actually produce such an impact? In this chapter, we outline a number of measures used in practice to cause a variety of effects on organizational performance and behavior.

The needs and means to target and adversely impact organizations are not new to human conflicts. In particular, warfare has *always* targeted the perception, decision-making performance, operational effectiveness, and ultimately the will of the organizations of national governments, their security services, and militaries. The advent of ubiquitous communication, computation, and sensing used by adversary organizations has enabled new methods and mechanisms to influence these targets—providing access to human decision-makers at all levels of the organization through the sensing and information systems that provide perception, the communication nets that enable collaboration and shared awareness, and the networks that distribute organization intent and commands. These channels and the dependencies of organization on the information they provide offer the *potential* for more sophisticated technical means to impact adversary organizations.

Operations to impact an adversary governments, for example, have traditionally been conducted by covert operations that targeted organizations within government, seeking to destabilize or overthrow them while influencing their popular support. Such large-scale counterorganization political-military operations often included activities to undermine the performance of the organizations of government while degrading popular support to ultimately overthrow the government [1]. The planning of activities for such organizational operations included the orchestration of four lines of effort:

- *Defection*: Recruit, train, and establish individuals within the organization and the supporting population capable of conducting resistance and dissent activities leading up to, if required, a *coup d'etat*.
- *Division*: Conduct psychological operations to disrupt the unity of purpose, discipline, and agreement between government bodies and between the government and population groups.
- *Deception*: Protect the true intentions of the counterorganization operations by operational security while revealing selected opposition activities and simulating false activities that cause the target to believe a different adversary plan and approach is being implemented.
- *Diversion*: Divert or misdirect attention from the true sources of the opposition; secure third-party support to divert attention from the primary source of the attack.

The advent of communication and information processing networks (and the increasing dependence of all categories of organizations on these to function, including governments, military and paramilitary units, and terrorist groups) has enabled much more refined methods of access to organizations and entire social populations at all levels of their network structures. This and subsequent chapters focus on the practical means to impact organizations using organizational intelligence (gathered by means described in earlier chapters), and the planning and access means required.

Earlier in Chapter 1, we introduced the role of intelligence in collecting information about organizations to model their structure and behavior in each of these areas. In this chapter we introduce the practical means to conceive, evaluate, and carry out operations to influence organization using this knowledge. The first section introduces the general planning process for creating impacts on organizations by targeting decision-makers at all levels of the organization or entire social processes. Specific methods to induce effects in organizations, based on decision-making and organizational vulnerabilities and pathologies, are introduced in the following section. The final sections develop the targeting process and describe the rationale for using dynamic organization models to perform effects-based analysis and the methods to model organization behavior and assess the potential effects (desired, undesired, and unintended consequences) of candidate actions. Where Chapter 1 was oriented toward the organizational intelligence analyst, this and subsequent chapters are oriented toward the operations planner who will use the intelligence to bring about change in a targeted organization.

Planning Operations for Organization Impact

The pioneering work of RAND researchers John Arquilla and David Ronfeldt on the potential effects of global networks introduced concepts of new forms of competitions among organizations by exploiting the impact of attacks on the networks of people via their networked information systems. In *Networks and Netwars*, they identified five levels of abstraction and practice at which these new forms of competition will occur [2]. At the *organizational level*, people are related in both formal and informal structures that make up the organization design. The *social level* describes the informal relationships not found on formal organization charts: personal ties, shared values, common experiences, and culture that give rise to confidence and trust. At the *narrative level*, the organization has a mission that defines its behavior—it is the story that is being told by the organization. The *doctrinal level* describes the policies and strategies (decision-making, information sharing, collaboration, synchronization, and the like) that guide the operation of the organization. The *technological* level describes the information systems and physical infrastructure that support the organization. Each of these levels must be considered by the operations planner seeking to influence an organization. Notice that the first four levels correspond to the cognitive domain of the organization introduced in Chapter 1, and the technology level corresponds to the domains of abstract information and elements of the physical world.

Planning to create impacts on adversary organizations across all of these domains must first consider the desired intermediate effects and end states (final effects) and then identify the set of feasible means to achieve those ends. Classical attrition-based warfare focuses on reducing the capability (and therefore, ultimately, the will) of the adversary by combat operations that attrite the adversary's human forces and their weapons (e.g., "killing people and breaking things"). By contrast, organization warfare impact planning generally falls in the domain of information operations that focus on attrition of the will by other means, including a combination of physical and nonphysical mechanisms to [3] "deter, discourage, dissuade and direct an adversary, thereby disrupting [the adversary's] unity of command and purpose while preserving our own."

The alternative military operations (means of impact) across the cognitive, information, and physical domains and competitive areas defined by RAND researchers (Table 4.1) encompass both physical and nonphysical means that are directed at the information and the minds of the actors in the organization.

The focus of operational planning is the defeat of the adversary organization, targeting centers of gravity of the organization by creating effects in the mental and emotional states of members of the organization to change operational capability. These nonphysical centers of gravity are an interacting set of mental faculties that, although not a linear causal chain, can be ordered in a general decision-making sequence:

Table 4.1 Means of Attack for Impact

Domain of Attack	Competition Areas	Typical Operations
Cognitive-emotive	Organizational: Formal organization structures Social: Informal, cultural relationships Doctrinal: Process, procedural Narrative: Mission explicitly articulated	Psychological operations (PSYOP) Military deception (MILDEC)
Information	Technological: Abstract information; tacit and explicit knowledge	Computer network operations (CNO) Electronic operations (EW)
Physical	Technological: Information, support, and combat systems; the physical bodies of people	Physical combat operations (kinetic) Electronic operations (directed energy)

- *Perception*: This is the awareness of the evolving situation at three levels: perception of the elements of the current situation, comprehension of the meaning of the current situation, and projection of the future status, as a function of alternative actions or inaction [4].
- *Emotion*: This is the subjective psychological-physiological experience expressed in terms of feelings (e.g., courage, fear, joy, sadness, love, hate, pleasure, and pain); it is a response to the perception of the current situational state of mind.
- *Cognition*: This is understanding and reasoning applied to a decision calculus, including a range of factors: cognitive capabilities, beliefs, cost-benefit, risk, and other decision criteria.
- *Will*: This is the conscious choice, decision, commitment, and intention to act.
- *Relation*: This is the properties of connections to others in the group (e.g., group cohesion or agreement).

Consider two distinct and complementary approaches to targeting the organization for impact (Figure 4.1), one focused on the critical nodes of the organization (human decision-makers, automated processes making decisions) and the other focused on the entire organization as a social entity. One method achieves success by knowledge of the critical causal links from decision to operational effects; the other achieves success by knowledge of the behavior of the entire organization acting as a fully interconnected organism. The methods are not mutually exclusive or competitive; they are complementary and strongly related. Subsequent chapters, in fact, do not differentiate between these two approaches. Social influence methods can set the

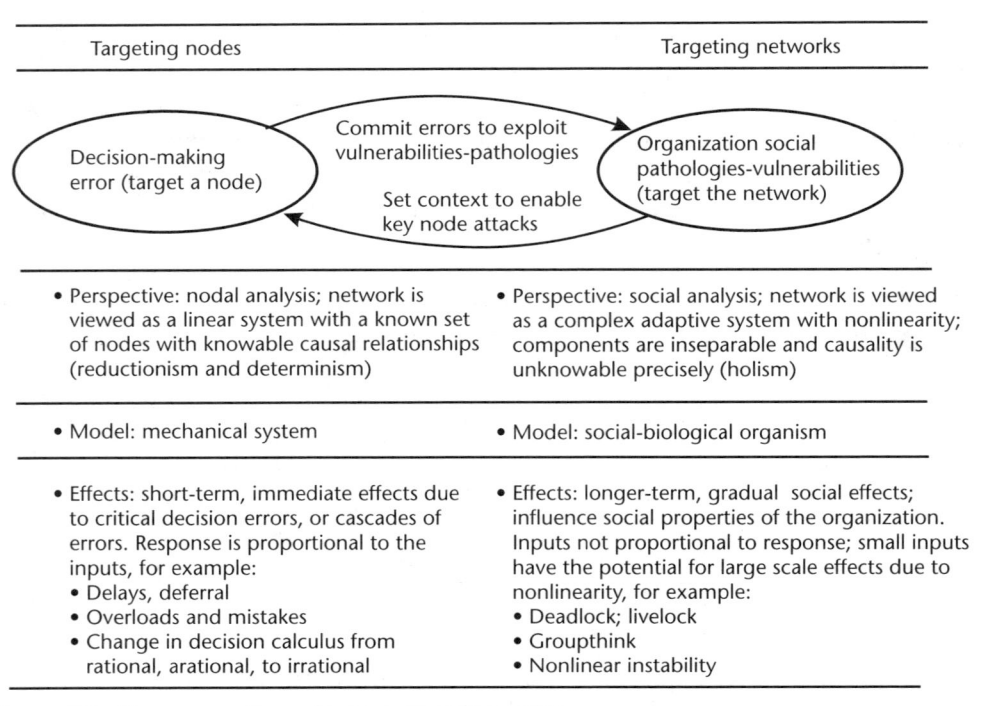

Targeting nodes	Targeting networks
Decision-making error (target a node) ⇄ Commit errors to exploit vulnerabilities-pathologies / Set context to enable key node attacks ⇄ Organization social pathologies-vulnerabilities (target the network)	
• Perspective: nodal analysis; network is viewed as a linear system with a known set of nodes with knowable causal relationships (reductionism and determinism)	• Perspective: social analysis; network is viewed as a complex adaptive system with nonlinearity; components are inseparable and causality is unknowable precisely (holism)
• Model: mechanical system	• Model: social-biological organism

• Effects: short-term, immediate effects due to critical decision errors, or cascades of errors. Response is proportional to the inputs, for example:
  • Delays, deferral
  • Overloads and mistakes
  • Change in decision calculus from rational, arational, to irrational
	• Effects: longer-term, gradual social effects; influence social properties of the organization. Inputs not proportional to response; small inputs have the potential for large scale effects due to nonlinearity, for example:
	  • Deadlock; livelock
	  • Groupthink
	  • Nonlinear instability

Figure 4.1 Two perspectives of organizational targeting.

context in which critical decisions are made, and individual decisions have networkwide influences.

Targeting Decision-Making

The *nodal* approach focuses targeting on critical nodes that can individually or collectively change the behavior (performance and effectiveness) of the organization. As introduced earlier in Chapter 1, critical nodes may have high betweenness, centrality, prestige, or other network properties that are evidence of their relatively high influence within the network. The nodes of attack may be humans (e.g., decision-makers at all levels, computer administrators, or network dispatchers), nodes of their supporting information networks, or nodes of essential physical supporting infrastructure (e.g., electrical power generation or communication switching). The objectives of nodal attacks are generally direct (denial, disruption, or destruction), the decision effects sought are discrete (delay, confusion, or error), and operational effects are short term (failure to act or error in acting). Degrading the decision-making performance of one critical individual, a portion of, or an entire organization is the principal avenue of attack. The planner seeks to influence the decision-making process by the perception of decision-makers—current or future leaders and their supporters. (Note that even when an adversary leader is removed, the influence is on the perception of those that take over leadership and subordinates who may perceive an uncertain future [5].)

The process of managing the perception of an adversary is defined as the actions to convey and/or deny selected information and indicators to targeted audiences to influence their emotions, motives, and objective reasoning as well as to intelligence systems and leaders at all levels to influence official estimates, ultimately resulting in opponent's behaviors and actions favorable to the originator's objectives. In various ways, perception management combines truth projection, operations security, cover and deception, and psychological operations [6].

We can define six conceptual levels of decision performance degradation (Table 4.2) to illustrate the increasing reduction of the of a target's decision-making capability relative to an objective. With confidence in the information available provided (by humans or technical information systems) and perceiving that the truth is known, a human can be decisive, making an optimal choice from among alternatives (the decision that yields the maximum utility for a given objective being pursued). As an attacker degrades the information available (by reduction, overload, or manipulation), the human loses confidence in the information, due to unavailability, uncertainty in content, apparent errors, or conflicts; these factors cause decisions to be delayed (deferred). Under these circumstances, the target may focus attention on incorrect information, waste time reviewing an overload of data, become confused, and ultimately be forced to choose between alternatives without confidence in their decisions. The sixth level of degraded decision-making occurs when a deceiver has effectively presented sufficient information to create a false perception, and the human places confidence (incorrectly) in the false alternative over the true; the result is an optimally incorrect decision—the target makes a decision chosen by the attacker.

Table 4.2 Levels of Decision Performance

Decision Performance	Confidence	Perception
Correct and accurate: knowledge of the truth; optimal and timely choice among alternatives	Trust in information	Belief in the truth; accurate knowledge
Decisive, but suboptimal choices		
Indecisive (delayed, deferred decision)	Uncertainty, ambiguity, and distrust in information	Increasing degrees of uncertainty about the truth; unreliable information
Distracted, believing in false sufficiently to pay attention		
Misdirected to focus on the false as reality		
Deceived: choosing the decision directed by the attacker	Trust in information	Belief in the false; false knowledge

Impacting the rational elements of decision-making is not the only means to degrade performance or the will to perform—avenues to the emotion may also be employed. Sticha et al. have demonstrated an analytic method to represent the influence of psychological and situational variables in predicting the behavior of decision-makers using Bayesian networks to provide a probabilistic prediction of a leader's decisions. The Bayesian network provides a means for the analyst to explicitly represent estimates of the psychological and situational variables and explore sensitivities of predicted decisions to estimates of personality and leadership factors [7].

Targeting the Entire Organization

While decision-making targeting generally focuses on the nodes (decision-makers) of an organization, the entire organization network may also be targeted in a holistic manner to create a shift in the overall dynamics by gradual and distributed changes across the organization network. This approach views the target as a large-scale organism, in which all nodes are critical and highly interrelated. Subtle shifts in the entire organization are sought, rather than sharp, immediate effects. In contrast to the nodal approach, the objectives of network attacks are generally indirect (deception or subtle influence of many nodes), the decision effects sought are more continuous (organization perception shift, gradually modified social behavior, or conditioning to selected stimuli or situations), and operational effects are over a longer term (change in policies, reduction in confidence, or desensitization). The recent applied research in business process reengineering and organizational change management has focused on such organizationwide social change. For example, methods for the analysis of organizational trust networks have been reported that apply social network analysis and simulation tools to identify organizational vulnerabilities and explore remedies [8]. In order to carry out organizational change, intelligence on the political dynamics and social context of the organization is essential. The U.S. Commission studying the intelligence available on Iraq in 2003 noted that an understanding of the political dynamics of the Iraqi government was an important context-setting factor that was necessary for analyzing its critical national decision-making, yet was lacking. The Commission noted [9]: "But the failure even to *consider* how the political dynamics in Iraq might have affected Saddam's decisions about his WMD programs was a serious shortcoming that resulted in an incomplete analytical picture."

We provide the distinction in of *targeting nodes* and *targeting networks* to emphasize two complementary perspectives rather than two competing theories. Subsequent chapters in this text illustrate the nodal approach that brings rapid and dramatic effects by introducing a precise combination and sequence of impacts of several or multiple nodes to cause broad-ranging propagation of effects, often rapid and drastic, through the entire organization. Large-scale organizations, including entire social populations, however, pose great challenges to the nodal approach, where critical nodes are not known, network dynamics are not understood, and the scale of interactions between actors prohibit a deterministic model useful for nodal attack. In these cases, the networkwide targeting perspective is appropriate, and exploratory computational social science models are useful to understand possible networkwide shaping strategies.

Effects-Based Targeting

The process of implementing and integrating both targeting approaches (nodal or holistic) applies operational *actions* (methods or mechanisms) across access *channels* to *targets (*nodes or large portions of the organization*)* to achieve specific organization *effects*. This process, illustrated in Figure 4.2, requires the planner to identify each specific element of a candidate impact plan:

- *Actions* coordinate a variety of operational, technical, and psychological methods to perform physical events, create and transmit information, and conduct technical operations over channels of access to the target. Actions can be further distinguished by their desired effect:
    - *Probes* are exploratory actions intended to elicit a response from the targeted organizational network to gain or refine knowledge about its structure, behavior, strengths, or vulnerabilities. Sun Tzu referred to probing actions when he directed the military commander to [10]: "Rouse him, and learn the principle of his activity or inactivity ... Force him to reveal himself, so as to find out his vulnerable spots."
    - *Attacks* are actions taken against an organizational network to achieve an operational end state, usually characterized by organizational coercion to cooperation, subornation, or defeat. (In the same sense that artillery fires for effect are preceded by aiming rounds, organizational probing actions generally precede attack actions.) Figure 4.2 illustrates a range of attacks including CNA, MILDEC injection into sensors, and EW attacks against radio frequency data links.
- *Channels* to the targets include people (e.g., diplomats, cooperative third parties with access to the organization, and agents in place), network channels opened by CNOs, electronic channels opened by EW, and other channels that allow insertion of data through open public media (e.g., the Internet, broadcast news reports, or newsprint) or sensors (e.g., open physical activities

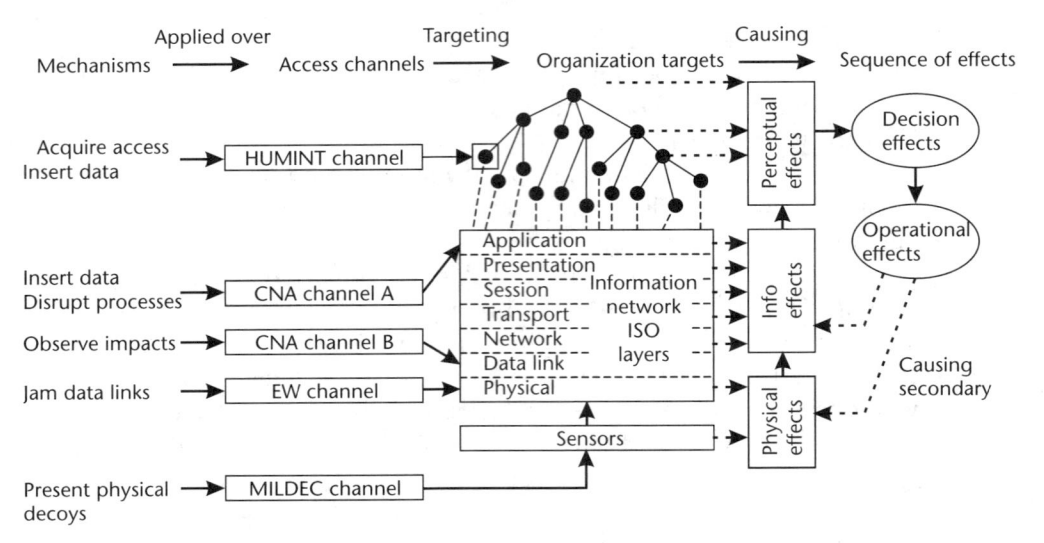

Figure 4.2 Organizational impact methods, channels, and targets.

prepared for observers or military deception or operations that are observed by adversary sensors).

- *Organization targets* may include nodes (people or computer network nodes such as switches, relays, or computer terminals) or the entire network as a whole. The figure illustrates that targets include the abstract layers of the organization, ranging from physical sensors and the physical layer of the network, through the layers of the organization's computer network, and then up through the layers of the social and formal network structure of the human organization.

- *Effects* may occur in the physical domain, the domain of the information systems, or in the perception, reasoning, or emotions of the members of the organization. The effects include effects on physical objects (sensors, switches), on information artifacts (data in messages, databases, processes), or in the perceptions of human users of information systems. These perceptual effects influence decision-making effects, which influence operational effects.

It is important to recognize that this model of channels to targets applies to nontechnical social systems as well as technology-based information and infrastructure systems. The basic mechanisms have been applied since ancient times; the tactics of Sun Tzu and Machiavelli can be described by their actions, channels to targets, and effects, just as the more technical methods are described. This model applies equally across military, law enforcement, business, political, and other competitive-conflict domains.

The planner synchronizes specific actions across channels to create desired effects on targets following an effects-based approach to planning. This approach, developed and refined by the U.S. Joint Forces Command (JFCOM), proceeds to derive effects from a high-level objective before developing actions to achieve those effects in the adversary organization [11]:

Planning Operations for Organization Impact

- The high-level objective is translated into the desired transitional and end states of the adversary systems.
- Desired effects are defined and related to corresponding states, and end states are the goals of plans and operations.
- Organization impact plans and operations contain courses of action (COAs) or policies that are implemented by actions that target specific nodes of the target system.
- The actions change the state of the targeted system (to intentional and predicted, unintentional-unexpected states), producing effects and observable state indicators [12].
- These indicators can be compared to desired states (transitional and end states) to determine if the COAs and policies are producing the desired effects.

The effects-based approach is illustrated in Figure 4.3, which adopts the key terminology of the NATO effects-based operations concept model (top of figure) and describes the causal flow of a simple plan to attack a C2 organization, with the desired effect of causing the system to fail to detect or track a penetrating aircraft [13]. The planned action is the disruption of a critical surveillance sensor (system node 1) that provides detection data to a target dispatcher (human node 2), which, in turn, delays a reliable target handoff to fire control tracking sensors (system node 3), degrading the accuracy of targeting data, causing the end state: a failure to

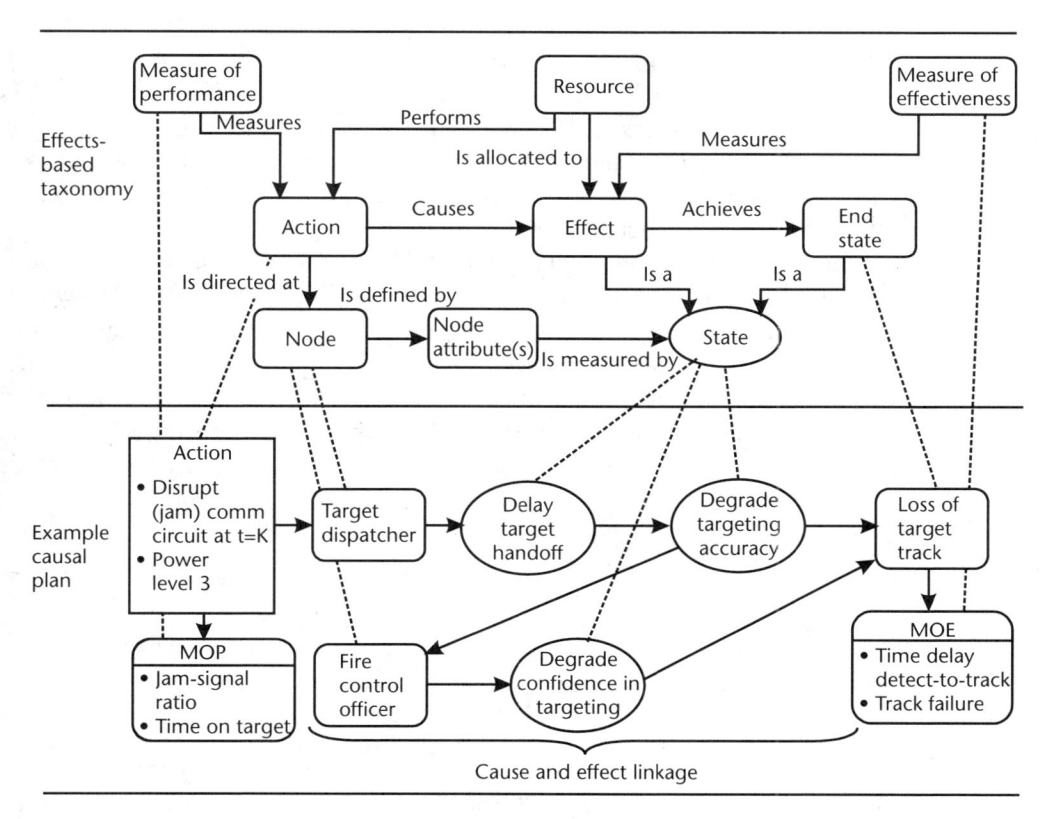

Figure 4.3 Effects-based approach to organization impact planning.

acquire or loss of target track. The action is quantified by a measure of performance—the required jamming energy and duration to deny or disrupt node 1; the effect is quantified by a measure of effectiveness—the degree to which target acquisition is delayed or denied and tracking is lost. Notice that a secondary chain of causality is planned; the sustained failure of the dispatcher to handoff tracks is expected to cause the fire control officer (human node 4) to lose confidence in the targeting data that is provided by the dispatcher and in the system.

The intended effects of actions may be applied in increasing intensity and force to produce a range of representative effects (Table 4.3). Effects may also be categorized by their temporal characteristics (transient and steady state behavior) or by temporal state changes.

Effects are also categorized as direct or indirect; indirect effects may be causally related by their causal distance from the direct effect (e.g., the second, third, or higher orders). Strategic military planning and decision-making training has long considered these effects-based perspectives; for example [14]:

> Some decision effects are *indirect*, often unforeseen, and therefore *unintended*. A given policy decision may set into motion a string of cause-and-effect events that play out over a number of years. Some of these *second-* and *third-* and even *fourth-order effects* may be unanticipated, and undesired. Effective strategic decision making requires planned responses to *second-* and *third-order* effects; it is more like chess than checkers. There are more options, the game is not linear, and the plethora of potential outcomes often is unanticipated.

Effects-based planning is based on the premise that the nodes and causal relationships between actions on nodes and organizationwide effects can be understood and represented to a sufficient degree to plan and conduct organization influence operations. This requires that abstract models be constructed to adequately represent the behavior of real-world "soft" social-psychological systems and related "hard" physical infrastructure and information systems. We do not argue that this is an easy task. The effects-based planner is challenged first by the difficulty in describing causality in social systems. Unlike the faithful descriptions of causality in physical systems (at least at the Newtonian, if not the quantum, level), causal relations and descriptions in the social sciences have been more difficult to demonstrate with predictive accuracy. The computational social sciences provide exploratory methods to discover causal relationships and hold the promise to improve our understanding of causality, if even in a statistical sense. The planner must also consider the

Table 4.3 Basic Action-Effect Relationships

Basic Action	*Representative Effects*
Encourage, reassure	Continue and reinforce the target's current behaviors
Influence or coerce, prevent, deter	Persuade the target to take or dissuade the target from taking an action
Disrupt, isolate, degrade	Temporarily disable a functional capability, the flow of information, or the human will of the target
Disable, destroy	Permanently disable a functional capability, the flow of information, or the human will of the target

difficulty of modeling the behavior of systems that have a high degree of *complexity*—measured by the scale of its independent actors and the scale and effects of interactions between those actors. Some organizational systems with even a relatively small number of actors and relationships produce nonlinear behaviors that produce unpredictable (from the component behaviors) or emergent behaviors. These *complex adaptive systems* are so characterized because their behavior is only understood in terms of the full interaction of the components, not by the behaviors of the actors themselves. Casti has pointed out that such systems are further characterized by behaviors that defy causal explanations and prediction in the traditional sense [15]:

1. They are *unstable*, subject to large effects being produced by small inputs or changes;
2. They appear random, though they are not—they are *chaotic*, in that relatively simple deterministic process rules produce *apparently* random behavior;
3. They are *irreducible* into their independent components and relationships between components—they defy decomposition;
4. They are *paradoxical* in the sense that there may be multiple independent solutions or explanations for behavior;
5. Their behavior is *emergent*—it arises from the interaction of all of the independent actors in the system. The emergent behavior patterns form out of interactions and self-organization, not strategic plans.

Jervis has pointed out how the high degree of interaction between international policy-making actors, for example, confounds linear analysis and causal prediction; in such cases, (1) results of the system cannot be predicted from separate actions of individuals, (2) strategies of any actor depends upon the strategies of others, and (3) the behaviors of interacting actors even changes the environment in which they interact [16]. Due to *complexity*, the emergent property of the organization's social behavior caused primarily by the interactions of its independent actors, rather than on the properties of the actors themselves, the organization behavior cannot be predicted by models of the properties of the actors nor by a simple linear combination of them. The approach to study such organizations that exhibit such complexity is not analytic (decomposition of the organization to reduce its behavior to a closed-form solution); rather it requires a synthetic approach, whereby *representative* models synthesize (simulate) behavior that may be compared to the observed world and refined to understand behavior in a more holistic manner.

In this text, approaches are described to develop synthetic simulations that can account for uncertainty in the organization models and produce a range, or envelope, of possible future behaviors to explore the set of outcomes of planned actions. The planning methodology using these simulations is *exploratory* in nature—developing models, probing the organization, and refining models to develop acceptable envelopes of response for planned actions.

The effects planning methodology to engage organizations must consider four alternative perspectives (or avenues of influence) to manipulate a targeted organization, each with appropriate tools for analysis and planning (Table 4.4). At the highest level of abstraction, the organization's mission and strategy must be understood, including its perspective of goals and payoffs, alternative moves, and risks. At the next level, the organization that is structured to carry out the mission provides an avenue for influence by manipulating its network structure (nodes or links). The decision and emotive processes within the organization, represented by high- and low-level decision-makers throughout the organization, provide the lowest level of influence at the human nodes of the network.

In the next section, we introduce the planning methodology to implement plans that address these four perspectives of influence.

Inducing Effects for Defeat

Organizational defeat can be achieved by a number of alternative means, ranging from attrition of the organization's physical capacity (e.g., attrition of people or resources) or functional information capacity (e.g., operational functions such as sensors, communications, or computation) to the attrition of abstract capacities of the reasoning and will. These traditional categories of military defeat (Table 4.5) are related to the three abstract domains of attack introduced in Chapter 1 and are causally related; for example, attrition of physical resources (e.g., data links) causes reduction or distortion of information available to the target, and the loss of or errors in that information can cause cognitive-emotive effects within individuals across the organization. Beyond classic attrition and maneuver warfare in the physical domain (targeting physical weapon systems, computers, communications links, and the physical bodies of people), organization warfare emphasizes abstract (nonphysical) information and cognitive-emotive defeat mechanisms:

Table 4.4 Alternative Approaches to Organization Influence Planning

Method	*Basis of Influence Method*	*Analysis and Planning Tools*
Strategy (attack strategy vulnerabilities)	Presumes rational actor competition—defined goals, options, and payoffs; presumes opponent knows options and payoffs	*Game and hypergame analysis* of opponents' moves and relative payoffs; strategy simulation to explore effects of moves
Organization (attack organization social structural vulnerabilities)	Presumes knowledge of the target organization structure (relationships) and behavior (dynamics); presumes a causal model of the impact of higher order node-relationship properties (decisions, emotions) on network behavior	*Social network analysis* to identify critical and vulnerable nodes; network simulation to explore effects of nodal attacks (remove or degrade nodes, links)
Decision process (attack the cognitive aspects of decision-making)	Presumes rational, repeatable decision process (e.g., based on doctrine or optimal performance); presumes knowledge of causal link between decisions and their operational effects	*Decision analysis* to assess individual and propagation of effects of decision errors; *network simulation* to explore decision errors, the propagation of errors and operational effects
Emotive process (attack the emotive aspects of decision-making)	Requires knowledge of organization culture and psychological profiles of leaders (decision-makers) and representative members; presumes knowable psychological links between emotions (panic, fear, distrust) and decision-making behaviors	*Psychological analysis* of human actors to identify vulnerabilities to emotive influences

Table 4.5 Military Defeat Mechanisms and Effects

Domain of Attack	Defeat Mechanisms	Typical Military Effects Sought
Physical defeat	Attrition of adversary's physical resources that are sources of power; maneuver of attacker's physical resources to exceed the rate of target response	Annihilation of physical resources; inability to track and respond to attacker's physical assets
Information defeat	Attrition of adversary's information resources that are sources of power; maneuver of attacker's information-based attacks to exceed the target's ability to recover and restore networks	Degradation and loss of data sources, communication links, and ability to process information; degradation of networks to reduce information sharing, shared situation awareness, and ability to self-synchronize
Cognitive-emotive defeat	Attrition or exhaustion of the mind and will; dislocation of state of mind of leadership; disintegration of the state of mind of combatants and their cohesion	Uncertainty, distracted, misdirected, and deceived state of mind; loss of situation awareness; indecision, delayed, and erroneous decisions; loss of morale and will to fight

- *Information mechanisms* attack the sources of information power, specifically the information available to make decisions, seeking to degrade the sharing, awareness, and synchronizing properties that are the fundamental enablers of an agile organization's ability to conduct effective operations [17].
- *Cognitive-emotive mechanisms* attack the leaders (decision-makers) and combatants (other organization members) to degrade organizational effectiveness and will to the point of achieving a breakdown in organizational structure and operation; when the organization cannot adapt and respond to the sustained attrition and maneuver of the attacker, it is poised for capitulation and defeat [18].

Organizational warfare is inherently nonlinear because the organization target is inherently reactive and adaptive, requiring the defeat mechanisms to be dynamic, tracking the organization responses to probes and attacks, and adapting the defeat tactics to the target. Current research in nonlinear tactics, including swarming, focus on convergent defeat mechanisms that emphasize superior situational awareness, stealth, standoff, encirclement of the target, and simultaneity of attacks [19]. Glenn has developed a conceptual approach to describing defeat mechanisms in terms of changes to a complex adaptive network of actors (decision-makers), information artifacts, and their relationships. Glenn's four complexity-based targeting mechanisms (Table 4.6) can be illustrated by representative actions across the three domains described earlier [20].

The following sections introduce four fundamental levels of effect-inducing actions, each requiring increasing sophistication and difficulty to conduct; all can be orchestrated to accomplish these operational or system-level defeat mechanisms. These actions also move from relatively discrete (e.g., turn on–turn off) actions with relatively coarse effects (e.g., start-stop) to more continuous actions (e.g., sequences of coordinated actions) with greater control of precision effects.

Denial or Destruction

The denial of critical decision-making information has the potential to reduce the timeliness, effectiveness, and correctness of adversary decisions, and, under certain

Table 4.6 Complex System Targeting Mechanisms

System Attack Mechanisms	Representative Actions Across Three Domains		
	Physical Domain	Information Domain	Cognitive-Emotive Domain
Decrease system variety	Destroy physical components of computer nets that contain information artifacts (e.g., databases)	Deny, destroy, or corrupt information artifacts available to actors; deny sources of information artifacts (e.g., reports)	Eliminate (deny or destroy) organization actors; eliminate variety of goals achievable by actors
Decrease system interactions	Deny, disrupt, or reconfigure communication channels between actors	Disrupt or destroy critical nodes between actors	Increase simultaneity; speed up decision time scale to reduce interactions
Decrease energy available to the system (to increase entropy and disorder)	Deny or disrupt external sources of physical energy (fuel, electrical power)	Deny or disrupt external sources of economic, intelligence, or support; disrupt high energy sources (critical technology, intelligence sources)	Disrupt recruitment of new actors; disrupt political-social support; disrupt or deny critical political power support, credibility, and so on
Alter system feedback and control	Deny or corrupt critical control channels	Deny or deceive sensors or processors to commit control errors	Deceive decision-makers to make control errors

circumstances, may even direct a target to a specific (incorrect) decision. Information denial to a target may be caused by mechanisms that range from causing a temporary pause in information to physical destruction of a capability (e.g., a radar sensor or a data link) that causes more permanent denial until the capability can be restored.

Denial operations target the following major target types:

- *Sensors:* Denial of information to sensors can be performed by *passive* means (e.g., by camouflaging a signature) or by *active* means (e.g., by concealing a radar signature with jamming or camouflaging an optical signature with paint).
- *Communication links:* Denial of information links can be achieved by "cutting" a link outright, overloading the link so access to the information is delayed or limited, or jamming the link with disruptive information (noise) to render the channel ineffective.
- *Services:* Information-processing services or services that support processing (e.g., electrical power and communication switches) can be denied.
- *Personnel:* Key personnel, such as administrators, critical support, or decision-makers themselves can be denied access or removed from the organization by arrest, detention, or other means.

The denial process is rarely covert; the target of denial is generally aware that the loss of information is attributable to an attacker's actions. The effects of denial on a human are situation- and target-dependent, and they include arousal of suspicion of an attack, blinding or disorienting effects that result in indecision, decision deferral, and delay. The effects of denial are also dependent on the timing, duration of dwell on target, and degree of information denied.

Disruption

Beyond the discrete denial of information, *disruption* includes the delay, interruption of continuity (discontinuity), or outright manipulation of information as well as the disruption of actors' abilities to function within a targeted organization. Disruption can include diplomatic or political activities that interrupt a policy-making organization's ability to conduct normal operations, overt law enforcement surveillance, search and seizure to upset a criminal organization's momentum, or military actions to degrade critical communications of an adversary (e.g., brute force or selective jamming, computer data link corruption by man-in-the-middle attacks).

The U.S. Joint Warfare Analysis Center (JWAC) has reported the results of disruption simulations against organization network models, noting that the continuous, sustained, and steady removal of the cognitive leadership (direction, decision-making, and so on) is more effective at degrading organization performance than efforts to remove multiple critical nodes simultaneously (e.g., decapitation attacks) that force the organization to reorganize and regenerate [21]. The study, reported by Mahncke, identified the following guidelines for disrupting networks:

1. *Required network understanding*: Full and complete knowledge of the network's structure is not required to take disruptive actions; the basic network structure and associated vulnerability classes allow probing and disruption to begin. Probing actions should disrupt high betweenness actors first, observing the responses that will reveal more network links as participants respond to adapt to the loss.
2. *Disruption timing*: Multiple and sustained disruptions are necessary to develop cumulative netwide effects that are observable. As the sustained attack proceeds, the net will attempt to reorganize and recruit new actors; these responses must also be identified and disrupted. Covert networks recover more slowly than overt networks due to the friction of their security mechanisms.
3. *Disrupt at the seams of the network*: The seams of an organizational network is where internal subnets connect and the network interacts with the external world; these boundaries are key points of vulnerability and offer the greatest potential impact of disruption on the net, depending on its structure.
4. *Disrupt social cohesion and trust*: The social cohesion (a capital measure of shared values, norms, and goals among organizational actors) and associated trust is vital to the effective operation of the organization. Cohesion and trust are extremely critical to covert organizations and are therefore a key target if vulnerable; disruption (and deception by net manipulation) operations that erode these properties can breed insecurity, paranoia, and suspicion that undermine operational effectiveness.
5. *Disrupt network resilience*: Disruption of the structural and functional (e.g., ideological, technical, political, financial) properties of the organization to adapt and innovate reduce the potential for emergence of new behaviors and recovery from other disruptions. Strategies such as isolation (reducing the ability to find and form new links), intimidation (reducing the ability to explore alternative threat avenues), resource degradation (removal of

redundancy and reserve), and decapitation (remove of creative and inspirational actors that provide conceptual innovation) may degrade emergence and the ability to adapt and recover from sustained disruption.

Carley et al. have affirmed and refined the basic disruption concepts in network simulations, demonstrating attack approaches that reduce the flow of information through the net that degrade the ability of decision processes to arrive at a consensus, resulting in operational breakdown (inability to accomplish tasks) [22].

Deception

The actions to deny or disrupt information to a target may be selectively orchestrated with the introduction of false information to deceive the target. This deception process misleads the target by mechanisms of manipulation, distortion, or falsification of information; the objective is to induce the target to react in a manner prejudicial to the enemy's interests [23]. Military planners trained in Sun Tzu's classic *Art of War* recognize that deception is at the core of warfare [24]: "Warfare is the way of deception. Therefore, if able, appear unable, if active, appear not active, if near, appear far, if far, appear near …" Deception is employed against sensors, computer processes, and humans to guide adversary decision-makers away from one set of decisions and toward another set of decisions favorable to the attacker [25].

The fundamental deception mechanisms are illustrated in the deception matrix (Figure 4.4) that distinguishes the two principal dimensions of deception: (1) what is true and false, and (2) what is revealed to the target of deception and what is withheld. The most critical interdependent components that always work in tandem include concealing the true (dissimulation) and revealing the false (simulation). The dissimulation-simulation cells include the principal methods described in Barton Whaley's theory of deception, organized in order of increasing complexity (and

	Reveal	Conceal
True	• Reveal truthful but selected information to create false perceptions (e.g., weakness or strength)	• Mask: conceal or eliminate a pattern; blend into the background to be undetectable Repackage: add or subtract characteristics to modify an existing pattern to appear as if another • Dazzle: obscure characteristics to distort or blur a pattern, reducing its detection or recognition
False	• Mimic: simulate information that copies or imitates the pattern of something not present, but known to the target (and expected) • Invent: simulate the characteristics of a new pattern not known to the target • Decoy: create an alternative pattern to draw the target away from the real to choose the false	• Conduct operational security activities to conceal the deception plan, mechanisms, and channels

Figure 4.4 The elements of a deception matrix.

decreasing effectiveness) [26]. According to Whaley, the most effective dissimulation-simulation pair is the mask and mimic approach, which directs the target away from the correct perception to a single, incorrect perception that results in a decision chosen by the deceiver. In Whaley's classic treatise on deception, *Stratagem*, the basic process of deception is described as [27]: "a decision-making procedure for the protagonist that will induce dysfunction in his enemy's decision process." According to Whaley, the most elegant stratagem, the *baited gambit*, simply confronts the victim with ambiguity and supplies at least one plausible alternative; the deceiver biases the alternative by masking (the truth) and mimicking (the false) and the victim makes a decisive (and incorrect) decision.

Supporting dissimulation-simulation are efforts on the other diagonal of the matrix to selectively reveal truthful information and carefully conceal the deception itself (the existence, plan, mechanisms, channels, and targets). Selective revealing of truthful information is performed in a manner that supports the desired perception of the target; for example, a limited number of forces may be exposed to verify the presence of forces but with decoys to exaggerate their numbers. Deception planning requires careful application of multiple methods across channels to limit the target's ability to compare multiple sources for conflicts, ambiguities, or uncertainties that may provide cues to the presence of a deception.

Direction and Reflexion

While denial, disruption, and deception actions are generally distinct actions or operations that occur at single points in time, or of short duration, the continuous application of these mechanisms over time to direct or control the target are described as *reflexive control*. This process provides the most refined level of influence on organizational targets—modeling and monitoring the opponent's cognitive-emotive process in an effort to convey specially prepared information to the target to produce mental states that will cause the target to voluntarily make decisions in favor of the manipulator.

In Russian researcher V. A. Lefebreve's mathematical approach to psychology, the process of *reflexion* attempts to construct a mental representation, first of the self, then of a target's thoughts and feelings (their "image" of reality) with multiple representations [28]. Reflexive control is the formal method to quantify the influence on a target's perception, decisions, and goals, based on an understanding of the target's awareness of reality and a mathematical procedure to guide that perception by "transferring" an image to the target that may have several effects:

- Transfer an incorrect image of the situation (e.g., weakness where there is strength or incomplete presented as complete) to induce specific decision errors.
- Transfer an image of the situation that guides the target to select goals that can be exploited.
- Transfer an image of the deceiver that leads the target to incorrectly infer the deceiver's goals, situation, and doctrine.

Reflexive control, organized in typical control system configuration (Figure 4.5), manipulates the target using all four methods of the deception matrix and tracks the resulting behavior to refine the target model and update control actions. The model is an effects-based approach that follows four steps in a competition between two parties, A and B:

1. Party A defines the desired goals and the effects (perceptions, attitudes, decisions, behaviors) within the target, B, that will support those goals.
2. Party A develops multiple-representation reflexion models of itself, of B, and of B's perception of A's self-model. Using these models, A conducts assessments of alternative control actions to influence B to make decisions favorable to A.
3. Once a deception plan is accepted and information is presented to B, the effects are observed and compared to the expected responses by the reflexion model of B.
4. Differences in anticipated and actual responses may be used to refine the deception plan in a typical feedback control manner and refine (or choose among alternative) reflexion models of B.

Thomas has summarized practical applications of reflexive control in anecdotal applications by the Russians and alternative analytic methods developed by Russian leaders in the field [29]. Most important to the reflexive control concept is the recognition that the target (individual or group decision-making) is regarded as a complex and adaptive system with issues of stability, response, and emergence.

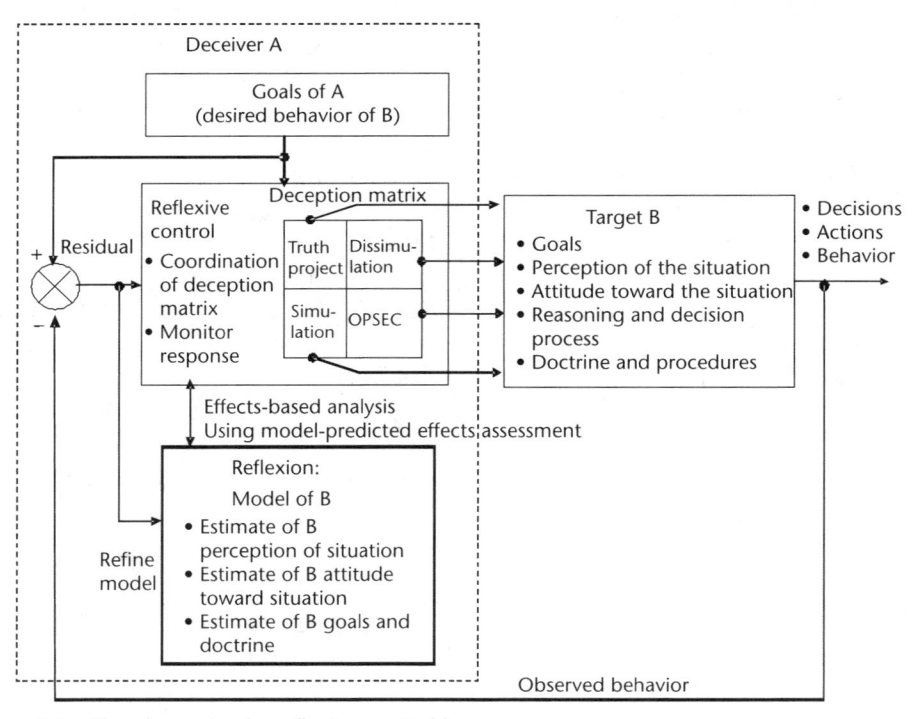

Figure 4.5 The elements of a reflexive control loop.

McCloskey has developed a theoretical concept for a cognitive campaign plan that attempts to understand the adversary's decision space (a tree structure of decision branches and sequels), then creates conditions to guide the adversary over a trajectory of decisions that will result in defeat [30]. Much like a chess game, the cognitive campaign requires the attack planner to maintain a significantly greater awareness of the decision space of the adversary and approaches to induce the victim's choices. Based on the reflexive theory, McCloskey's concept identifies vulnerabilities in the victim's perception (orientation or framing) process that provides the context for understanding the situation.

Targeting for Effects

The process of targeting the organization includes selecting and prioritizing targets of the organization while selecting and assigning attack mechanisms to achieve a desired strategic effect within a defined time period. The planning and targeting process can now be integrated with the organizational intelligence analysis workflow introduced in Chapter 1 (Figure 1.4). The combined workflow (Figure 4.6) proceeds from the model of the organization target provided by analysis, per-

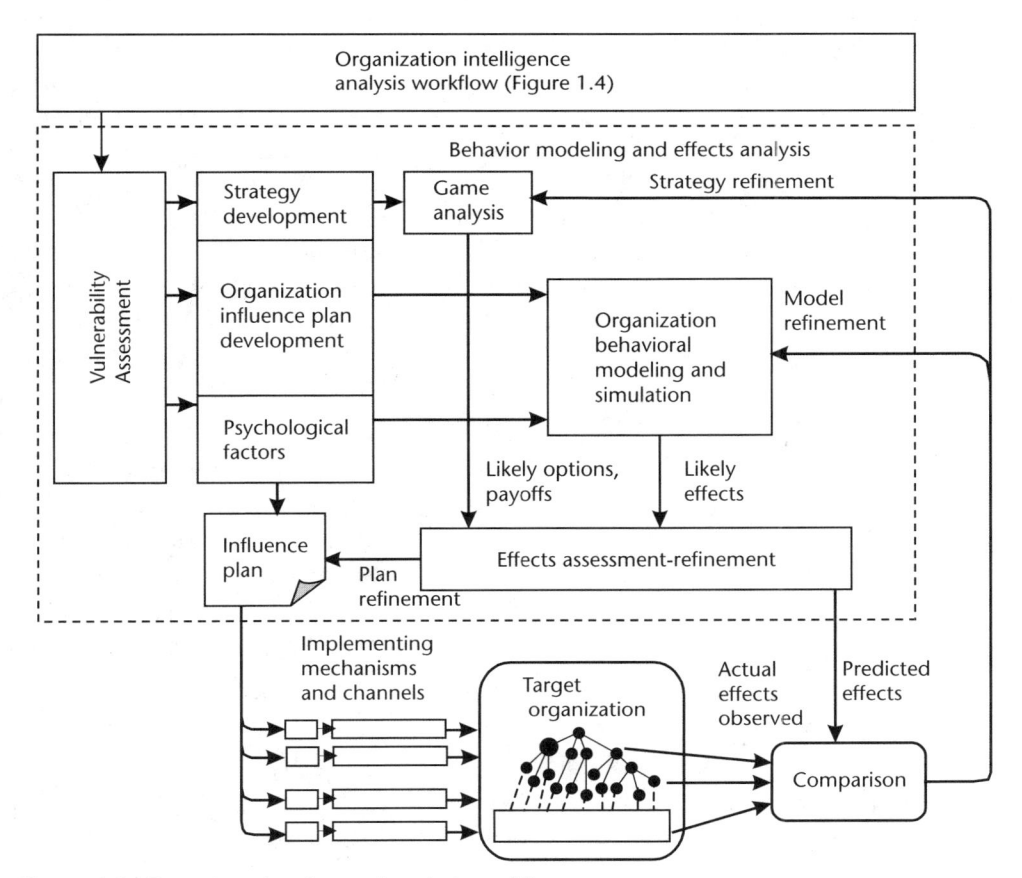

Figure 4.6 Targeting planning and analysis workflow.

forming a vulnerability analysis before plan development, and behavior modeling and simulation at each of the strategic, organization, and personal-psychological levels described earlier in this chapter. The result of the process is an effects-based influence plan that creates probe and attack actions that are implemented by multiple mechanisms across multiple channels. The observed indicators of effects on the target are compared to predicted effects to refine the behavior models and plan subsequent actions.

Vulnerability Assessment

The assessment of organizational vulnerabilities considers a range of adversary reasoning factors (cognition) [31], trust in automation [32], and emotion. Studies of decision-making pathologies for military command and control organizations [33] and business organizations [34] have enumerated the basic vulnerabilities to be catalogued and considered by the planner. A summary of the basic decision-making effects and operational effects resulting from these pathologies (Table 4.7) distinguishes those errors that are aimed at critical nodes (decision-makers) and those aimed across the network. The following chapter details the vulnerabilities and effects sought by attackers in each of the categories.

Plan Development

Based on the effects desired, the available target vulnerabilities, and channels of access, the planner synthesizes candidate courses of action that identify actions, their timing, and selected access channels to the target. In the planning phase, the resources required are identified as well as measures of performance for each action and observable indicators of effects that can provide a measure of effectiveness of the course of action as it plays out. The plan is depicted on a *synchronization matrix* that depicts the relationships and dependencies between desired effects, planned actions, and their associated resources as a function of time and space.

Table 4.7 Categories of Targets and Effects

Target	Decision-Making Effects	Representative Operational Effects
Nodes (critical decision makers)	Delay or defer decisions	Reduced targeting effectiveness
	Change decision threshold	Missed detection
	Overload	Loss of context; missed detection
	Self-reinforce	Nonproportional responses
	Impact emotions	Timid or aggressive response to stimuli
Network (entire social system)	Misallocate authority	
	Cascading decision errors	Decision breakdown; major op errors
	Lock condition	Indecision; failure to respond
	Deadlock; livelock	Delayed decisions; delayed operations
	Induce social pathologies (e.g., groupthink, group confusion)	Suboptimal and biased decision making; nonlinear effects; irregular operational behavior

Strategic Analysis

The planner must also consider the alterative strategies available to (or perceived by) the adversary organization and the relative payoffs for each competitor. This provides a necessary context for planning actions and may apply game theoretic analysis that quantifies the alternative choices that each competitor has from among available strategies under conditions of imperfect information and interdependence of the competitors.

Organization Behavior Analysis

Static modeling tools for representing the structure of the organization (nodes and links) were introduced in Chapter 1, and now organization dynamics are explored by models and simulations that represent the temporal interactions between actors with goals-directed behaviors. Operations researchers refer to *models* as the physical, mathematical, or otherwise logical representations of systems, entities, phenomena, or processes (including dynamics and time dependencies, as in differential equations) and refer to *simulations* as those methods to implement models over time [35]. For example, in Chapter 6 dynamic models are solved analytically to illustrate dynamic, time-dependent effects; in Chapter 8 probabilistic simulations are used to synthesize the range of behaviors of an organization. These tools provide a means for analysts and planners to be immersed in the modeled organization, its structure, dynamic behavior, and responses to probes and attacks. It is a tool for experimentation and exploration, providing deeper understanding of the factors that influence behavior.

A variety of simulation approaches may be applied to represent organizational dynamics at varying levels of fidelity (Table 4.8). The static social network models introduced in Chapter 1 and these dynamic simulation methods are inherently collaborative because they explicitly represent the assumptions, elements, and behavioral relationships (versus mental models), permitting planning and analysis teams to collectively assemble, refine, test, and develop plans with a best estimate of expected effects.

Representative examples of applications include simulations of covert networks and national leadership. Carley et al. have pioneered the use of social network analysis to understand the structure of large-scale covert networks and agent based simulation (DyNet) to explore vulnerabilities and responses to alternative attacks [36]. Taylor, Vane, and Waltz have described the application of agent-based simulations to represent the decision-making dynamics of foreign leadership to assess the effects of decision-making stresses under political, social, economic, information, and military conditions [37]. Clark has described the particular issues associated with methods to model covert networks [38]. In addition to computer simulation tools, red team or red cell events may be conducted to simulate organization behavior by employing subject matter experts to role play the targeted organization and assess organizational dynamics under a variety of planned conditions (situations and actions) [39]. In this book, subsequent chapters apply simulation approaches to explore organization dynamics and to plan probes and attacks.

Table 4.8 Representative Target Modeling and Simulation Tools

Simulation Approach	Description and Characteristics	Application to Organizational Impact Analysis
General causal modeling	Static Bayes networks represent chains of actions to nodes and resulting effects; dynamic Bayes nets add a representation of complex states and transitions at nodes to represent the aggregate dynamics of a causal networks	Supports general nodal-based analysis where causality is known (e.g., doctrinal behavior); subject matter experts estimate transition probabilities (example tools: Norsys Netica, Bayes Net Toolbox for Matlab)
Continuous or discrete-time simulation	Time-based simulation of continuous or time-discrete processes defined by differential equations; represent continuous processes by state-machine simulation of all processes for each discrete-time increment	Simulates organization supporting infrastructure; may simulate discrete organizational decision processes (example tools: Imagine That, Extend)
Discrete event simulation	Simulate event-based systems using queuing models of queues—servers, Petri nets, Markov, and other models that define nodes, links, and resources to simulate process interactions, synchronization, and scheduling of discrete events	Simulate stochastic flow within sequential decision-making processes—for example, C2 and automated decision processes (example tools: MathWorks MATLAB SimuLink and SimEvents, University of California at Berkeley Ptolemy)
System dynamics simulation	System dynamics flow models are based on the principle of accumulation, representing the flow of resources to accumulate stocks of products; system dynamics causal models account for circular feedback across processes and represent nonlinear behavior in complex systems	Simulate rational and optimizing behaviors of integrated physical-information and social systems; extensively applied to business modeling (example tools: ISEE Systems iThink, Powersim Software PowerSim Studio, Ventana Vensim)

Application Considerations

To apply the planning and operations concepts in this and subsequent chapters, the planner must consider three crucial issues: assumptions, effects, and legality. First, it is critical that planner consider fundamental assumptions in three areas:

- *Available intelligence:* We assume the ability to observe, collect information on, and describe a specific targeted organization across multiple levels of abstraction with *sufficient* fidelity to faithfully represent its behavior. The degree of fidelity required is unique to each situation (number of organization actors, their degree of influence, the sophistication of interactions, and so forth); analysts and planners must have a means to justify what level of understanding is sufficient for planning. In addition, intelligence is desired to observe the results of probes and attacks to refine organization models and confirm that desired effects are achieved, and unintended effects are not.
- *Sound organization theory:* We also assume *sufficient* knowledge about human decision-making and free will, social behavior within organizations, and information processes of organizations to describe the regions of predictable causality and behavior—and the regions where they are unknown. While this text focuses on technical methods using computational models to study organizations, the requirement to understand the dynamic behavior of organizations is not new. Sun Tzu acknowledged this need for an understanding of

organizations [40]: "If you know the enemy and know yourself, you need not fear the result of a hundred battles ..."

- *Representative explicit modeling:* We assume an ability to explicitly represent the structure and behavior with *sufficient* fidelity, granularity, and accuracy to provide an acceptable degree of behavior and response to understand the effects of planned actions, including unintended, collateral effects.

Each of these assumptions requires the analyst and operator to identify and describe the uncertainties in collected data, behavior models applied, and the explicit representation in tools. It is the responsibility of the intelligence analyst to estimate the uncertainty in information about the target; the operations analyst (planner) must estimate the uncertainty in predicted effects of planned actions.

The second issue that the planner must consider is the effects of actions and the uncertainty in the range of these effects. Small actions against organizations' complex social systems have the potential of inducing large unexpected organizational and societal effects (desired, undesired, and potentially unintended catastrophic consequences). Quantitative risk management tools provide a means to assess the relative risks of alternative plans, compelling the analyst to explicitly represent and compare the utility of each candidate against the potential for risk.

Next, while the impact mechanisms in this text focus on targeting human reasoning (decision-making), there remains the question of how to influence an individual's will. William Murray has articulated the issues related to quantifying and measuring the adversary's will; on the causal link from human reasoning to the will, he cautions [41]:

> The answer is that no one really knows. The question has no easy answers. The makeup of an enemy's will is so complicated, and varies so much with each scenario, that claims of universal applicability of any given means of countering an enemy's will are immediately suspect.

Finally, and not of least importance, the planner must consider the ethical and legal implications of direct effects on organizations and the indirect effects on other social organizations, civil populations, institutions, information, and infrastructure. The planner must consider the legality of organization influence operations and the potential collateral effects on civil populations to remain compliant with directly relevant international legal protocols [42].

Endnotes

[1] See, for example, *Power Moves Involved in the Overthrow of an Unfriendly Government*, Washington D.C.: CIA Historical Release Program 4-15-2003, January 5, 1970.

[2] Arquilla, J., and D. Ronfeldt, (eds.), *Networks and Netwars: The Future of Terror, Crime and Militancy*, Santa Monica, CA: RAND, 2001; see also Arquilla, J., et al., *In Athena's Camp, Preparing for Conflict in the Information Age*, Santa Monica, CA: RAND, 1997.

[3] DOD Information Operations Roadmap, October 30, 2003, p. 8.

[4] Endsley, M. R., "Design and Evaluation for Situation Awareness Enhancement," *Proc. Human Factors Soc. 32nd Ann. Meeting*, Santa Monica, CA, 1988, pp. 97–101.

[5] For an analysis of the approaches, issues effects, and consequences of leadership, or "decapitation" attacks, see Hosmer, S. T., *Operations Against Enemy Leaders*, Santa Monica, CA: RAND, MR-1385-AF, 2001.
[6] "Definition of Perception Management," Joint Publication JP 1-02, *DoD Dictionary of Military and Associated Terms*, U.S. Joint Chiefs of Staff, April 12, 2001.
[7] Sticha, P. J., D. M. Buede, and R. L. Rees, "It's the People, Stupid: The Role of Personality and Situational Variables in Predicting Decisionmaker Behavior," *Proc. of 73rd MORS Symp.*, 2006.
[8] Krackhardt, D., and J. R. Hanson, "Informal Networks: The Company Behind the Chart," *Harvard Business Review*, Reprint 93406, July–August 1993.
[9] *The Commission on the Intelligence Capabilities of the United States Regarding Weapons of Mass Destruction*, Washington, D.C., March 31, 2005, p. 174.
[10] Giles, L., *Sun Tzu on the Art of War: The Oldest Military Treatise in the World*, translated from the Chinese with Introduction and Critical Notes, 1910, Chapter VI, "Weak Points and Strong" para. 23.
[11] For an introduction to effects-based approaches, see: *Effects Based Operations Concept Primer*, U.S. Joint Forces Command, November 2003, and *Doctrinal Implications of Operational Net Assessment (ONA)*, The Joint Warfighting Center Joint Doctrine Series, Pamphlet 4, February 24, 2004.
[12] Notice that some states are directly observable (e.g. physical bomb damage and functional destruction of electrical power production capability) while other states are hidden from direct observation (e.g. adversary leaders' states of mind). Indicators provide insight to reveal the hidden states, by inference.
[13] *EBO Business Model*, Version 0.5 DRAFT, Brussels: NATO C3 Agency, March 18, 2005.
[14] *Strategic Leadership and Decision Making*, National Defense University, Chapter 1, Overview, accessed online September 14, 2003, http://www.au.af.mil/au/awc/awcgate/ndu/strat-ldr-dm/pt1ch1.html.
[15] Casti, J. L., *Complexification: Explaining a Paradoxical World Through the Science of Surprise*, New York: Perennial, 1995.
[16] Jervis, R., *System Effects: Complexity in Political and Social Life*, Princeton: Princeton University Press, 1997; see also Jervis, R., "Complex Systems: The Role of Interactions," in Alberts, D. S., and T.J. Czerwinski, (eds.), *Complexity, Global Politics and National Security*, Washington, D.C.: National Defense University, June 1997, pp. 45–72.
[17] These are the core elements of network-centric warfare; see Alberts, D. S., J. J. Garstka, and F. P. Stein, *Network Centric Warfare: Developing and Leveraging Information Superiority*, Washington, D.C.: DoD C4ISR Cooperative Research Program, 2nd ed., 1999.
[18] For an overview of defeat mechanisms that recognize the inherent complexity of the organization systems, see Brown, M., A. May, and M. Slater, *Defeat Mechanisms, Military Organizations as Complex, Adaptive, Nonlinear Systems*, report prepared for the Office of the Secretary of Defense, Net Assessment Contract No. DASW01-95-D-0060, 2000; See also DeLancey, Douglas J. (Maj. USA), *Adopting the Brigadier General (Retired) Huba Wass de Czege Model of Defeat Mechanisms Based on Historical Evidence and Current Need*, Fort Leavenworth, KS: U.S. Army Command and General Staff College, 2001.
[19] Edwards, S. J. A., *Swarming and the Future of Warfare*, Santa Monica, CA: RAND, 2005.
[20] Glenn, K. B. (Maj. USAF), *Complex Targeting: A Complexity-Based Theory of Targeting and Its Application to Radical Islamic Terrorism*, Maxwell AFB: Air University, 2002; see Chapter 4.
[21] Mahncke, F. C., "Network Disruption: Project MacDuff," *Proc. ASD/C3I Conference on Swarming*, January 13, 2003.
[22] Carley, K., et al., "Destabilizing Dynamic Covert Networks," *Proc. of the 8th International Command and Control Research and Technology Symposium Conference*, National Defense War College, Washington, D.C., 2003.
[23] Based on the definition of deception in JP 1-02, military deception (MILDEC) is further defined as those actions executed to deliberately mislead adversary military decision-makers as to friendly military capabilities, intentions, and operations, thereby causing the adversary to take specific actions (or inactions) that will contribute to the accomplishment of the friendly forces mission.

[24] Sun Tzu, *The Art of War*, Chapter 1, "Calculations"; this often-cited passage paraphrases the principles of deception and illustrates both simulation and dissimulation.

[25] This and the following section are adapted by permission from the author's chapters in Bennett, M., and E. Waltz, *Counterdeception Principles and Applications for National Security*, Norwood, MA: Artech House, 2007.

[26] Whaley, B., "Toward a General Theory of Deception," in Gooch, J., and A. Perlmutter, (eds.), *Military Deception and Strategic Surprise*, Totowa, NJ: Frank Cass & Co, 1982. p. 178.

[27] Whaley, B., *Stratagem: Deception and Surprise in War*, Center for International Studies, Massachusetts Institute of Technology, 1969, p. 139.

[28] See the overview: Lefebvre, V. A., "Sketch of Reflexive Game Theory," *Proc. Workshop on Multi-Reflexive Models of Agent Behavior*, 1998, pp. 1-42; original concepts were published in Lefebvre, V. A., "Basic Ideas of the Logic of Reflexive Games," in *Problemy Issledovania Sistemi Struktur*, Moscow: Academy of Sciences of the USSR Press, 1965.

[29] Thomas, T. L., "Reflexive Control in Russia: Theory and Military Applications," *International Interdisciplinary Scientific and Practical Journal, Issue on Reflexive Process and Control*, July–December 2002, Vol. 1, No. 2, pp. 60–76. See also Thomas, T. L., "Dialectical Versus Empirical Thinking: Ten Key Elements of the Russian Understanding of Information Operations," *CALL Publication #98-21*, Fort Leavenworth, KS: U.S Army Foreign Military Studies Office, 1998, and Thomas, T. L., "Russian Information-Psychological Actions: Implications for U.S. PSYOP," *Special Warfare*, Vol. 10, No. 1, Winter 1997, pp. 12–19. For a description of reflexive control principles, see Reid, C., "Reflexive Control in Soviet Military Planning," in Dailey, B., and P. Patker, (eds.), *Soviet Strategic Deception*, Stanford, CA: The Hoover Institution Press, 1987, pp. 293–312.

[30] McCroskey, E. D. (MAJ USAF), "Decision Space Operations: Campaign Design Aimed at an Adversary's Decision Making" School of Advanced Military Studies United States Army Command and General Staff College, Fort Leavenworth, KS, 2003.

[31] For a description of cognitive shortcomings that may be exploited, see: Richards, H., *The Psychology of Intelligence Analysis*, CIA Center for the Study of Intelligence, Washington, D.C., 1999; Reyna, V., "How People Make Decisions That Involve Risk," *J. of the American Psychological Society*, Vol. 13, No. 2, pp. 60–66; Gilovich, T., *How We Know What Isn't So: The Fallibility of Human Reason in Everyday Life*, New York: Free Press, 1991; Gilovich, T., D. Griffin, and D. Kahneman, *Heuristics and Biases: The Psychology of Intuitive Judgment*, New York: Cambridge University Press, 2002.

[32] For a study of the potential dependence of organizations on trust in automation, see: Bisantz, A. M., et al., *Empirical Investigations of Trust-Related System Vulnerabilities in Aided Adversarial Decision Making*, Center for Multi-Source Information Fusion, State University of New York at Buffalo, January 2000; Llinas, J., et al., *Studies and Analyses of Aided Adversarial Decision-Making: Phase 2 Research in Human Trust in Automation*, State University of New York at Buffalo, April 1998.

[33] Rasch, R., A. Kott, and K. D. Forbus, "AI on the Battlefield: An Experimental Exploration," *Proc. of 14th Innovative Applications of Artificial Intelligence Conference*, Command and Control text; see Chapter 3, "Pathologies in Control: How C2 Systems Can Go Wrong," July 2002, pp. 95–122.

[34] Nieves, J. M., and A. P. Sage, "Human and Organizational Error As a Basis for Process Reengineering: With Applications to Systems Integration Planning and Marketing," *IEEE Trans. on Systems, Man and Cybernetics, Part A—Systems and Humans*, Vol. 28, No. 6, November 1998.

[35] Definitions from the Defense Modeling and Simulation Office Modeling and Simulation Glossary.

[36] Carley, K. M., "Estimating Vulnerabilities in Large Covert Networks," *Proc. CASOS 2004 Conference*; Carley, K. M., et al., "Destabilizing Dynamic Covert Networks," *Proc. of the 8th International Command and Control Research and Technology Symp.*, National Defense War College, Washington, D.C., 2003; Dombroski, M. J., and K. M. Carley, "NETEST: Estimating a Terrorist Network's Structure," *Proc. CASOS 2002 Conference*; Tsvetovat, M., and K. M. Carley, "Knowing the Enemy: A Simulation of Terrorist Organizations and Counter-Terrorism Strategies," *Proc. CASOS Conference*

2002; Prietula, M. J., K. M. Carley, and L. Gasser, (eds.), *Simulating Organizations: Computational Models of Institutions and Groups*, Menlo Park, CA: AAAI Press/The MIT Press, 1998.

[37] Taylor, G., R. Vane, and E. Waltz, "Agent-Based Simulation of Geo-Political Conflict," *Proc. I AAAI Innovative Applications of Artificial Intelligence 2004*, July 2004.

[38] Clark, C. R. (Captain, USAF), "Modeling and Analysis of Clandestine Networks," AFIT/GOR/ENS/05-04, Air Force Institute of Technology, March 2005.

[39] Murdock, C. A., "The Role of Red Teaming in Defense Planning," Defense Adaptive Red Team (DART) Working Paper #03-3, August 2003.

[40] Giles, L., *Sun Tzu on the Art of War: The Oldest Military Treatise in the World*, translated from the Chinese with Introduction and Critical Notes, 1910, Chapter III "Attack by Stratagem," para. 15.

[41] Murray, W., "A Will to Measure," *Parameters*, Autumn 2001, p. 147.

[42] Relevant authorities include Protocol I addition to the 4th Geneva Convention of 1949, particularly Article 36 (New Weapons) and Article 49 (paragraph 3) that addresses, "...any land, air or sea warfare which may affect the civilian population, individual civilians or civilian objects on land"; also United Nations Resolution 3384, November 10, 1975.

CHAPTER 5
Breakdown of Control: Common Malfunctions of Organizational Decision-Making

Alexander Kott

When attacking an adversary decision-making organization, one expects that such attacks would cause the target organization to malfunction—to operate in a way that was unintended by the original designer of the organization and is probably undesirable for the organization. But how and why exactly would it malfunction? We have discussed what can be done to an organization. But what happens with the organization, within the organization? Why can't a properly designed and trained organization avoid being impacted by such an attack?

In this chapter, we pursue two objectives. First, we explore the possible mechanisms and behaviors of organizational malfunctions. Second, we consider how such behaviors can be intentionally caused or exacerbated by means identified in the previous chapters (e.g., by deception or denial).

To understand the mechanisms and behaviors of malfunctions, we take a relatively uncommon approach: the malfunctions are culled from a variety of technical disciplines rather than from organizational and management sciences. This chapter takes a look at what can be learned from discrete event system theory, and classical and model-predictive control literature focused on chemical and power plant control. There are at least two major benefits in taking such an approach. First, drawing on the fields of control theory and related disciplines, we are able to quarry a wealth of concepts, phenomena, insights, analytical methods, and design techniques. Second, we discover that much of organizational malfunctions are not necessarily products of human (individual or organizational) cognitive limitations, psychology, training, or culture. Even though the significance of such factors must not be minimized (see, for example, [1]), the malfunctions are also likely to originate in the systemic limitations of an organization's design (i.e., its structure and processes). Our special focus is on malfunctions that arise specifically because of the large, distributed nature of decision-making organizations and attending systems. In other words, these are undesirable system behaviors that occur despite the fact that the individual decision-making elements are functioning correctly (i.e., as they were intended to function). The behaviors are illustrated with a few examples from the domains of military command, as well as political, administrative, and business decision-making.

Analysts often attribute failures in decision-making organizations to human error [2]. In many cases, however, such failures have less to do with human error than with the inherent complexities of large-scale dynamic systems. Because failures in complex systems are themselves the subject of major studies (e.g., in control theory, in discrete event system theory, and in computer science), it is reasonable to assume that these other disciplines can offer insights into—and solutions for—problems in organizational performance. After all, damage mitigation is precisely what a well-designed control system is intended to do—to take corrective action when errors and disturbances happen—before they produce unacceptable ramifications or grow to unacceptable proportions. Yet even conventional control systems, relatively simple and relatively easy to understand, produce complex and remarkably counterintuitive behaviors that require sophisticated analysis and synthesis techniques. Therefore, it is only logical to presume that there will be more of this type of behavior for organizations that, by definition, are large and complex, and involve ill-understood human dimensions and often an intelligent adversary.

Although inspired to a large extent by technical analogies, the discussion here makes no assumption that a human organization necessarily involves computerization, automation, or high-technology communications. Since the presumption is that errors and other malfunctions are endemic in organizations, technology per se does not bear on the basic thesis. In fact, such behaviors can occur in organizations that employ purely human decision-making and venerable means of information gathering and communications. While it is presumed that the ongoing introduction of sophisticated technologies into organizations will only increase the likelihood that malfunctions will occur more frequently, such pathologies are inherent in the nature of organizational decision-making—whether operated via smoke signals or supercomputers. Nevertheless, it is certain that with the ever-increasing role of technology, there will be a concomitant need for increasingly rigorous approaches for understanding, predicting, and avoiding malfunctions in decision-making organizations.

Because decision-making in most organizations is at least partly hierarchical, this structure is reflected in many of the examples in this chapter. We use the following abbreviations: decision-maker (DM), higher-level decision-maker (HLDM), and lower-level decision-maker (LLDM). The usual assumption is that LLDMs are subordinates of an HLDM. In this chapter, LLDMs are presumed to direct or command an operational unit (e.g., a military force that exerts direct impact, positive or negative, on other units, friendly or enemy, or a business entity that performs direct actions affecting the market).

The malfunctions discussed in this chapter include, for example:

- Allocation of decision authority—too little or too much local authority—leads to tradeoffs between safety and agility.
- Loss of coordination and synchronization is common in distributed control architectures as a well as in management and command organizations.
- Deadlock is a deadly embrace in which several entities hold up each other's resources.
- Thrashing and livelock—repetitive patterns of unproductive actions—are common in computing systems and in management.

- Excessive gain in control systems is analogous to an overly aggressive decision-maker, while a controller with low gain is like an overly cautious one.
- Time delays between an event's occurrence and a corrective action can cause instabilities—counterintuitive and violent divergence from normal behavior.

We also link these malfunctions with the specific techniques of information operations (discussed in the previous chapter). For example, we note that deception is a form of positive feedback. Deception is also a way to increase uncertainty of signal and increase data acquisition and processing load at selected decision-making nodes, thereby causing such malfunctions as time delay (with potential instability), loss of coordination and synchronization, reduced gain (sluggish actions), and so forth.

Tardy Decision

Perhaps the most common explanation of a decision-making failure is something of the sort "by the time we got the information, it was too late," or its more positive version "given the information we had at the time, our decision was correct." The main point of such an argument is that the information necessary to make the right decision arrived too late. Indeed, for many years, time delays have been studied as one of the classic problems in feedback control theory. Long before the advent of computer control, so-called transport delays—those delays in sensing the effects of control actions—had to be mitigated in process control systems. Typically, transport delays occur when sensors are downstream from the point of control influence, thereby introducing a delay in the information received by the controller. For example, such delays are experienced every time one takes a shower: the effect of a change in the hot or cold faucet position (a control action) isn't felt until the water flows through the pipe, out the showerhead, and to the skin (the sensor). In this case, if the delay is not taken into account, the target temperature will be overshot and undershot several times, making water-temperature adjustment a tricky and uncomfortable procedure. Time delays are a key source of instabilities—often frightening and counterintuitive system behaviors that quickly diverge from normal, sometimes with violent oscillations.

In organizations, delays typically occur in the communication channel or in the decision-making process—and the effect is the same: the decision-maker takes too long to assess the effects of decisions. Organizations must contend with many sources of time delays, including the time required to:

- Collect information about the current situation;
- Assess and aggregate the information into a form suitable for presentation to an HLDM;
- Transmit the information through the layers of decision-makers;
- Process and evaluate the information;
- Collect additional data and verify all the information;
- Perform and coordinate all decisions;

- Issue orders to LLDMs;
- Make the necessary decisions and preparation at the LLDM level.

All these factors add to the significant delay between the time a situation is observed and when the control action is executed. Indeed, by the time the action is executed, the situation may have changed, resulting in such possibilities as missing an opportunity to gain an advantage, not countering threats on time, or not lending timely support to a suborganization or an ally.

An adversary can further aggravate the problem of time delays by applying the techniques of information warfare. Concealment, denial, and disruption force the decision-makers to use alternative, multiple, often slower means of collecting and distributing information, and to take additional, time-consuming steps to formulate and verify accurate estimates from limited, less complete information elements. When the opponent disrupts one's decision-making by imposing additional urgent tasks (e.g., an attack on the military command post compels the decision-makers to attend to the immediate defense task, or a law suit by a competitor forces a business to dedicate efforts to legal defense), the decision is further delayed.

The likelihood of deception imposes the burden of collecting additional information from multiple sources, verifying the information, and analyzing it with greater scrutiny. It also poses a more difficult problem with complex risks and mandates a more complicated and involved (and thus usually slower) decision-making process. A deception that induces distrust within an organization also may compel the organization to institute additional compartmentalization and verification measures that slow down the decision-making (see Chapter 7).

In all ages, business and military practitioners have attempted to reduce time delays—by organizational mechanisms, training, and technology. However, overall time delays can remain substantial, in spite of greatly improved means of communications. The advantages of faster communication can be more than offset by increased complexity and demands in other aspects of the overall decision-making chain. Still, there are ways to minimize such delays. These include, for example, procedural and technical means to enable shorter cycle, continuous dynamic action replanning and rescheduling, or reduction of delays through the radical decentralization of decision-making.

Traditional feedback control systems deal with time delays in two principal ways. One approach is simply to detune the controller (i.e., making the control action far less aggressive). Such an approach makes it possible to assess the effects of control actions before the actions go too far in compensating for perceived errors in the controlled variables. The second approach is to base the control action on a prediction of its effect rather than to wait for information from the sensors [3]. In this approach, the signal used to evaluate the controller's effect is actually the output of a model inside the controller that gives an estimate of the effect immediately. This way, the process can be controlled as if there were no delay, and the model predictions are adjusted appropriately by actual sensed values. Returning to the shower analogy, both of these approaches are familiar. When in an unfamiliar setting, such as a hotel, one makes small adjustments to bring the temperature slowly to the desired level. This is a cautious detuned control strategy. However, in a familiar home shower, one makes more bold adjustments before the effect is actually sensed because one has a good mental predictive model of what the effects will be when the faucets are turned a certain amount.

In organizations that face an intelligent adversary, using a detuned approach to deal with time delays—that is, taking small, incremental actions until the desired result is achieved—is usually not viable. An adversary will find a way to recognize and exploit such cautious actions. However, the other approach—predictive modeling—is viable and widely used in human decision-making. Humans often use models—both mental and computerized—of one's own organization and of the adversarial factors to predict the effects of decisions. Predictive modeling allows the decision-maker to be more agile and aggressive; as long as he trusts the predictions of his model, he does not need to wait until sensors (or subordinates) bring reports on the actual events.

Of course, decision aggressiveness must be balanced against the degree of credence that one puts in these models. When there is uncertainty about how well models can predict effects, commands may have to be detuned (i.e., caution is used so that real data can be received before too many commitments are made). Yet the degree of caution must be balanced against the concern that it will provide the adversary with additional time to deduce one's intent and take counteractions.

An example of a delayed decision-making is the collapse of Arthur Andersen, at one time among the largest accounting and business consulting companies in the United States. In early October 2001, Arthur Andersen CEO Joe Berardino learned about irregularities found in recent reports of Enron, one of the firm's most important clients. Partners (top-level owners and managers of the firm) met and discussed the attendant risks. Given the enormous fees paid by Enron to Arthur Andersen—on the order of $100 million a year—the partners were unable to decide on any specific action. Merely a couple of weeks later, Enron was forced to announce a shocking loss, and six weeks later, on December 2, 2001, Enron filed for bankruptcy, the largest at the time in American history. Alarmed by the developments, David Duncan, the Andersen partner responsible for the Enron account, executed a massive campaign of document shredding, in late October and early November 2001. Then the U.S. Security and Exchange Commission subpoenaed all Enron-related information from Andersen.

In public speeches and articles, Berardino continued to insist that Andersen had done nothing wrong. Although in early January 2002, the firm was forced to admit to massive document shredding and to fire several key management figures, it still tried to place all blame on the rogue destroyer, Duncan. Attempting to stem the ensuing flight of key customers, Andersen created an outside group headed by Paul Volcker, the former chairman of the Federal Reserve Board, presumably empowered to change the Andersen's fundamental business model. Volcker's team made extensive recommendations for changes intended to reassure Andersen's clients and to restore its credibility. Still, the partners remained deeply divided and unable to accept the recommendations. Instead, they were exploring other solutions, such as a sale of assets to a competitor. They also refused to settle with the U.S. government. On March 14, 2002, infuriated by Andersen's delays and evasions, Assistant Attorney General Michael Chertoff indicted the firm. Soon, the history of the 89-year-old organization ended [4].

The underlying reasons for the firm's inability to make timely decisions were similar to those we already discussed. The partners had difficulties in obtaining full information about the unfolding events and accurately assessing their potential impact. The adversarial nature of relations with other key players (U.S. government investigators, competitors, potential buyers, public critics) involved a degree of information concealment and perhaps even deception—intentional or otherwise—and exacerbated the decision uncertainty. Assessing the available information, making estimates of the potential future outcomes, and arriving to a consensus decision was also a slow process. For example, some of the partners believed that Chertoff was unlikely to proceed with the threatened indictment, while others were confident that a merger with a competitor would be easy to accomplish.

Low and High Threshold

Suppose the information does arrive in a timely manner. Is every new piece of information a good reason to make a new decision? Under what condition should a decision-maker undertake and execute a new decision? This is a difficult question. Often the decisions of a decision-maker are triggered by signals exceeding specified thresholds. When the threshold for a deviation signal is set too low, excessively frequent change of orders may occur. Alternatively, a system may have too high a threshold for a deviation signal, causing the system to stay the course until it enters a danger state.

Some of the reasons for excessively high or low thresholds are related to information warfare. Knowing that the adversary practices concealment, denial, and disruption, the decision-maker recognizes that she may have to act on a very limited amount of information (i.e., she is compelled to set a very low threshold). Conversely, concerns about a possible deception lead the decision-maker to wait until a sufficient amount of evidence is accumulated and verified—in effect constituting a high threshold for initiating a decision.

Consider for example the behavior of Spanish decision-makers during the 1587 campaign of English admiral Drake against Spanish ports. Industrious and detail-oriented, Spain's Philip II continuously monitored reports about Drake's actions and kept sending a stream of frequent orders to his administrators and military commanders. The contemporaries complained that Philip's new orders were promulgated in response to news of even slightest movements in Drake's disposition. The relentless flood of daily orders far exceeded the ability of Spanish subordinates to execute them effectively [5]. Each order would supersede or modify the one issued a day earlier, confusing the execution and preventing a meaningful coordination between the subordinate decision-makers. In effect, Philip's threshold for making and issuing each new decision was set too low. We do not know whether Drake intentionally made a multitude of his relatively insignificant actions known to his Spanish enemy in order to confuse its command chain, but it certainly could be an effective form of disruption.

Excess of Timidity or Aggressiveness

Suppose that the information is arriving on time, and the decision-maker has the means to determine the right time to decide and to act based on the newly arriving information. Then another challenge for the decision-maker is to choose the correct extent or the strength of the response to the new situation. In control theory, this is called gain, and the fundamental problem in feedback control theory is selecting the gains in the controller [6]. In the simplest controller, a single controller gain is applied to the system error (the deviation of the system response from the desired response), defining the control action that is supposed to drive the system error to zero. Selecting the magnitude of the controller gain is a classic design tradeoff. On the one hand, a higher gain usually means a faster, more aggressive response to quickly eliminate undesirable deviations in the variables being controlled. On the other hand, a high gain can cause the system to overshoot the target, making it necessary to take corrective action in the other direction. When the controller is too aggressive (that is, the gain is too high), the system becomes either marginally stable (e.g., it goes into sustained, undamped oscillations) or even unstable (e.g., the oscillations begin to grow). High gains can also lead to control commands that exceed the control hardware's capabilities, so limits are hit and signals saturate.

In an organization, a high gain problem manifests itself when the HLDM's orders cause an LLDM's actions to exceed desired limits. Such a problem is especially prevalent when the system is forced to operate close to its limits. In an organization operating in adversarial environments, it is almost always the case that one or both opponents operate near the limits of their capabilities. The troubling results of a high-gain decision could include placing the decision-maker's own assets in an untenable situation, endangering other units, causing political complications, expending resources that could be better used elsewhere, and so on. In such cases, historians speak of lack of caution, insufficient planning or intelligence gathering, misinterpretation of intelligence, poor judgment or arrogance, underestimating the adversary, and so on [7].

The Bible (Joshua 8:3–9) describes how Joshua and his army attacked the town of Ai. He selected a large contingent of quality troops and sent them under the cover of night into an ambush position behind the town. With his remaining troops Joshua demonstrated in front of the town, and then he performed a feint attack. When the town defenders responded with a sortie, he simulated a disordered retreat. The Ai defending forces, lured from their town fortress by the apparent flight of the enemy, engaged into confident pursuit. Joshua's ambush force, meanwhile, entered the undefended town and set it on fire. Now, realizing the unfolding disaster, the Ai pursuing force turned back to their burning town. Then they were caught in the middle between the two parts of Joshua's forces and decisively defeated.

Here, the Ai defenders committed a fatal mistake: they executed an excessively aggressive move without adequate collection of pertinent information and assessment of risks. Note that Joshua took special measures to deny them the information—he used the cover of darkness and placed his ambush force in a well-concealed position. He also exploited the recognition-primed [8] or reflexive decision-making of the enemy: when the Ai commander recognized what clearly appeared to be a fleeing force, his experience and training dictated an apparently correct decision to exploit the precious opportunity, to pursue and finish off the disorganized attackers. The apparently correct decision turned out to be high gain: an overly aggressive move, an excessive response to the situation.

Similarly, an overly cautious or hesitant decision-maker is analogous to a controller with low gain. Here, the results of low-gain control in a warfare environment could be a failure to exploit a time-critical opportunity, an inadequate response to a threat, the tardy achievement of a desired objective, or allowing the adversary to gain an advantageous position.

The gain challenge lends itself to exploitation in information warfare. For example, the adversary can manipulate the information available to the organization in a way that causes the organization to form a wrong estimate (model) of the adversary's capabilities or operational procedures. Then, at a decisive moment, the organization will end up applying a wrong gain. Also, in a deception-prone situation, the decision-maker is likely to act with less certainty and with greater caution (i.e., with lower gain than the situation might require). Looking at this issue from the opposite perspective, one can estimate the gain preferences of the opponent, use this knowledge to infer the likely adversary responses, and to design one's own course of action in a way that exploits the estimated responses.

Self-Reinforcing Error

In the case of an incorrect gain, at least the decision is made to take an action in the right direction. But there exists a class of severe decision-making malfunctions where even the direction of the corrective action is entirely wrong. In control theory, the term positive feedback covers a broad class of instabilities in which errors get magnified rather than attenuated in the feedback loop. Or, to put it somewhat differently, the controller's corrective action is applied in the wrong direction, making things worse rather than better. Here, the system undergoes a self-reinforcing cycle of deterioration. Sometimes such a cycle is caused by a gain in the feedback loop that has the opposite sign from what was assumed in the system design. As a simple but common example, positive feedback can occur in industrial control systems when the polarity of a connection is reversed by inadvertently switching the wires for a voltage or current signal. Consequently, control actions are in the wrong direction—and, as errors are made larger rather than smaller by the feedback loop, the system becomes unstable. This example is in contrast to the instabilities described in the previous section, where it is the magnitude, rather than the direction, of the control action that causes instability.

A subtler source of positive feedback occurs when a feedback loop develops that was not anticipated or intended in the system design. One example of an unintended—and detrimental—feedback loop is the familiar squeal of an auditorium's sound system. In this case, the microphone picks up sound from the speaker, which in turn painfully reamplifies the stray sound.

Consider an example in the domain of business decisions. In early 1995, the management and shareholders of Apple Computer, Inc., the manufacturer of Macintosh (Mac) personal computers, watched with alarm the continuing reduction of its market share. The slippage of the market share was caused largely by the shrinking availability of software for Mac computers as compared to that for "Wintel" computers based on Intel processors and Microsoft's Windows operating system. This was a part of the self-reinforcing cycle: dropping market share led to declining stock prices, leading to dismissals of top management and growing chaos within the company's sales and engineering organizations, leading to inadequate support to independent, third-party software developers. Further, because of inadequate support and because of Apple's declining market share, fewer software developers were willing to build or upgrade software for the Mac. With less software available, fewer consumers wanted to buy the Mac, further reducing market share. That in turn led to further instability within Apple's management and engineering decision-making [9].

Overloaded with multiple problems, such as Apple cash flow and failure to develop the next generation of the Mac operating system, the senior management was unable to address the long-standing deficiencies in availability of support to the application development community. It was unlikely that there was intentional information warfare against the Apple management. However, the unrelenting onslaught of competition, bad business news, and public criticism produced a massive overloading and disruptive effect on the management's ability to assess and process the relevant information.

A form of positive feedback can also occur within decision-making process itself. For example, the HLDM issues orders to LLDMs. However, the orders happen to be erroneous and cause undesirable results. Faced with an onslaught of feedback—demands from above to explain and fix the situation, and requests from LLDMs for guidance and support—the HLDM is pressed to plan and issue new sets of orders. Conceived under growing pressure, and in increasing haste, the new orders are likely to contain even more errors, in turn causing even greater deterioration of the situation, and so on. In the next chapter, we analyze this phenomenon in detail. Many other factors—political, psychological, and environmental—can also introduce similar positive (that is, self-reinforcing) feedback loops.

In an adversarial situation, one of the most important sources of potential positive feedback is a deception executed by an opponent. For example, when one executes an action against the opponent and the action does not produce the desired effect, the opponent may choose to display signs that the action was quite effective. One will then commit even more resources to the ineffectual approach. In the series of deceptions supporting the Operation Overlord of World War II, the British planners were able to induce the German intelligence into a self-reinforcing cycle of misinterpretations: the more intelligence the Germans collected, the more efforts they applied to reassure themselves, and the more comfortable they grew with the misleading information supplied by the British [10].

What can be done to mitigate the possibilities and effects of positive feedback? By identifying and modeling potential sources and channels of positive feedback, some can—and should—be prevented at the design stage. Most, however, are run-time problems (i.e., problems that arise only in implementation rather than design). For run-time problems, it is necessary to have mechanisms for comparing what is happening in the system with an expectation of what should be happening. When humans perform control, and if the right information is presented in the right way, such run-time problems are often detected immediately. However, when computers—or at least some of them—make control decisions, it is vitally important not to lose diagnostic capabilities. Monitoring and diagnostic functions—human or computerized—are critical to detection and mitigation of positive feedback. The same functions are also important targets for an adversary who wishes to exploit a positive feedback loop within an organization.

Overload

Why would an intelligent, well-trained decision-maker not notice that he is committing an error like selecting a wrong gain, following a poorly defined threshold, or staying trapped in a self-reinforcing vicious circle like we discuss earlier? One common reason could be overload. Even if a system were properly sized for normal operating conditions, under certain transient conditions, an organization can overload its decision-making or execution capacity of some components or links. This phenomenon—of a dynamic variable hitting a limit—is called saturation; it is one type—perhaps the simplest yet most common type—of what is known as nonlinearity. Saturation arises when actuators, process variables, and sensors reach physical limits. Thus, when a system is linear, it is easy and intuitive to think about the effects of decisions: results (outputs) scale with actions (inputs). If, for example, one doubles the input, the output naturally doubles. In the case of saturation, however, some variable in the system has hit a limit. When it occurs, increases in the control input—which try to make the saturated variable further exceed the limit—simply have no effect. Here, the system is nonlinear, because doubling input does not double output. Although its workings are simple, saturation is subtle—because as long as variables are not at their limits, the system behaves in a perfectly linear and intuitive way.

As an illustration of saturation in a military organization, an HLDM sees an impending enemy attack in area B and hastens to bring in assets from area A. Yet unless circumstances are favorable, he may not be able to succeed—the redeployment takes time, his logistics are already stretched to the limit, the troops are already engaged, and so on. In this case, the system's execution elements are saturated.

Saturation can also occur in a system's decision-making element. It is not too difficult to imagine a case in which an HLDM has too many subordinate LLDMs (excessive span of control), which in turn will overwhelm him both with information and requests for decisions (probably in the most critical situations). Likewise, a decision-maker may be required to have too much peer-to-peer coordination, which will similarly overwhelm her with information and requests for decisions. This sort of eventuality is also likely to happen at the most unfortunate moment, just when the decision-maker is called upon to make critical decisions. Here, if saturation is not detected, and decisions continue to be made as if variables can go beyond their saturation limits, the overall decision-making system will malfunction. Other decision-makers, unaware that saturation has occurred, will continue to form their plans and expectations under the assumption of normal operations. As a result, they will not receive timely or accurate inputs from the saturated element.

Practical manifestations of such decision-making saturation or overload in commercial and industrial organizations—typically at the middle-management levels—were described by Galbraith and others [11] decades ago and simulated in more recent work (see Chapter 8). Recognition of overload-related difficulties in military decision-making is a large part of the recent drive in the U.S. military toward a highly decentralized, self-synchronized style of command [12].

Clearly, information warfare increases the likelihood of overload in an organization under attack. Opponents' denial, disruption, and deception tend to increase the uncertainty in the information available to an organization. This causes an increased information collection and processing load. One way an organization can deal with a threat of saturation is to avoid it—never driving the system into regimes where the limits are active. However, not taking a system to its limits also can mean that the system will not be operated to its full potential—often unacceptable in an adversarial environment. In Chapter 6, we explore effects of overload on decision-making in detail and with some mathematical rigor.

Cascading Collapse

Once a decision-maker is overloaded, it is possible that the resulting wrong decisions, indecision, or spillover of the overload will propagate to other decision-makers within the organization and cause a cascading collapse. An example of cascading collapse in traditional feedback control systems is the well-known power system blackout. Typically, blackouts occur when a single event causes a severe overload at some point in the power grid; that overload, in turn, causes equipment to fail or be taken out of service by protective relaying. Such equipment failure then causes the overload to spread to other parts of the system, thereby causing more equipment to be removed from service. The process then cascades through a large part of the power system—hence a blackout. Other networks—including telecommunications, computer, and traffic—also experience cascading failure when various arteries become blocked by congestion, leading to gridlock spread throughout the entire system. In military settings, a breach in a line of defense is one example of cascading collapse. Although only a small fraction of the overall force is defeated, the units next to the actual breach experience a great increase in the pressure applied to them. Frequently, then, the entire line begins to unravel [13].

A similar phenomenon occurs in the decision-making network. When one decision-maker is overloaded, the effects spill over to other decision-makers in the organization (particularly through an increased number of erroneous decisions made by the overloaded element) and cause the deterioration of their performance as well. We explore such a propagation of malfunctions in the next chapter. A computational technique capable of simulating a cascading collapse in an overloaded project team is discussed in detail in Chapter 8. Note that the same techniques of information warfare that cause an overload of a decision-maker (we discussed them in the preceding section) would help an opponent induce a cascading collapse in a decision-making organization. One of the most effective ways to cause such an overload in an organization is to induce a local spike of concerns about the trustworthiness of a decision-maker or a source of information (see Chapter 7).

Consider how an intentional inducement of cascading collapse incapacitated a terrorist organization of the early twentieth century, the Russian Party of Socialist-Revolutionaries (PSR), an effective perpetrator of multiple terrorist acts against Russian government officials and business leaders. In spring of 1908, Russian revolutionary Vladimir Burtsev delivered a startling notification to the Central Committee of the PSR. The notification referred to Evno Azef, a highly respected member of the Central Committee and a key liaison with the party's combat organization, the armed wing responsible for execution of terrorist acts. Burtsev offered evidence that Azef was an agent of the Russian police. After a formal investigation, in late 1908, the frustrated Central Committee was compelled to announce that Azef, one of the PSR's heroes and leaders, was in fact a police agent [14].

The impact of the exposure was enormous: each revolutionary cell of the PSR, even each individual member, were now forced to suspect each other, verify all credentials, investigate all connections, reexamine all suspicions. There was, in effect, a cascading effect where the regional organizations, cells, and subcells were increasingly consumed by investigations of each other, by professing innocence in face of suspicions of others and defending themselves in interrogations and proceedings by PSR's secret courts. Indeed, the wave of investigations revealed several other police agents within the party, propagating the wave of suspicions and investigations further through the organization. To the extent that a cell had a connection to another cell, both had to investigate each other and to defend against the other's accusations. A serious suspicion about a cell caused concerns about reliability and trustworthiness of all connected cells, and neither of the affected cells was able to collaborate further in their revolutionary activities.

The collapse of organizational effectiveness propagated from cell to cell. The normal functioning of the organization was curtailed. Its elements were overloaded by this additional, massive psychological, logistical, and decision-making burden. The party splintered, accusations flew in all directions, members feuded and melted away, and the number of terrorist acts dropped dramatically. The agents planted by the Russian police caused a massive cascading collapse of a major revolutionary organization.

The most effective method for avoiding cascading collapse is to keep the initial overload from spreading—for example, by absorbing its impact with local losses. In power systems, analyses of major blackouts indicate that catastrophes could have been completely circumvented: at the time of the first local overload, systems should intentionally shed a small amount of load (that is, cut off service to some customers). The difficult problem, of course, is to know exactly when such measures need to be taken. Another way of keeping the initial spike from spreading is to always keep the connections between systems as limited as possible. In a revolutionary organization like PSR, or in an intelligence organization, this often means a highly compartmentalized organization, where the links between individual cells are extremely limited, controlled, and easily severed. This, however, is an interesting example of a measure that may have both negative and positive impacts on organizational performance: compartmentalization within an organization increases time delays but may also help minimize the problems associated with propagation of overload within one's organization.

Misallocation of Authority

One effective way to prevent a decision-making organization from a local or a cascading overload is to distribute the decision-making authority among multiple decision-makers with well-defined responsibilities. However, allocating control authority is known as one of the most challenging problems for designing distributed, hierarchical control systems [15]. A large manufacturing plant, for example, has thousands of local controllers for machinery and material handling systems. While many decisions and control actions might be based entirely on local information, such as bar codes on containers, to achieve optimal performance, the various processes need to be coordinated by higher level centralized controllers. At the same time, when emergency situations occur—due to equipment failure, for example—time may not be available to communicate the situation to a centralized control system so that it can decide what to do. Indeed, it may be most expedient to respond to these situations immediately, with local decisions; communication with the central controller could occur later, so that any effect on the overall operation could then be assessed. To design these systems, many factors need to be evaluated to determine when local controllers should make decisions and report results, as opposed to when they should report information and wait for instructions.

Designers of robust organizations face similar challenges. A hierarchical organization may give the LLDMs too little authority, making it impossible for them to respond to unanticipated situations in a timely manner. For example, an LLDM sees an opportunity to acquire a very advantageous position that would enable faster achievement of the goal; however, following that course would require abandoning preplanned actions that would clearly lead to a much less attractive position. What does he do? In order to deviate from the plan, the LLDM will have to request permission. To do so of necessity would delay the planned action. In turn, delaying the planned action would result in dire consequences. Therefore, the LLDM proceeds as planned—and foregoes a better course of action.

Perhaps an unforeseen situation arises that puts the LLDM in grave danger. To avoid disaster, he must take action beyond his allotted authority. While waiting for approval, time runs out. Clearly, then, allocation of authority is an age-old concern about finding the right balance between giving subordinate decision-makers enough autonomy while retaining the desired degree of control and safety.

Considerations of information warfare must be a key concern in determining a suitable scheme for allocation of authority. On one hand, when one's opponent has strong capabilities in denial and disruptions, the communications between HLDM and LLDM are less reliable, and LLDMs should be given a greater authority to decide and act in accordance with their local knowledge of the situation. On the other hand, when the opponent is likely to execute a deception, the HLDM is often in a better position to collect more complete, verified information and to make a deception-resistant decision. Thus, the adversary gets a vote in the allocation of authority.

In fact, allocation and misallocation of authority is a classic organizational dilemma, and authority is always misallocated whenever an organization faces a mission different from the one for which it was originally designed, as we discuss in Chapter 9. Jay Galbraith, for example, discussed these challenges [16] and offered business and technological arguments (particularly the advances in information processing) for emerging dominance of nonhierarchical, lateral, distributed, network-oriented business organizations and processes. Decentralized networked organizations are less likely to find themselves in situations of misallocated authority and are more likely to redistribute such authority in an agile and efficient manner.

The story of the Betamax video recording system offers an example of challenges involved in making proper decisions about allocation of authority. In late 1970s, Sony's American division was embroiled in a law suit regarding the Betamax [17]. Harvey Schein, the successful president of Sony America, delighted in the publicity generated by the trial and considered it a great sales booster. Yet Akio Morito, the Japanese founder and president of Sony, was worried. He demanded that much more money be spent on Betamax advertising. Schein detested this demand on his operating budget and pointed to the marketing polls that showed the wide popularity of Betamax among American consumers.

Morito, however, had a broader view of the situation. He learned that a number of Sony's Japanese competitors had already adopted the VHS model, a video recording system that required fewer parts, was cheaper to manufacture, and could play a tape longer than Betamax could. He knew that the competition was about to enter the market. Morito literally forced Schein to commit funds to a major advertisement campaign, which Schein continued to resist as unnecessary. Morito's model of the situation, however, was more on target than Schein's: by 1980, the rapid rise of VHS drove Betamax off the market. For years before that incident, Schein's extensive authority and independence have been crucial factors behind the success of Sony America. Schein was empowered to make agile independent decisions in order to respond to competitive threats or to exploit opportunities. In this case, however, Morito's in-depth understanding of the global competitive situation warranted reallocation of authority. Although Morito attempted to execute this reallocation, it was bitterly resisted by Schein and remained ineffective.

One could debate whether a better approach to allocation of authority between these two decision-makers was indeed possible, or whether it could help in the Betamax case. What is clear, however, is that in a complex distributed organization, it is challenging to design an effective allocation of authority, especially if the allocation is expected to change dynamically depending on circumstances. Recent research on so-called reconfigurable control systems [18] has begun to consider strategies for using distributed controllers designed for normal operation to handle new and possibly unanticipated situations. For decision-making organizations, such capability would be highly desirable. For example, while local decision-makers certainly need the ability to act according to proscribed protocols and limits under normal conditions, they also need the agility and freedom to respond more autonomously when required by extreme conditions. In the next chapter, we consider an approach to such reconfiguration in a decision-making team.

Lack of Synchronization and Coordination

Even when the responsibilities for decision making are effectively distributed, it is difficult to keep decisions properly aligned. They must be synchronized in time and coordinated in purpose. Consider large-scale control processes, such as chemical plants, which are usually controlled by many distributed controllers, each responsible for regulating a particular local process variable [15]. As such, a distributed architecture can have several advantages over a centralized scheme. Dedicated, single-variable controllers are easier to install, tune, and maintain. Information is obtained and used locally, avoiding the need for a high-speed communication network. The system is robust against single-point-of-failure outages—if one controller goes down, the whole process may still operate acceptably because other controllers are still performing their tasks. Such a system is not unlike an organization, where controllers are multiple and distributed, and coordination and synchronization between multiple operational units is necessary to achieve a common objective.

When there are multiple controllers acting on a system, however, it is possible that they can begin to work at cross-purposes, thereby leading to instabilities. An obvious example is when two units, commanded by different decision-makers, initiate an attack against the enemy. Unless the two attacks are coordinated in time and space, the desired impact on the enemy is lost. In all likelihood, the enemy will be able to defeat the attack, often by concentrating his assets first on one unit, then on the other.

In general, then, multiple independently executing tasks must be coordinated to achieve a synergistic effect. An example could be a series of independent attacks on a power grid. Each attack will produce a local effect—but if the time can be properly synchronized, the total effect will be a massive collapse of the entire grid. The question, then, is how can we control these actions across many organizational hierarchies?

In decision-making organizations, multiple-unit coordination and synchronization is achieved in part by an HLDM who plans synchronized actions, issues orders to LLDMs, and provides coordination instructions. In addition, the HLDM often issues instructions for coordination between units on a peer-to-peer basis. None of his actions are foolproof, however, because in a dynamically changing world with inevitably incomplete information, LLDMs often must act only on information they have from their relatively local—and, of necessity, myopic—view of events. As a military example, LLDM A sees an opportunity to take advantage of an adversary's local weakness, thereby assuming a better position. LLDM A therefore leaves position X for position Y. At the same time, however, LLDM B, assuming it can call on LLDM A at position X if his forces need additional help, decides to take an aggressive action toward position Z. Sadly, by the time word reaches LLDM A (now at position Y) that LLDM B needs help, it is too late.

Synchronization of actions must occur not only in time but also in space. As another example, a military unit is busy erecting a tent city for refugees in location C. Meanwhile, a nongovernmental medical organization builds a medical facility in location D. Heavy traffic of ill refugees now ensues between C and D, precisely across location E—which another military unit has begun to use for logistics facilities. As this example demonstrates, real-world LLDMs not only do not necessarily report to the same HLDM, they also often have a variety of inconsistent—if not conflicting—agendas and objectives.

In fact, not merely the presence or absence of coordination, but even the specific arrangements of coordination can have a critical impact on the effectiveness and timeliness of an organizational process. For example, different coordination schemes between engineering design teams and activities (e.g., sequential or parallel) can have a dramatic and nonobvious impact on the overall product development time [19].

Most commonly, failures of synchronization and coordination have to do with conflicting allocations of a resource in time and space. Consider the historical event when several decision-makers failed to coordinate the use of one critical resource, the warship Powhatan, at the same time but at two different locations [20]. In April 1861, several Southern states of the United States threatened to take possession of the federal military installations located within the states. Among the installations were Fort Sumter, South Carolina, and Fort Pickens, Florida. Navy Secretary Welles championed a naval expedition to reinforce Fort Sumter, which would include several supply boats protected by a powerful warship Powhatan. Simultaneously, the Secretary the State Seward advocated reinforcement of Fort Pickens using the same Powhatan. Concerned about numerous Southern spies among the Navy officers, the two teams planned their respective expeditions in secrecy and with little awareness of each other. Overworked and unable to verify all the details, President Lincoln signed orders for both expeditions without realizing that both relied on the same resource—the warship Powhatan. On April 11, Powhatan left with orders to proceed to Fort Pickens, while several supply ships steamed to Fort Sumter expecting to meet Powhatan there. Shortly thereafter, Welles, Seward, and Lincoln discovered the error and attempted to reroute Powhatan. It was too late. On April 12, Confederates attacked Fort Sumter, and the Civil War started.

Given the importance of effective communications for proper synchronization and coordination, techniques of denial or disruption of an opponent's communications are effective in reducing her ability to synchronize and coordinate the actions of distributed decision-makers. Deception is also effective in forcing the opponent to adopt a more complex decision-making process that requires a greater degree of coordination. For example, the ability to detect and interpret a deception sometimes requires a culture of accepting and exploring multiple hypotheses within the decision-making process [21], open communications upward, and the freedom to voice an alternative interpretation of the available information. But this in turn requires a more complex, multipath process that calls for more intricate, harder to achieve coordination.

It is interesting to look at how such challenges are handled in control theory and engineering practice. There, a key method to prevent distributed controllers from working at cross-purposes is to introduce coordination, through either peer-to-peer communication between distributed controllers or a supervisor in a hierarchical control structure. Hierarchical schemes are analogous to human hierarchical organizations. A peer-to-peer communication scheme is where each controller broadcasts its measurements and actions to all other controllers—thereby giving all controllers global information. However, even with global information, local computations have to be designed to take into account ways that other controllers will use the information. This could lead to such extensive computations at each node—equivalent to solving the global control problem, or worse—that the advantages accruing to a distributed architecture are at least partially lost. One might argue that in human organizations much of coordination functions are handled by the ubiquitous informal networks. However, as illustrated in our examples, such informal coordination networks are not always effective, and in fact may exacerbate the lack of coordination as often as they mitigate it.

Deadlock

Sometimes a disconnect between decision-makers grinds the decision-making process to a halt. The organization may stop functioning entirely. An example is deadlock, a classic problem that has been studied extensively in the theory of concurrent discrete event systems [22]. It occurs whenever there is a circular wait for resources, a so-called deadly embrace. In a computing system, deadlock occurs when a cycle of processes uses shared resources—and each process holds a resource needed by the next process in the cycle. Therefore, any given process in the cycle cannot proceed because it is waiting for another process to release a resource it needs, and at the same time it is holding a resource needed by another process.

As an organizational example, decision-maker A requests information that must come from decision-maker B, who in turn requests information from decision-maker C, who requests information from decision-maker A. However, decision-maker A cannot respond to decision-maker C because he is waiting for information from decision-maker B. And so on. Here, the cycle is clearly deadlocked. A similar case would be that decision-maker A needs resources X and Y to accomplish his mission—while decision-maker B also needs resources X and Y to accomplish his mission. Decision-maker A acquires resource X, decision-maker B acquires resource Y—but neither can progress farther. The system is deadlocked.

A tragic instance of this example occurred when on August 29, 2005, the hurricane Katrina inundated New Orleans and southeast Louisiana, killing 1,307 people and causing more than $150 billion in damage along the Gulf Coast. While the full details and explanations remain disputed, it appears that the Homeland Security Secretary Michael Chertoff and Michael D. Brown, the head of the Federal Emergency Management Agency (FEMA) at that time waited for several days before bringing federal resources to help the devastated city of New Orleans. Further, on numerous occasions, the federal agencies waited for requests from the local agencies before taking decisive actions. Their explanation was that according to the existing policies, federal agencies had to wait for state and local agencies to request specific kinds of assistance [23]. Meanwhile, the local agencies, overwhelmed by the disaster and unsure of the procedures, waited for instructions and help from the federal agencies. This "waiting for request" phenomenon was so pronounced and so central to the malfunction of the overall system that the congressional report regarding the hurricane [24] was titled "A Failure of Initiative." The deadlock persisted for several days and caused unnecessary deaths and suffering. Clearly, the hurricane imposed a form of information attack on the decision-making systems: disruption of communications and overload by tasks and information flows. These probably exacerbated the deadlock by preventing officials from realizing that a deadlock had in fact occurred.

In a case like this, although many are tempted to blame the incompetence of the officials or their lack of initiative, we must also consider the role of a systemic organizational failure. When the investigating committee questioned the mayor of New Orleans regarding his failure to request the federal assistance, he argued that the press immediately and widely reported the distressed conditions in New Orleans, and that should have been a sufficient notification to FEMA. Clearly, the mayor did not recognize that the existing policies required him to issue a specific request for help. It is also possible that the head of FEMA did not recognize at the moment that his subordinate organizational layers were waiting for formal requests required by the governing policies. Even when obvious in hindsight, a deadlock within numerous, complex organizations and policies is difficult to notice in the midst of a large-scale crisis.

It is known that the problem of detecting whether a deadlock can occur in a given collection of processes and resources is computationally extremely difficult. One way to prevent deadlocks is to implement an appropriate protocol for reserving and releasing resources. In fact, FEMA's system of requests for assistance was an instance of such a protocol. In addition, one can implement deadlock-detection mechanisms, accompanied with procedures for eliminating deadlock conditions. However, all these countermeasures rely on effective communications. If communications are denied or compromised by the opponent's information attacks, or by a natural disaster like the hurricane Katrina, the deadlock-handling techniques may not work. For example, if a deadlock resolution between decision-makers A and B is dependent on a protocol managed by decision-maker C, then the opponent might isolate C and leave A and B deadlocked.

Thrashing and Livelock

Even when an organization is not deadlocked and is proceeding toward its intended goals, it can do so in an inefficient manner, spending much effort on unproductive activities. The ability of such an organization to resist an opponent is greatly diminished. Thrashing describes a number of different phenomena that manifest themselves in a similar manner: a system undergoes multiple unproductive cycles of behavior.

Chattering, one type of thrashing found in traditional switching control systems, arises when an action taken after a switching condition is reached (typically when some signal crosses a threshold) drives the system quickly back to the condition before the switch occurred. Sometimes, cycling between two control modes is desired—for example, when a furnace thermostat detects that the temperature has gone below the specified set point, the action (turning on the furnace) returns the temperature above the threshold (where the furnace shuts off). Although a thermostat is designed to cycle, it would be inappropriate if it switched the furnace on and off every 10 seconds, as might happen if the thermostat is too sensitive to temperature changes. In this case, it might be said that the thermostat controller is chattering.

Generally, an organization may perform unnecessary switching—from task to task or from plan to plan—to the point that excessive resources are used up simply by switching and not by any productive effort. Even though the causes of such continuous changes of plans or tasks may be perfectly reasonable—the arrival of new information, for example—such thrashing can be extremely disruptive and disorienting, particularly for lower level units [5, 25].

One measure that can be taken to reduce thrashing's potential harmful effects is to introduce a cost for switching, an approach applicable to controllers making decisions based on optimization. If switching does have a cost, a controller will not switch until other factors have changed sufficiently to compensate for the switching cost. Indeed, in many applications undesirable thrashing occurs because the cost of switching has been overlooked in the controller design. For example, in a traditional control system, excessive switching might cause harmful wear to the equipment. In human organizations, thrashing will probably be minimized if the true cost of a switching decision—moving resources, for example—is assessed correctly through formal or cultural disincentives to the decision-maker responsible for the change. By incorporating such a cost, one achieves a dwell time that reflects the system's true operating costs and objectives.

Livelock, another thrashing behavior, occurs when processes appear to be making progress at each decision point but are actually stuck in an unproductive cycle of steps. For example, in state A, the decision-maker sees path A-B-C as the most appropriate toward the goal, and so he moves to B. From that vantage, path B-C-D looks optimal, so he moves to C. From there, path C-D-A looks optimal—and so on. Here, either cyclical or noncyclical nonconvergent behavior is possible.

In a system of multiple decision-makers, a process of negotiations between decision-makers can also cause a pattern of unproductive, thrashing cycles. In a sequence of events between two decision-makers, (a) decision-maker A recommends decision X1 to decision-maker B, (b) B responds with recommendation X2 to A, (c) A recommends X3 to B, (d) B recommends X4 to A, then (e) again, and the chain repeats—livelock [26].

In general, like all large systems, an organization is capable of both mitigating and magnifying its own follies. Indeed, the key purpose and advantage of building an organization is to increase the information-processing and decision-making capabilities by combining multiple decision-makers. Yet a key drawback of any organization is that combining multiple decision-makers, each of whom inevitably, with some probability, makes errors, combines and magnifies those errors, and increases the error-producing potential of the organization. Clearly, a robust organization design must mitigate the probability errors (e.g., by cross-checks, coordination, or specialization). In other words, a good organization is an error-reducing system. Still, many organizations are not effective error-reducing systems; instead they are error-inducing systems [27]. An organization's adversary must look for ways to minimize the organization's error-reducing qualities and to exploit its error-inducing ones.

Acknowledgments

This chapter draws in part on the ideas developed in earlier work in collaboration with Professor Bruce Krogh of Carnegie Mellon University.

References

[1] Cooper, J. R., *Curing Analytic Pathologies: Pathways to Improved Intelligence Analysis,* Langley, VA: CIA Center for the Study of Intelligence, 1995.

[2] Cohen, E. A., and J. Gooch, *Military Misfortunes: The Anatomy of Failure in War,* New York: Vintage Books, 1991.

[3] Mayne, D. Q., et al., "Constrained Model Predictive Control: Stability and Optimality," *Automatica,* 2000, pp. 789–814.

[4] Toffler, B. L., *Final Accounting: Ambition, Greed, and the Fall of Arthur Andersen,* New York: Broadway Books, 2003, pp. 209–219.

[5] Hanson, N., *The Confident Hope of Miracle,* New York: Knopf, 2005, p. 85.

[6] Kailath, T., *Linear Systems,* Englewood Cliffs, NJ: Prentice-Hall, 1980.

[7] Russell, M., "Personality Styles of Effective Soldiers," *Military Review,* January–February 2000, pp. 69–74.

[8] Klein, G. A., "A Recognition-Primed Decision (RPD) Model of Rapid Decision Making," in G. A. Klein, et al., (eds.), *Decision Making in Action: Models and Methods,* Norwood, NJ: Ablex Publishing, 1993, pp. 138–147.

[9] Carlton, J., *Apple,* New York: Random House, 1997, pp. 252–253, 345, 372–373.

[10] Latimer, J., *Deception at War,* Woodstock, NY: The Overlook Press, 2001, pp. 222–231.

[11] Galbraith, Jay, "Organization Design: An Information Processing View," *Interfaces,* Vol. 4, May 1974, pp. 28–36.

[12] Hayes, R. E., and D. S. Albert, *Power to the Edge: Command and Control in the Information Age,* Washington, D.C.: CCRP Publications, 2003.

[13] Pfaff, C., "Chaos, Complexity and the Battlefield," *Military Review,* July–August 2000, pp. 83–86.

[14] Geifman, A., *Thou Shalt Kill: Revolutionary Terrorism in Russia, 1894–1917,* Princeton, NJ: Princeton University Press, 1993, pp. 232–237.

[15] Dullerud, G. E., and R. D'Andrea, "Distributed Control of Heterogeneous Systems," *IEEE Trans. on Automatic Control,* 2002.

[16] Galbraith, J. R., "The Business Unit of the Future," in Galbraith, J., and E. Lawler, III, (eds.), *Organizing for the Future: The New Logic for Managing Complex Organizations,* San Francisco, CA: Jossey-Bass Publishers, 1993, pp. 43–64.

[17] Nathan, J., *Sony: The Private Life,* Boston, MA: Houghton Mifflin, 1999, pp.106–111.

[18] Patton, R. J., "Fault Tolerant Control: The 1997 Situation (Survey)," *Proc. of the IFAC Symposium SAFEPROCESS '97,* Hull, U.K., Vol. 2, August 26–28, 1997, pp. 1033–1055.

[19] Carrascosa M., S. D. Eppinger, and D. E. Whitney, "Using the Design Structure Matrix to Estimate Product Development Time," *Proc. of the 1998 ASME Design Engineering Technical Conferences,* Atlanta, GA, September 13–16, 1998.

[20] Goodwin, D. K., *Team of Rivals: The Political Genius of Abraham Lincoln,* New York: Simon & Schuster, 2005, pp. 343–345.

[21] Elsaesser, C., and F. Stech, "Detecting Deception," in Kott, A., and W. M. McEneaney, (eds.), *Adversarial Reasoning,* Boca Raton, FL: CRC Press, 2006.

[22] Cassandras, C., and S. Lafortune, *Introduction to Discrete Event Systems,* Boston, MA: Kluwer Academic, 1999.

[23] Newman, J., "Report Blames Katrina Response on Chertoff," *Los Angeles Times,* February 2, 2006.

[24] U.S. House of Representatives, *A Failure of Initiative: Final Report of the Select Bipartisan Committee to Investigate the Preparation for and Response to Hurricane Katrina,* June 2006.

[25] Van Creveld, M., *Command in War,* Cambridge, MA: Harvard University Press, 1985, pp. 203–218.

[26] Meyer, L., "The Decision to Invade North Africa," in *Command Decisions,* Center of Military History, U.S. Army, 1987, pp. 185–187.

[27] Perrow, C., *Normal Accidents: Living with High-Risk Technologies,* Princeton, NJ: Princeton University Press, 1999.

CHAPTER 6
Propagation of Defeat: Inducing and Mitigating a Self-Reinforcing Degradation

Paul Hubbard, Alexander Kott, and Michael Martin

Until now, we presented a strictly qualitative discussion of organizational malfunctions. It is time to take a more quantitative look at some of the malfunctions introduced in the previous chapter by exploring the potential of computational models for describing their mechanisms and manifestations. The key practical questions to be answered with such models concern how one would identify and plan a course of action that could either induce or mitigate organizational malfunctions.

All remaining chapters of this volume describe and apply computational models of organizational behavior. Computational analyses of organizational behavior employ either intellective or emulative models [1]. The models described in this chapter and in Chapter 7 are intellective models. Intellective models are usually abstract, small, relatively simple, and incorporate only a few parameters. They help modelers make general predictions about trends and the relative benefit of organizational changes, identify the range of likely behaviors, and qualitatively compare the impact of different types of policies or technologies on expected behaviors. Chapters 8 and 9 describe emulative models. Emulative models are much more detailed, incorporate a large number of parameters or rules, are more difficult and expensive to construct, and require large volumes of input data. They help modelers make specific predictions about quantitative characteristics of organizational behavior for specific organizations under specific conditions, and compare quantitatively the expected impact of alternative organizational designs or policy changes.

Recalling the malfunctions described in the previous chapter, the models in this chapter demonstrate how self-reinforcing error due to positive feedback can lead to overload and saturation of decision-making elements, and ultimately the cascading collapse of an organization due to the propagation of overload and erroneous decisions throughout the organization.

We begin the chapter with an analysis of the stability of the decision-making aspects of command organizations from a system-theoretic perspective. A simple dynamic model shows how an organization can enter into a self-reinforcing cycle of increasing decision workload until the demand for decisions exceeds the decision-making capacity of the organization. In this model we consider only two components corresponding to two layers of an organization—an upper command layer and a subordinate execution layer. We show that even this simple model offers useful insights into conditions under which an organization can experience a rapid decrease in decision quality.

We then extend the model to more complex networked organizations and show that they also experience a form of self-reinforcing degradation. In particular, we find that the degradation in decision quality has a tendency to propagate through the hierarchical structure (i.e., overload at one location affects other locations by overloading the higher level components, which then in turn overload their subordinates).

But how would one devise measures that actually induce such a malfunction in an enemy organization? Conversely, how would one devise a set of actions that mitigate a malfunction in one's own organization? Our computational experiments suggest several strategies for mitigating this type of malfunction: dumping excessive load, empowering lower echelons, minimizing the need for coordination, using command-by-negation, insulating weak performers, and applying online diagnostics. Further, a suitable compensating component (e.g., a brokering mechanism that dynamically redistributes responsibilities within the organization as it begins to malfunction) can dramatically increase the envelope of stable performance. We describe a method to allocate decision responsibility and arrange information flow dynamically within a team of decision-makers for command and control. We argue that dynamic modification of the decision responsibilities and information-sharing links within a decision-making team can either degrade or improve (depending on the intent of the one who performs the modifications) stability and performance in terms of quality of decisions produced by the team.

A Simple Model for Self-Reinforcing Decision Overload

Let us consider a network of decision-making entities, perhaps individuals, teams of individuals, information-processing tools, or artificial agents, operating jointly in accordance with organizational procedures and protocols. Such a decision-making organization acquires, transforms, generates and disseminates information in order to acquire, allocate, and deploy its resources so that its objectives can be accomplished efficiently and effectively. A decision-making organization's ultimate product is the commands it issues to those operational elements that execute direct effects on the environment of the organization. We label the totality of these executing elements as the *field*. The field may include salespeople who are trying to affect the behavior of the buyers in the market; workers who assemble the products; trading floor clerks who execute the transactions; pilots of military aircraft who fly to bomb their targets. To illustrate the kinds of malfunctions we wish to model here, consider the following scenarios.

- *Scenario A.* After a major financial loss, a corporation forces one of its underperforming divisions into a major restructuring. Several new senior managers and advisors are brought into the divisional operations. The existing personnel dedicate a large fraction of their time to explaining and justifying their decisions to the new managers, and modifying their procedures and plans according to the new guidance. The day-to-day decisions receive less attention and their quality suffers. Mistakes are made more often. The field personnel resent the erroneous guidance, and morale and discipline decline. Performance of the division suffers even further. The corporate management decides to step up the restructuring ... and the vicious cycle continues.

- *Scenario B.* A corporation faces a new, unexpected tactic employed by its competitor. The tactic is successful and rapidly makes the business plans and procedures of the corporation inapplicable. Management attempts to introduce new ideas and approaches. Field personnel are bewildered and call for explanations and support. Decisions with new unfamiliar approaches become harder just as the attention of management is distracted by the competitor's new tactics. The quality of decisions deteriorates. Management's confidence plummets and decisions take even more effort. The competition exploits the errors and continues to succeed in altering the market position, which in turn requires more adjustments in corporate business tactic ... which in turn causes more confusion and errors.

Qualitative discussions of challenges and phenomena in decision-making organizations are numerous (e.g., [2–4]). Related issues have been studied previously in a variety of fields including organizational design [5], distributed and group decision-making [6, 7], human-automation interaction [8], and manufacturing systems [9]. Here, we focus on a quantitative analysis of the stability and overall performance of the decision-making aspects of an organization from a systems-theoretic perspective. We model the organization as a dynamic system of multiple decision-making models.

The representation of human decision-making is a crucial aspect of our intellective models of decision-making organizations. For our purposes, we require highly abstract representations that reasonably approximate human behavior without reference to the semantic content of particular decision-making tasks. The representation we choose is based on the fact that decisions (and all other cognitive processes) take time. Thus, decision quality (e.g., accuracy) decreases when decision-making environments dictate that decisions be made before decision processes can be fully executed. More specifically, the accuracy of human decision-making for a particular decision-making task decreases in a nonlinear fashion as the rate at which decisions must be made increases. Demonstrations of this tradeoff between decision speed and decision accuracy are widespread in experimental psychology. Furthermore, the behavioral research base overwhelmingly shows this tradeoff to be S-shaped [10, 11], as shown in the notional plot of decision accuracy as function of decision workload (i.e., rate) in Figure 6.1. There is nothing magical about this S-shaped function. It merely shows a soft threshold for the impact of time-pressure on decision quality (i.e., accuracy). The negative acceleration at the tails of the curve simply indicates that the effect of time pressure on decrements or increments in decision quality diminishes as decision accuracy approaches the limits of 0 and 1. It should be noted, however, that conclusive evidence of speed-accuracy tradeoffs in more complex decision tasks (e.g., team decision-making) is scarce. Given the methodological difficulties associated with demonstrating speed-accuracy tradeoffs in basic laboratory tasks, the lack of evidence in complex tasks is not surprising, but one may conjecture that the S-shape tradeoff also applies.

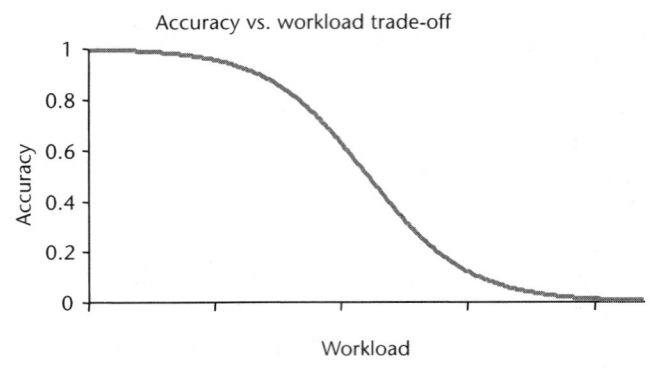

Figure 6.1 Accuracy-workload tradeoff curve for human decision-makers is commonly described by an S-curve.

The rate of requests for decisions is only one measure of decision workload. Other factors contributing to greater workload may include, for example, the complexity of decisions, criticality or risk associated with decisions, uncertainty in the available data, and latency of the available data. Arguably the impact of these factors should be accounted for in models of decision-making organizations. However, the veracity of such arguments must be considered with respect to the goals of the modeling endeavor. Decisions in dynamic environments do indeed engage a variety of interrelated cognitive processes, ranging from monitoring, recognition, and information search to planning, judgment, and choice. Incorporating these processes into a simulation to account for the variety of factors that may influence decision workload requires a commitment to a particular model of decision-making processes—a topic of continuing debate. The indisputable fact that remains is that each of the cognitive processes engaged by dynamic decision-making tasks takes time. Thus, in dynamic environments decision-makers are placed in a situation where they must control one time-dependent process (i.e., the evolving business situation) with another time-dependent process (i.e., the cognitive processes underlying dynamic decision-making). Decision-makers face this situation regardless of the complexity of a decision and any risks associated with it or the quality of the data on which that decision is based. For our purposes, therefore, we focused on the simplest and arguably the least contentious measure of decision workload—the rate of the decision requests per unit time.

Consider a model of a decision-making organization consisting of a *headquarters* (HQ) component and a *field* component, as illustrated in Figure 6.2. The HQ component receives an input flow of orders from a higher authority (u—the number of orders per unit time) as well as a flow of requests for decisions from the field component ($x2$). The HQ component produces a flow of commands ($x1$) and sends them to the field. In general, some of the commands may be erroneous. A workload-accuracy tradeoff function $f(x)$—an example of the S-curve discussed earlier—governs the fraction of the errors. If a command is correct, it is assumed that the field component executes it successfully. If the command is erroneous, it results in problems in the field. The problems manifest themselves in the number of requests for decisions generated by the field component and sent back to the HQ component. A constant coefficient, K, relates the number of erroneous commands to the number of new

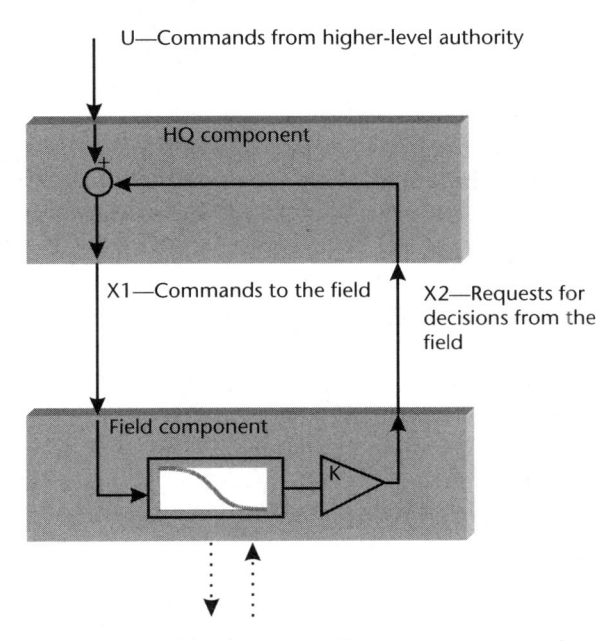

A Simple Model for Self-Reinforcing Decision Overload

Figure 6.2 A decision-making organization consisting of an HQ component and a field component.

decisions that must be made as results of the errors. A greater value of K corresponds to a greater confusion caused by an erroneous command within the field component and to a greater ability of the adversary to exploit the error. Here, a discrete-time approach is used; the accuracy, at a particular time instant, of outgoing decisions is a function of the decision requirements at the previous time instant.

If we use $x1$ and $x2$ as internal states, the dynamics of the system are given by

$$x_1(k+1) = x_2(k) + u(k)$$

$$x_2(k+1) = x_1(k) \cdot K \cdot f(x_1(k))$$

where

$$f(x) = 1 - \frac{1}{1 + e^{(x-a)/b}}$$

Linearization yields a sufficient condition for stability of the original nonlinear system:

$$K < 1 \text{ or } -1 < \left[1 - \frac{(1 - x_1/b) \cdot e^{(x_1-a)/b} + 1}{\left(1 + e^{(x_1-a)/b}\right)^2}\right] \cdot K < 1 \qquad (6.1)$$

and equilibrium points $\bar{x}_1, \bar{x}_2, \bar{u}$ must satisfy

$$\bar{u} = \bar{x}_1 \cdot (1 - K \cdot f(\bar{x}_1))$$
$$\bar{x}_2 = \bar{x}_1 \cdot K \cdot f(\bar{x}_1)$$
(6.2)

Numerical computations using Matlab [12] yield the results depicted in Figure 6.3. The lines marked "20%" and "80%" show where the fraction of the erroneous commands issued by the HQ component stays at the levels of 0.2 and 0.8, respectively. We observe that for $K < 1$ the system remains theoretically stable, but higher values of u can lead to a rapid increase in the fraction of the erroneous decisions, in essence supplying a domain-specific type of instability. For $K > 1$, the system can exhibit unstable behavior at higher values of u. As K increases, the instability occurs at progressively lower values of u. In domain-specific terms, this means that if one lowers the ability of the field component to correct for erroneous commands, then the combined decision-making organization becomes unstable at lower values of input commands (u).

Clearly, this result is qualitatively consistent with the realistic scenarios A and B we introduced earlier. It is also reminiscent of the discussions of overload and self-reinforcing error in the previous chapter, including the examples of real-world challenges faced by organizations (e.g., the self-reinforcing failure of Apple to recognize the importance of support to third-party software developers). The value of the model is not in its ability to predict specific quantitative behavioral characteristics of a specific organization. Rather, it offers easily comprehended suggestions of possible phenomena to explore with more in-depth analyses (e.g., with high-fidelity emulative models). So, what does this model suggest?

First, it points to the importance of modeling feedback channels, particularly those associated with erroneous decisions. In fact, the creators of the emulative model described in Chapter 8 found it important to model such a link. In particular, that model pays great attention to the feedback channel from a decision-maker or a suborganization when receiving a product of unacceptably poor quality from an upstream suborganization that must be rerouted for rework to the originating suborganization. Depending on the structure and tasks of the organization, the impact on its productivity can vary widely. The predictions of that emulative model

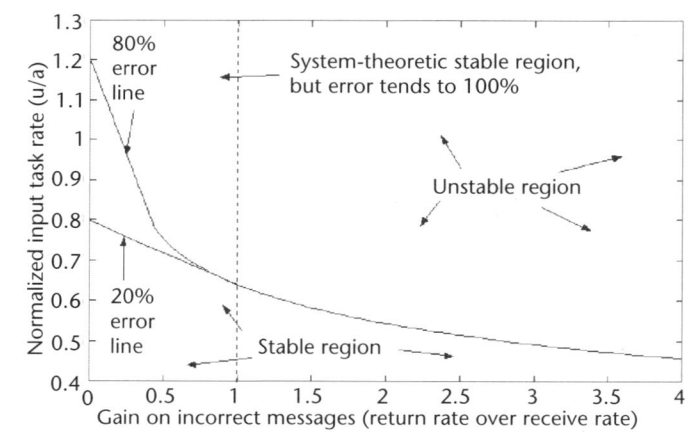

Figure 6.3 Stability regions as a function of gain on incorrect messages and of the input task rate.

have been extensively validated in practical applications to real organizations (see Chapter 8).

Second, our model highlights the importance of modeling the reduction in decision quality due to time pressure. Again, the emulative model described in Chapter 8 converges on this finding. There, it was found that modeling the drop in quality of decision-making due to the complexity of a task can have a massive impact on the predicted performance of an organization, a prediction supported by empirical findings.

Third, our model allows us to hypothesize useful directions for further study in terms of the design of experiments that employ high-fidelity emulative models. In particular, the susceptibility of an organization to self-reinforcing error, according to this simple intellective model, depends mainly on two key parameters: a, the extent to which the decision-maker is able to absorb the flow of higher authority commands (recall the example of the unfortunate subordinates of Phillip II in the previous chapter), and K, the measure of how well the field operators are enabled and empowered to handle locally erroneous or late decisions coming from their superiors. Knowing these major influencers can be valuable guidance for a designer of an organization who models its performance in order to verify its robustness under dynamic conditions.

Propagation of Disruptions in Organizations

Now let us extend the simple model of the previous section to consider networks of decision-making units. In order to do this, we first build up a generic component model that can be inserted in a standard hierarchical authority structure where tasks or messages are received from a superior and distributed among subordinates. Such a model is illustrated in Figure 6.4.

In this model, message flow down through the hierarchy is characterized by the rate and accuracy (percent error), while message flow up is characterized only by the message rate. We emphasize that this model is also intellective, as in the previous example. The incoming message load to this component is $u_2(k) + u_3(k)$, and we assume that the errors created by this component are based on the same speed-accuracy tradeoff in the first example—that is, the incurred error percentage is

$$f(u_2(k) + u_3(k)), \text{ where } f(x) = 1 - \frac{1}{1 + e^{(x-a)/b}}$$

This error percentage is in addition to that already in the messages from the superior, $u_1(k)$. The rate of messages to the superior is meant to capture the need for clarification, and therefore we assume this is the product of the current incoming rate from the superior, the total error percentage of those messages $u_1(k) + f(u_2(k) + u_3(k))$, and a gain factor K representing the susceptibility to confusion at this component. We assume the percent error in the messages that flow through to subordinate nodes is the sum of the received error rate from superior nodes, $u_1(k)$, and the incurred errors by this component in the correct messages, $(1 - u_1(k))f(u_2(k) + u_3(k))$. And, finally, the message rate to the subordinate nodes is set equal to the received

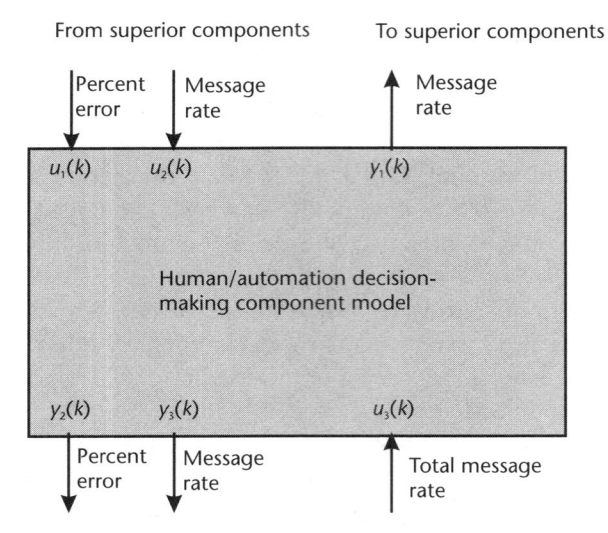

Figure 6.4 An intermediate component in a network of decision-makers.

message rate from the superior node at the previous time step. Note that the message rate received from subordinates, $u_3(k)$, may also include exogenous inputs representing additional requests for decisions directly from the field.

This results in the following set of difference equations.

$$y_1(k+1) = K\big[u_1(k) + f(u_2(k) + u_3(k))\big]u_2(k)$$

$$y_2(k+1) = u_1(k) + (1 - u_1(k))f(u_2(k) + u_3(k))$$

$$y_3(k+1) = u_2(k)$$

where again

$$f(x) = 1 - \frac{1}{1 + e^{(x-a)/b}}$$

This model of an intermediate component is used to construct networks of decision-makers by linking inputs and outputs. We consider first the network illustrated in Figure 6.5. This is a purely hierarchical network in which components receive and interpret messages from their superior components and then direct messages to their subordinate components (as well as respond to their superiors), as expected from the previous description.

The stability of the hierarchical network was analyzed in simulation, subject to the rate of exogenous high-level inputs from above—that is, $u_1(k)$ in the top component—and the rate of exogenous inputs from the field—that is, $u_3(k)$ in the four bottom components. Figure 6.6 shows the average error rate in commands to the field components—that is, the average of $y_2(k)$ in each bottom component, as the exogenous input from above is increased. We note that the system maintains its performance at a near-constant level and then rapidly collapses.

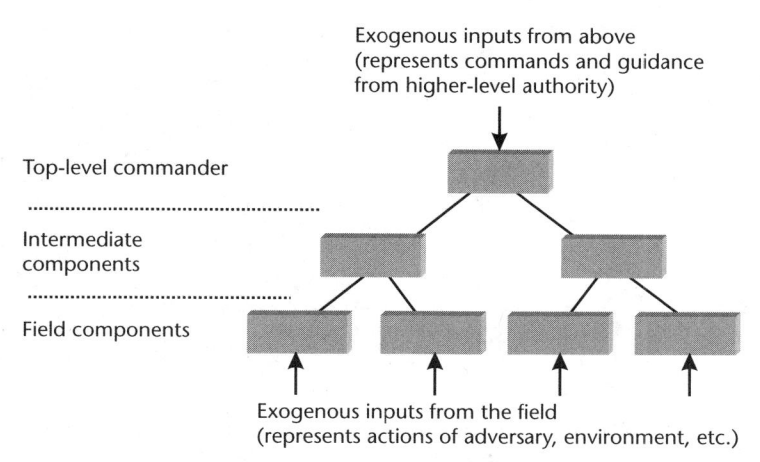

Figure 6.5 A hierarchical organization.

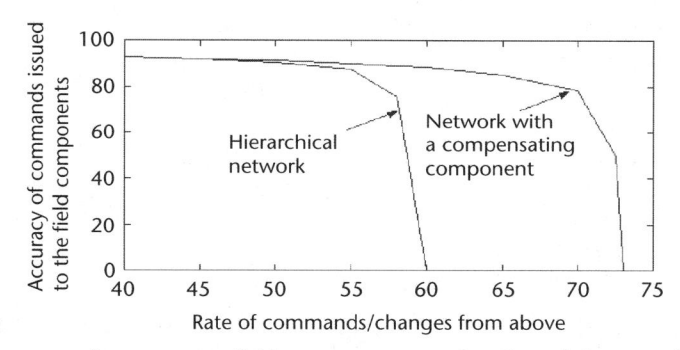

Figure 6.6 Accuracy of messages to field components as a function of the rate of high-level commands.

To evaluate the envelope for stability subject to variations in exogenous inputs from the field, we stimulated the inputs $u_3(k)$ in the far right and far left components with constant inputs. Figure 6.7 shows the stability envelope subject to the sum and difference of these constant inputs. The interpretation of *stability* in these graphs is not a standard systems-theoretic instability (i.e., bounded inputs result in bounded outputs) but rather a domain-specific meaning of instability alluded to earlier (i.e., feedback leading to the production of a high fraction of incorrect commands). In these simulations, systems-theoretic instability was excluded with the use of saturation devices. It was also observed during simulations of this network that the instability had a tendency to propagate through the hierarchical structure (i.e., an overload at one location affects other locations by overloading the higher level components, which then in turn overload their subordinates).

These results should remind us of two malfunctions we discussed in previous chapters. First, the *cascading collapse* in which the Russian terrorist organization collapsed in part because the flows of information each suborganization received from other organizational elements became progressively more unreliable and required progressively more investigation and verification. Second, the *failure of coordination and synchronization* in which the confusion about the warship Powhatan propagated and grew as it traveled through layers of authority, upward

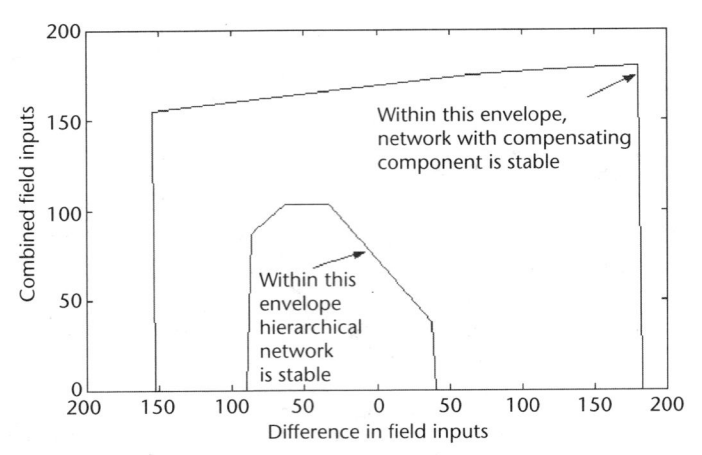

Figure 6.7 Stability regions with respect to combined field inputs and difference in field inputs at two locations.

and downward. In any event, such phenomena surely look very complicated and difficult to manage. Can something be done about them?

Active Compensation

In Chapter 5, we mentioned that monitoring and diagnostic functions—human or computerized—are critical to detection and mitigation of positive feedback. The intent of such monitoring is to introduce corrective measures when the system or organization approaches a danger zone, perhaps the limit of stability. We also mentioned the possibility of using reconfigurable systems to deal with undesirable and nonnormative situations. Such ideas lead us to consider a possible approach to mitigating degradations in decision quality based on an active compensating component.

Consider the network illustrated in Figure 6.8, in which a compensating component (e.g., a task allocator or broker that redistributes messages or tasks) is introduced to the network as an intermediary between superior and subordinates in order

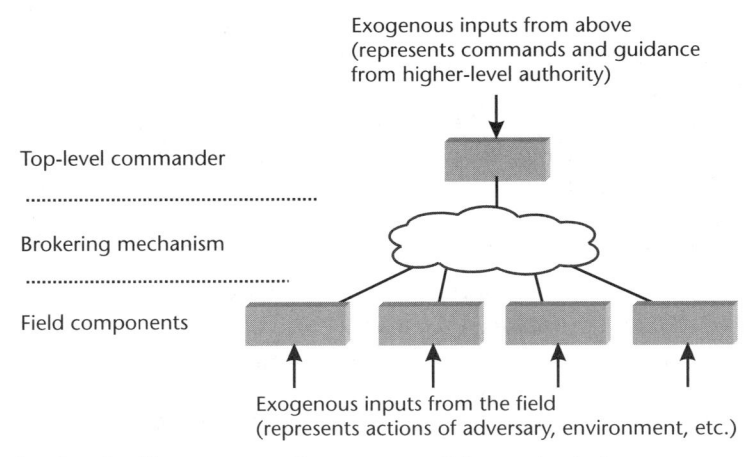

Figure 6.8 A network with a compensating component (e.g., a broker).

Active Compensation 145

to redistribute tasks or messages. The compensating component is illustrated in Figure 6.9. Although this component continues to propagate errors through the system, it attempts to drive the message rates to subordinates to a uniform distribution of messages (i.e., based on measured return rates from the subordinate components, future messages are directed away from components that have high levels of additional decision-making requests). It is assumed that the compensating component is itself subjected to the speed-accuracy tradeoff because any functions that were performed by the intermediate levels in Figure 6.5 remain. Hence, the redistributing component adds errors to the messages, depending on the total number of messages received from superior and subordinate components. This also allows us to focus on the impact of the redistribution of messages rather than the impact of the removal of the intermediate level of decision-making. The following difference equations are used to model the behavior of the compensating redistributing component.

$$y_1(k+1) = u_1(k)\left[f\left(u_1(k) + \sum_{i=3}^{n} u_i(k)\right)(1 - u_2(k)) + u_2(k)\right]$$

$$y_2(k+1) = f\left(u_1(k) + \sum_{i=3}^{n} u_i(k)\right)(1 - u_2(k)) + u_2(k)$$

$$y_i(k+1) = u_1(k)\left(\frac{n-4}{n-3}\right)\left(1 - \frac{u_i(k)}{\sum_{j=3}^{n} u_j(k)}\right), \quad 3 \le i \le n$$

Again, the stability of the network with a redistribution component was analyzed in simulation subject to the rate of exogenous high-level inputs from above—that is, $u_1(k)$ in the top component—and the rate of exogenous inputs from the field—that is, $u_3(k)$ in the four bottom components. Figure 6.6 shows the results,

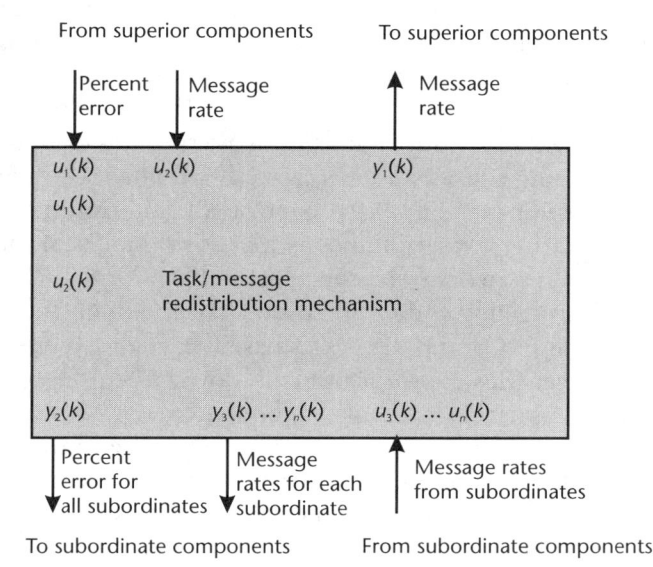

Figure 6.9 A compensating component.

along with the results for the hierarchical network. Again, the graphs show the average error rate in commands to the field components—that is, the average of $y_2(k)$ in each bottom component—as the exogenous input from above is increased. We note that with the redistribution component, the system also maintains its performance at a near-constant level and then rapidly collapses, though now at a higher total rate than the pure hierarchical network. Again, the stability was evaluated subject to variations in exogenous inputs from the field in the same way as for the hierarchical network—that is, inputs $u_3(k)$ in the far right and far left components were stimulated with different constant inputs. Figure 6.7 shows the stability envelope subject to the sum and difference of these constant inputs, alongside that for the hierarchical network. The stability envelope is increased with the inclusion of the compensating component, and therefore the system becomes more robust to variations in the difference in exogenous inputs at different components.

Dynamic Reorganization to Mitigate Malfunctions

The simplified models here indicate that an appropriately designed compensating component could significantly improve the performance of a decision-making organization over a broader range of operating conditions. But so far we've said nothing about how such a component would actually be implemented. We explore now a more specific scheme to such an active compensation and consider the internal workings of a compensating component that dynamically reorganizes the allocation of decision-making responsibilities and flows of relevant information.

In the traditional theory and practice of distributed and hierarchical control systems, a supervising controller issues commands to lower level controllers. These commands become inputs or set points for the control processes of the lower level controllers. The responsibilities and types of interactions between lower level controllers are static. They are fixed when the system is designed and remain constant while the system operates. When this constant definition of responsibilities and interactions becomes inadequate to adapt to the changing circumstances, as could be the case in an uncertain environment with adversarial threats and changing information requirements, the control system is unable to perform effectively. Additionally, in the context of a team of decision-makers that is composed of humans as well as automated control processes, it is typically a human senior leader (a military commander or a senior manager) who controls the allocation of decision-making and information-sharing responsibilities as well as the modes of such sharing. The structure of the team is also often fixed in a hierarchical format. A human leader typically adjusts decision-making and information-sharing responsibilities based on informal techniques, intuition, and best guesses based on prior experience and training [2, 5, 13, 14]. However, this control can break down when the complexity and speed of changes in the situation exceeds the cognitive and reasoning capabilities of the human controller. This problem may be exacerbated in those teams that include software agents and robots as decision-makers because these artificial decision-makers can observe, execute their decision-making algorithms, and act much faster than a human controller, potentially leading to a cascade of failures and poor team performance [15].

Here we explore a computer-assisted approach for managing decision-making, information-sharing responsibilities and modes of interactions between the team members that allows the team to effectively and rapidly adapt to changing circumstances, threats, and opportunities.

The predictive control scheme, illustrated in Figure 6.10, involves four main aspects that are described in the following sections. Predictive control is often employed, even if only implicitly, in planning and execution of military operations and industrial engineering applications. An internal model exists, even if only informally in the commander's situational awareness, and is used to determine an input that can then be applied to the real system. This is done either analytically if it is a mental model or through simulation if it is a computer model.

In Figure 6.10, we consider the input to the physical world or action space to be the *structure* of the decision-making team (i.e., the decision responsibilities and information channels). The epistemological content of the decisions themselves remains in the physical action space. The internal model within the predictive controller is only an abstraction that quantitatively captures the decision load and information channel loads. Similarly, as illustrated in Figure 6.10, the forecast decision requirements from the physical action space are abstracted to a set of time-varying parameters that are used as an open-loop input for the simulation of the decision-making process. The standard model-predictive control process is then enacted; different information structures are simulated with the given forecast decision requirements, and a best structure is chosen and used for the real decision-making team in the physical action space. This process of simulating a variety of information structures is captured by the internal optimization loop of Figure 6.10.

Modeling Team Decision-Making: Decision Responsibility and Information Structure

For illustrative purposes, we consider a simple example of a team of three decision-makers—a foreman, a scout, and a robot—involved in a mission whose objective is to extinguish a fire in a large chemical plant. We assume that an initial ad hoc plan

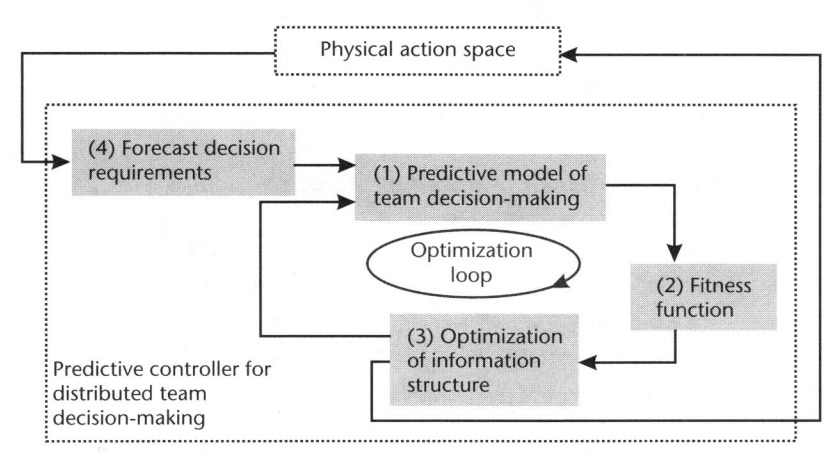

Figure 6.10 The control scheme for dynamic allocation of decision responsibilities and information sharing.

calls for the foreman to observe the fire from an observation post, the scout to identify a target location for dropping fire extinguishing material and then join the foreman, and the robot to drop the fire extinguishing material at the identified target location. An example of possible decisions required is as follows:

- U1: When and where to call a vehicle for the foreman escape;
- U2: When and on which target to drop fire extinguishing material;
- U3: Which flight path to take to reach the target;
- U4: When to egress and which egress route to choose for the scout.

We will revisit this example as notions are developed. The questions we wish to address in the context of this example are:

- Which of the three team-members should be responsible for deciding U1 through U4?
- How should the current decision outcomes be disseminated to other members in the team (noting, of course, the key fact that too much information may have negative consequences when an urgent decision is needed)?
- If we can predict the difficulty and urgency of decisions U1 through U4, can we manage the answers to the first two questions? That is, can we dynamically update who is responsible for which decisions and how the decision outcomes are disseminated?

Note that the decision-makers in this example are one and the same as the actors, but this is for convenience only. There may be entities in the decision-making teams that are physically removed from the field (e.g., an experienced foreman at another plant). Also note that here we are not considering the control of the observation process, again for convenience. Elsewhere [16], it is proposed that observation channels be controlled in much the same way as decisions. In our example, this could mean, for instance, appending the observations "Z1—Location of the fire" and "Z2—Types of target locations" to the list of decisions.

Let us employ a straightforward model to capture the traditional relationship between the task characteristics of complexity, urgency, and decision load and the resulting accuracy of the decision [10, 11, 17, 18]. A set of parameterized curves in which accuracy is inversely proportional to the time pressure is used for this purpose. An example of this relationship is shown in Figure 6.11. Unlike the simpler S-curve (Figure 6.1), here the additional parameters of normalized discriminability and the number of the options provide an upper and lower bound on the accuracy as illustrated. In a practical implementation of the scheme, characteristics of the decision-making entities could be used to quickly generate models from a predetermined virtual decision-maker coefficient database. Such characteristics might include rank and experience level for a human decision-maker or processor capability, and function and speed for an artificial decision-maker. As we are presently interested in the analysis of trends and the effectiveness of the approach, a relatively simple model suffices.

The relationship in Figure 6.11 is used to build dynamic input/output models for single decision-makers in the same way as was done in Figure 6.4 for the previous

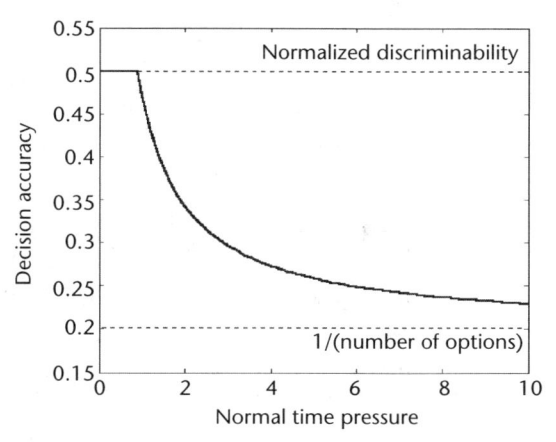

Figure 6.11 The decision-accuracy relationship for one decision-maker.

example. The normalized discriminability and number of options are exogenous inputs to the decision-making team, and we assume these are properties of the decisions themselves rather than the interaction processes in the team decision-making. Further, we assume the time pressure has two sources. First, there is an exogenous component that must be predicted, but additionally there is time pressure that results from the necessity of interactions with other team members. Clearly, if another team member is urgently soliciting information on a different topic, this reduces the amount of time that can be spent on the decision at hand and so increases the time pressure.

The input/output model for a single decision-maker is used as an atomic block to construct the team decision-making model. The connections between the atomic blocks are determined by what we will term the *information structure*. This term is not uncommon in the literature on decentralized and distributed control in more general settings [19–21]. We borrow the term here to refer specifically to the combination of three objects: an *observation structure*, which maps the incoming data streams to the decision-making entities or agents, a *decision-responsibility structure*, which partitions the outgoing decision variables among the agents, and a *decision-sharing structure*, which defines the information paths between agents.

A possible information structure for the firefighting example discussed earlier is partially illustrated in Figure 6.12. The decisions that each member is responsible for are indicated beside each agent (e.g., the foreman is responsible for U1), and this defines the *decision-responsibility structure*. The *decision-sharing structure* is illustrated with connecting arcs. The labels on these arcs show the decision that is being communicated and the *mode* in which it is communicated. Here, we consider only two modes: a *pull* mode, in which the first decision-maker communicates information only if the second requests the communication, and a *push* mode, in which the first decision-maker immediately communicates any new information. Though there is certainly a spectrum of possible interactions of this sort, we have identified these two modes for a preliminary analysis. The observation structure is again left out for convenience. Note that from a semantic point of view, there is a minimal necessary information transfer for the decisions to be made. For instance, decision U3 (the route taken to the target location) necessitates knowledge of decision U2

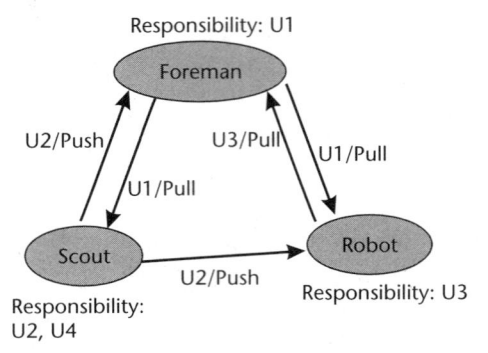

Figure 6.12 The decision-makers in the firefighting example communicate with one another and take responsibility for specific tasks as described by an information structure.

(selection of the location and timing for the target). Accordingly, we define a *decision dependency criterion* that underlies the dynamics of the model. Information structures that do not meet this minimal criterion will perform poorly because, for example, decisions concerning U3 made in the absence of knowledge of U2 are arbitrarily given an accuracy of 0 percent (we assume informally that two wrongs do not make a right). The decision dependencies for this example are that U1 requires knowledge of U2 and U3, U2 requires U1, U3 requires U1 and U2, and U4 requires U1 and U2. We assume that the minimal criterion can be met with either the pull or push mode communication. For simplicity, the observation process is not modeled in this example, and so the *observation structure* is not defined here. However, in general, the observation structure would define a set of observations (e.g., Z1 and Z2 for each decision-maker).

Measuring Decision-Making Performance

How do we know if one information structure is better than another? A quantitative measure of decision-making performance (e.g., overall accuracy or timeliness of decisions) is required as a measuring stick for the suitability of a given information structure. Clearly, the true intrinsic value of a given information structure is its effectiveness in the real world, but unfortunately without a high-fidelity simulation of the "physical action space" in Figure 6.10, this is not available. While a number of surrogate measures can be envisioned, we explore a particular one—a weighted average of the accuracies of all decisions being made by the team, integrated over the time horizon. This serves as an approximate indicator of performance. A potential implementation would couple of decision-making models with an action space simulation so that the predicted overall effectiveness could in fact be used as a fitness function.

As indicated in Figure 6.10, the value of the fitness function is used to fine-tune the information structure. Smoothness or monotonicity of the fitness function with respect to changes in information structure will not generally hold. More importantly, changes in the information structure are discrete, yielding a discrete (in state) and highly nonlinear optimization problem. In general, traditional continuous-state feedback control techniques are not directly applicable.

Repeated simulation for different alternative information structures can be performed to maximize the fitness function. This in effect requires a search of the space of information structures. The optimization problem can be stated formally in a form similar to the observation problem suggested in [21], and here, as there, it will generally bear no analytic solution. In practice, the optimization of the responsibility/information structure cannot be decoupled from the optimization of the physical actions, such as the execution of a military operation. However, an iterative or serial approach may provide reasonable solutions; current traditional operations planning tools do not explicitly consider the information structure as a quantity to be controlled; however, future operations planning may plan both physical operations and the information structure.

Forecasting Decision Requirements

In the decision-making accuracy curve in Figure 6.11, normalized discriminability, the number of options, and a time pressure characterize each decision. When this model is used in simulation within each dynamic model of decision-making, predictions for these characteristics are required in order to select the most appropriate information structure. The characteristics are predicted for a rolling time horizon (for the next 6 hours, for instance). Of course, these predictions should change due to action space events that are a result of the change in information structure. However, that is not considered here; the predictions are taken as fixed for the purposes of optimizing the information structure.

In practice, the raw information from which these characteristics could be predicted comes directly from the observations of the action space and other information and intelligence sources.

An example of predicted decision requirements for the four decisions for the firefighting scenario is shown in Figure 6.13 for a time horizon of 6 hours. For the purposes of our example, these were generated manually on the basis of the notional progression of the fire and the related actions of the firefighting team. For instance, the decision characteristics of U1 (corresponding to the decisions concerning the escape vehicle for the foreman) are such that there is little time pressure and many options at time zero, but as the fire progresses the time pressure increases (reflected by a decrease in the graph, which depicts time availability) and the number of options decreases steadily. This might be the case if escape vehicle options become scarcer as the fire approaches locations close to the foreman.

A Simulation of the Firefighting Example

Figure 6.14 shows a Matlab Simulink [12] implementation of the model in which the atomic blocks for the foreman, robot, and scout in the firefighting example are connected based on a given information structure (e.g., that in Figure 6.12). The forecast decision requirements in Figure 6.13 are used as exogenous inputs to this model. By simulating this model in time, we are able to generate, for any given infor-

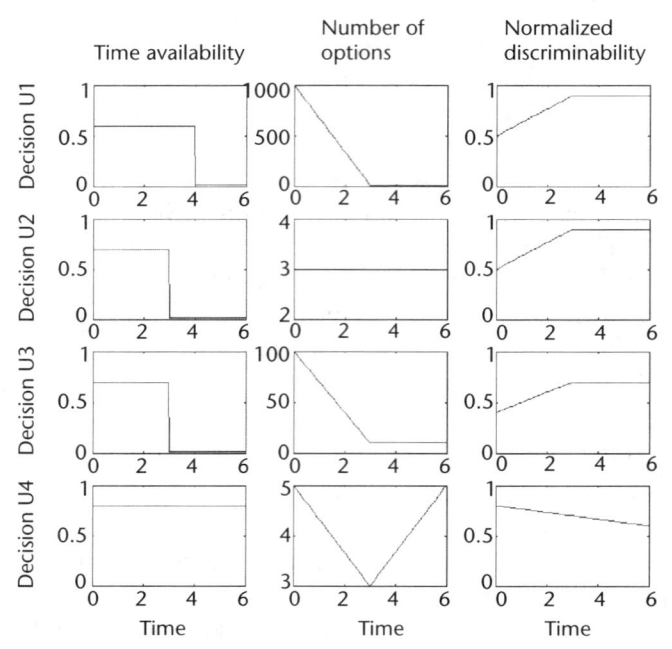

Figure 6.13 Future decision characteristics are predicted from information sources and used to fine-tune the information structure.

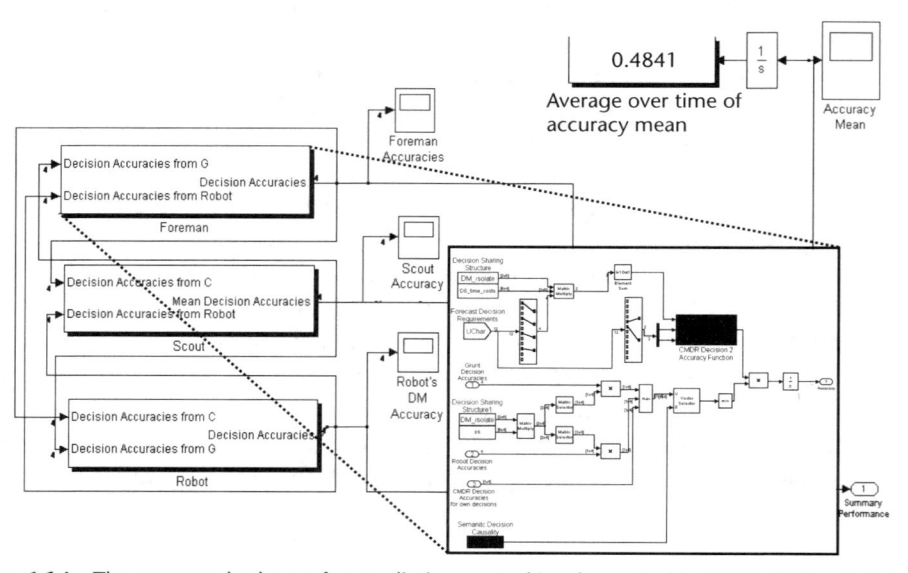

Figure 6.14 The proposed scheme for predictive control implemented in MATLAB Simulink for a small-scale example.

mation structure, an average accuracy of decisions over both time and decisions that provides a measure of performance for this information structure.

We investigated an optimization scheme based on a genetic algorithm, although a number of discrete optimization techniques could be potentially applicable. The

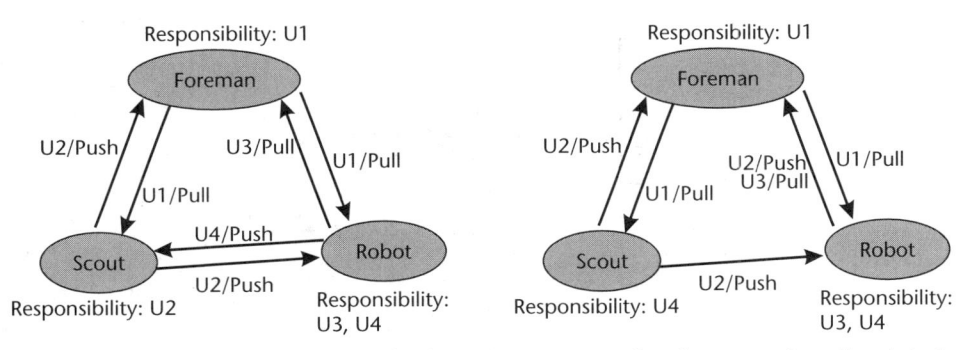

Figure 6.15 Examples of locally optimal information structures for given sets of predicted decision characteristics.

information structure is encoded as a chromosome without content loss, and the fitness of each chromosome is evaluated with a direct simulation of the team decision-making model in MATLAB Simulink. The Genetic Algorithm Optimization Toolbox [22] offers a large variety of mutation operators, crossover operators, and selection criteria and is well suited for this application.

Results for this optimization appear promising for the small-scale exploratory scenario with convergence to a (locally) optimal information structure with greatly improved fitness in realistic computational cost. A small variety of meaningful, common sense, information structures was produced (see Figure 6.15 for two examples), depending on how the forecast predictions were varied. This matches at least with intuitive notions of how teams should communicate and share tasks, and indicates the potential for the approach to be applied in real systems with higher fidelity decision-making models.

One might wonder how the effectiveness of such an approach depends on the characteristics of the environment in which the organization operates. Figure 6.16 conjectures a notional landscape of types of decision-making environments categorized by the accuracy of the predictions that can be made and the rate of change of the environment or operations tempo. The regions denote where we believe it is appropriate to use dynamically managed team decision-making rather than traditional team decision-making, where information structure is either rigid or evolves on an ad hoc basis. As the tempo of operations increases and the accuracy of predictions in the action space decrease, we believe it becomes more difficult to recognize the need for changes in information structure, and therefore it is in these cases that dynamic management of the information structure will improve performance.

Mitigating and Inducing Malfunctions

So, what do these models and simulations tell us about the susceptibility of an organizational decision-making process to an adversarial action? What can an organization do to mitigate its susceptibility to a malfunction? What can an adversary do to induce a malfunction? Based on the results of the simulations we described, one can suggest several approaches to architecting and operating an organization in a way that reduces its susceptibility to malfunctions.

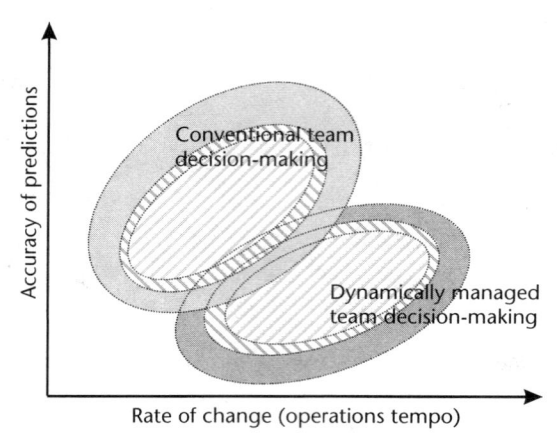

Figure 6.16 The regions of general applicability of conventional and dynamically managed team decision-making.

- *Empower subordinates via mission-type orders.* The onset of collapse can be delayed by reducing the number of requests sent from the field components to the decision-making components. This suggests that such a collapse can be mitigated by enabling the field components to operate as independently as possible in order to minimize the amount of cases in which they must call for the decision of the higher decision-making components. Providing the field components with maximum autonomy, using mission-type orders that specify the goals but not the specific ways to achieve them, and using command by negation (defined later) all contribute to the reduction of the K parameter and strongly improve stability of the overall system (see Figure 6.3). Conversely, an adversary may exploit the situations where field components within the target organization are compelled to operate under strict and detailed control by higher level decision-making components. An adversary may also attempt to induce this tight control by generating extensive negative publicity about any error committed by field components.
- *Prioritize and delegate.* Degradation can be avoided by dumping excessive messages (i.e., by ignoring some of them and insisting subordinate decision-makers handle problems autonomously). In practice, this is one of the mechanisms that organizations employ to reduce the effects of decision overload. Decision-makers learn to prioritize their decision-making load by ignoring what appears less important.
- *Use command by negation, not command by permission.* The fact that dumping of decision load is often necessary to avoid self-reinforcing degradation provides insight and support to the intuition that command by negation is advantageous as compared to command by permission. In the command-by-permission protocol, a lower level decision-making component detects a condition, formulates a plan for action, sends a request for permission to execute the action to the higher level component, waits for the permission (or denial) to arrive, and then executes the action. In the command-by-negation protocol, the lower level component does *not* wait for permission but proceeds to execute the action when the time is right, while

being prepared to abort the action if the higher-level component responds negatively. Clearly, avoidance of degradation by dumping at the higher-level component can be done more effectively in command by negation—if the higher level component ignores the message, it does not prevent the lower level component from executing the desired action. The same dumping of messages at the higher-level component in command by permission prevents the lower level component from executing the necessary actions to exploit an opportunity or to block a threat. In both cases, dumping enables the decision-making organization to avoid the degradation, but in the case of command by permission, this avoidance leads to greater rigidity and passivity.

- *Minimize the need for coordination.* Minimizing coordination loops, both vertical and horizontal, reduces susceptibility to self-reinforcing degradation. Although reduction in coordination may appear counterintuitive and controversial, some of the human factors literature has been calling attention to the potential negative impact of coordination requirements for a long time—for example, Morgan and Bowers [15] cite findings from Naylor and Briggs [23] as follows: "the performance of operators in a simulated air-intercept task was superior when the subjects worked independently of one another. Decrements in performance were observed when operators were placed in an organizational structure that encouraged interaction among the operators." Experimental findings (e.g., [13]) show that teams tend to perform better when they are able to communicate less under high-stress conditions. In the design of a decision-making organization, assigning tasks to minimize the need for coordination reduces the amount of knowledge the team members need to have about each other's roles and the amount they need to communicate, which can result in better overall performance [14]. An adversary, however, can compel increased coordination requirements by injecting misleading or conflicting information, or by presenting an organization with events that, by their very nature, require highly coordinated responses from multiple suborganizations.

- *Insulate the weak link.* Weaker decision-making components within the organization can accelerate the collapse of the entire system. It is advisable to insulate such a component from the rest of the system either by providing a greater degree of supervision or, if unavoidable, by allowing such a component to fail in its mission without expending excessive effort on the part of the superior component. This, however, may not be possible when the weak component is engaged in a critical task. This is also a situation that an intelligent adversary will attempt to create and exploit.

- *Diagnose online and compensate by dynamic reorganization.* In several experiments, we observed consistent symptoms of the onset of collapse manifesting themselves well in advance of the actual collapse. This observation suggests a possibility of introducing an online diagnostic mechanism. Further, diagnosis leads to a possible compensation by dynamic reorganization, such as we explored in this chapter (Figures 6.6 and 6.7). From the perspective of an adversary, however, the diagnostic and reorganization functions constitute excellent high-value targets.

Now, in spite of this rich harvest of practical insights, we must remind ourselves that intellective models are not intended to produce definitive recommendations. Their value is in generating useful suggestions: suggestions for high-fidelity modeling approaches, suggestions for the design of experiments, and suggestions for possible policy or structure redesigns. All these are to be explored and validated by other more concrete models or experiments. Despite their limitations, the strength of intellective models is that they are uniquely capable of identifying a phenomenon that would be difficult to detect in a pure and readily recognizable form in a real organization or even in an emulative model. By avoiding unnecessary detail, intellective models offer tools that help researchers visualize and explore the dependencies and the impacts of key factors in a manner that is relatively easy to grasp. Finally, intellective models are valuable in highlighting not only what they consider but also what they do not. Note that the series of related models we explored in this chapter all revolved around the measures of correctness or accuracy of information and decisions. They do not consider any measures of certainty, reliability, or trustworthiness of information and decision. They do not consider an important question: what happens when a decision-maker realizes that he received bad information or decision? And that brings us to the next chapter.

Acknowledgments

This chapter is based in part on the paper "Instability of Distributed Decision Making in Command and Control Systems," by A. Kott, P. Hubbard, and M. Martin, which appeared in *Proceedings of the 2001 American Control Conference*, Arlington, VA, June 2001, © 2001 IEEE. It is also based in part on the paper "Managing Responsibility and Information Flow in Dynamic Team Decision-Making," by P. Hubbard, A. Kott, and M. Martin, published in the *Proceedings of Command and Control Research and Technology Symposium*, 2002.

References

[1] Mavor, A. S., and R. W. Pew, (eds.), *Modeling Human and Organizational Behavior: Application to Military Simulations*, U.S. National Research Council, 1998, p. 275.
[2] Van Creveld, M., *Command in War*, Cambridge, MA: Harvard University Press, 1985.
[3] Cohen, E. A., and J. Gooch, *Military Misfortunes: The Anatomy of Failure in War*, New York: Vintage Books, 1991.
[4] Burrough, B., and J. Helyar, *Barbarians at the Gate: The Fall of RJR Nabisco*, New York: Harper and Row, 1990.
[5] Carley, K. M., M. J. Prietula, and L. Zhiang, "Design Versus Cognition: The Interaction of Agent Cognition and Organizational Design on Organizational Performance," *Journal of Artificial Societies and Social Simulation*, Vol. 1, No. 3, June 1998, pp. 1–19.
[6] Commission on Behavioral and Social Sciences and Education, *Distributed Decision Making: Report of a Workshop*, Washington, D.C.: National Academy Press, 1990.
[7] Sorkin, P. D., R. West, and D. E. Robinson, "Group Performance Depends on the Majority Rule," *Psychological Sciences*, Vol. 9, No. 6, November 1998.
[8] Sarter, N. B., D. D. Woods, and C. E. Billings, "Automation Surprises," in Salvendy, G., (ed.), *Handbook of Human Factors and Ergonomics*, New York: John Wiley & Sons, 1997.

[9] Kumar, P. R., and T. I. Seidman, "Dynamic Instabilities and Stabilization Methods in Distributed Real-Time Scheduling of Manufacturing Systems," *Proc. of the 28th IEEE Conference on Decision and Control*, Tampa FL, 1998.

[10] Luce, R. D., *Response Times: Their Role in Inferring Elementary Mental Organization*, New York: Oxford University Press, 1986.

[11] Louvet, A. -C., J. T. Casey, and A. H. Levis, "Experimental Investigation of the Bounded Rationality Constraint," in Johnson, S. E., and A. H. Levis, (eds.), *Science of Command and Control: Coping with Uncertainty*, AFCEA, Washington, D.C., 1988, pp. 73–82.

[12] MathWorks Web site, http://www.mathworks.com.

[13] Serfaty, D., E. E. Entin, and J. H. Johnston, "Team Coordination Training," in Cannon-Bowers, J. A., and E. Salas, (eds.), *Making Decisions Under Stress: Implications for Individual and Team Training*, Washington, D.C.: American Psychological Association, 1998, pp. 221–245.

[14] Entin, E. E., "Optimized Command and Control Architectures for Improved Process and Performance," *Proc. of the 1999 Command and Control Research and Technology Symp.*, Newport, RI, 1999.

[15] Morgan, Jr., B. B., and C. A. Bowers, "Teamwork Stress: Implications for Team Decision Making," in Guzzo, R. A., E. Salas, and Associates, (eds.), *Team Effectiveness and Decision Making in Organizations,* San Francisco, CA: Jossey-Bass Publishers, 1995.

[16] Kott, A., P. Hubbard, and M. Martin, "Method and System for Dynamic Allocation of Roles and Information Sharing Within a Decision-Making Team," Patent Application, BBN Technologies, 2002.

[17] Busemeyer, J. R., and J. T. Townsend, "Decision Feld Theory: A Dynamic-Cognitive Approach to Decision Making in an Uncertain Environment," *Psychological Review*, Vol. 100, 1993, pp. 432–459.

[18] Zsambok, C. E., and G. Klein, (eds.), *Naturalistic Decision Making*, Hillsdale, NJ: Lawrence Erlbaum Associates, 1997.

[19] Witsenhausen, H. S., "On Information Structures, Feedback and Causality," *SIAM Journal of Control*, Vol. 9, No. 2, May 1971.

[20] Teneketzis, D., "On Information Structures and Nonsequential Stochastic Control," *CWI Quarterly*, Vol. 10, No. 2, 1997.

[20] Ho, Y. -C., "Team Decision Theory and Information Structures," *Proc. of the IEEE*, Vol. 62, No. 6, June 1980.

[22] Genetic Algorithm Optimization Toolbox (GAOT) for Matlab 5 Web site, http://www.ie.ncsu.edu/mirage/GAToolBox/gaot/gaotindex.html.

[23] Naylor, J. C., and G. E. Briggs, "Team Training Effectiveness Under Various Conditions," *Journal of Applied Psychology*, Vol. 49, 1965, pp. 223–229.

CHAPTER 7
Gossip Matters: Destabilization of an Organization by Injecting Suspicion

Kathleen M. Carley and Michael J. Prietula

Accuracy of the information flowing through an organization is critical to its ability to function. We have seen that even a modest increase in the error rate generated in one node of an organization can induce a profound, far-ranging degradation of the entire organization's performance. However, another characteristic of organization's information flows—trustworthiness—can be at least as critical. Even with perfectly correct and useful information circulating through the organization's communication channel, suspicion or lack of trust with regard to the information can induce a dramatic collapse of the organization. Here, we explore such phenomena by focusing on a particular type of organization where the effects of trust and suspicion are relatively easy to understand and to model. We examine how smaller, more social types of organizations that exist on the Internet can be disrupted through loss of trust induced by deception. We show that a key mechanism that emerges to sustain cooperation in these organizations is gossip.

The remarkable communication capabilities of humans allows for symbolic representations and exchange of information, such as in written or spoken text [1–4]. A universal observation is that much of that information exchange, especially in relatively informal contexts and within informal groups, is comprised of *gossip*. Gossip, or its equivalent, "appears to be common to all mankind" ([5, p. 26]). Gossip (and related nontask social discussion) often makes up a remarkably large proportion of conversation, with estimates near 65% of conversational content [6]. Furthermore, children began to gossip about other children almost as soon as they "can talk and begin to recognize others" [7, p. 181; 8–11].

We explore what some of the theory and evidence tells us about gossip and its role in social settings, and how that impacts "veiled" groups that form to exchange information anonymously. Veiled groups are groups that operate with minimal social context cues, largely through message traffic such that the messengers are contextually anonymous; that is, they are known only by their nom-de-guerre and the messages they send. Online groups are a canonical example, and will be the specific subject of this investigation. By examining the impact of gossip in these groups, we will learn that, apart from what our teachers and our parents may have taught us, gossip is a fundamental component of who we are as social beings. It is part of our species-culture, and it is, in a literal sense, also a part of us. What might this tell us about our organizational decision processes and potential vulnerabilities? And what might this also tell us about insulating us against these vulnerabilities? To answer these questions, we take a simple, but surprisingly ubiquitous, type of organization—anonymous information networks—and build a computational simulation to explore the sensitivities of this type of organization to gossip. The specific type of anonymous information network that informs this discussion is the anonymous online group, such as an anonymous chat room. Gossip, as we shall see, matters—and it matters quite a lot.

Is Gossip Good or Bad?

From our own experiences, we see fundamentally different perspectives on the desirability and utility of gossip. On one hand, we are aware of the potential social damage that can be done through gossip and are familiar with various dictums that gossip should be avoided on moral. Consider the following observations by Kari Konkola, quoted in Shermer [12, pp. 45–46]:

> The Protestants of early modern England knew very well the habit to gossip and regarded it as a personality trait that absolutely had to be eliminated. Indeed, the commandment "thou shalt not give false witness" was believed to be specifically a prohibition of gossip.

Such guides are also found in tales told to children, as well as adults, as discussions of moral platitudes and passed down as cultural components of behavior. Consider the following (perhaps familiar) story [13, p. 143]:

> A man in a small village was a terrible gossip, always telling stories about his neighbors, even if he didn't know them. Wanting to change, he visited the Rabbi for advice. The Rabbi instructed him to buy a fresh chicken at the local market and bring it back to him (the Rabbi) as quickly as possible, plucking off every single feather as he ran. Not one feather was to remain. The man did as he was told, plucking as he ran and throwing the feathers every which way until not a feather remained. He handed the bare chicken over to the Rabbi, who then asked the man to go back and gather together all the feathers he had plucked and bring them back. The man protested that this was impossible as the wind must have carried those feathers in every direction and he could never find them all. The Rabbi said, "That's true. And that's how it is with gossip."

The moral is unequivocally related to the negative effects of gossip, which, in this example, is related to the spread and inability to retract "inappropriate" information. When we explore the historical role of gossip, we find the practical implications of this moral perspective. Spacks [14] describes the fundamentally important position that *reputation*, and hence gossip about reputations, played in early society. The social ecologies were such that gossip could (and often did) have grave consequences for the subject of the talk, such as seen in the accusations of witchcraft (e.g., [15]). Yet, societies sometimes established substantial punishments for the gossiping individual [16]. Spacks [14] notes that "the perception of derogatory talk as morally destructive continued unchanged from the Middle Ages to the Renaissance, as did the (negative) imagery associated with it" (p. 27) and that "one can hardly overstate the seriousness of sixteenth- and seventeenth-century attacks on malicious gossip" (p. 28).

On the other hand, emerging theory and research argue strongly that there is a social need to share information to establishing closeness in a group [17] and even an obligation to share such information [9–18]. Thus, on a more fundamental level, gossip is not only to be expected as a mechanism to support social groups, but is even seen as playing a critical role in the evolution of human society itself as a form of social grooming, bonding, and information maintenance [19]. Both the Social Brain Hypothesis [20] and the Machiavellian Intelligence Hypothesis [21] present strong arguments and evidence that intelligence and language emerged as it did through evolutionary pressures to successfully address complex social problems. Gossip is viewed as a critical behavior in controlling, and sometimes generating, these complex social problems.

Dunbar [22] argues that "without gossip, there would be no society" (p. 100) in support of Enquist and Leimar [23], who suggest that gossip evolved as a linguistic mechanism for handling free-riders (i.e., cheaters). Research has demonstrated in general that when exchange tasks are presented in a "social context," people are, in fact, more successful at detecting conditions for cheating [24–27] and that cheaters in social exchanges command more attention and are remembered more accurately than are cooperators [28, 29]. Furthermore, growing evidence suggests that there are neurological bases that help detect cheaters [30, 31]. Additionally, research has suggested that when humans deliberate about social contracts, they engage distinctly different reasoning and neural subsystems [32, 33].

The consequences of such detection can often result in stigmatization and ostracism of an individual from future interactions, and this exclusionary behavior has substantial evolutionary adaptive value for the group [34]. Moreover, there is evidence that such exclusion results in both social and physical pain [35], providing feedback for both individual and social learning of such consequences. Also, evidence is building for neurological bases for emotional responses to presumed perspectives of others—the "moral emotions" [36–39].

These apparently incommensurable views of gossip being "both good and bad" are reconciled when one sees gossip not as a construct that is either present or absent, or even as either good or bad, but a type of behavior that if somehow regulated toward some form and frequency can achieve a nondestructive equilibrium to the overall functioning of the group. Whether this arrangement is explained by an altruistic, community focus (e.g., [40]) or as a type of individualistic "information management" used to promote and protect individual interests by controlling dissemination [41], it is clear that gossip plays a variety of important functions that can be psychologically, physiologically, and sociologically beneficial or punitive. From a behavioral perspective, gossip appears in most, and probably all, cultures and in a wide variety of contexts [42–46].

In summary, in turns out that both research results and theoretical stances suggest that our brains and behaviors are indeed adapted to function in social groups in order to *facilitate sustained cooperative social exchange*. Strategically, the ability to detect cheaters and accurately associate prior interactions with these individuals is important for cooperation to emerge within a group [47]. But do these groups do better, given they have to engage in costly monitoring and punishment? In what are called common-pool resource problems (i.e., limited and important resources that require cooperative behaviors), groups of nonaltruistic agents that can maintain sustained cooperation (via punishment) have a substantial competitive advantage over groups that do not (e.g., [48, 49]). Gossip, then, can be a way to maintain cooperative social exchange, thus providing individual and group benefits [50, 51]. As we will argue, there are dynamic, emergent organizations in which gossip will be the *sole* mechanism by which cooperative social exchange is maintained. As we shall see, as long as there are consequences for targets of gossip, these organizations can be disrupted.

What Is Gossip?

This chapter is about gossip, but what *is* gossip? There is no widely accepted definition of gossip, and gossip is also confused with rumors. Though they do have elements in common, we view them as different, especially in their pure forms. First, let us define rumors. Three key properties of rumors are (1) the absence of demonstrable fact and presence of ambiguity, (2) some level of significant importance or interest on a scale larger than the current social context, and (3) a considerable duration and spread. In their classic work on the subject, Allport and Postman [52, p. ix] define rumor as a "specific (or topical) proposition for belief, passed along from person to person, usually by word of mouth, without secure standards of evidence being present" [53, 54]. Similarly, Rosnow and Fine [55, p. 10] suggest that "rumors are not facts, but hearsay: some rumors eventually prove to be accurate, but while they are in the stage described as 'rumor' they are not yet verified." Rumors address individuals, but also can address issues or events, that are "on a larger scale of relevance" than the immediate social context, though they may be (but not need be) relevant to the immediate social context. For example, Allport and Postman [54] studied rumors regarding World War II. Rosnow and Fine [55] explored a broad array of rumors ranging from the Beatles to the stock market. Knopf [56] studied the relation between rumors and race riots. Koenig [57] explored a wide variety of business rumors (e.g., the Proctor & Gamble rumors of associations with Satanic cults) and articulated the "three Cs" that define the substance of most rumor studies: crisis, conflict, and catastrophe. As rumors spread, they appear to evolve systematically and functionally, though there is no significant agreement on either form or function (e.g., [56, 58–60]), and rumors can be remarkable persistent [61]. Rumors can have varying (direct and indirect) impacts on social conditions on a wide scale, ranging from relatively innocuous discussions of celebrity events [62] to conspiracies [63, Chapters 3 and 4], to stock market fluctuations [64], to "mass hysteria" [65], to riots [56, 66], and even to more substantial conflicts, such as the French Revolution [67] or World War I [68].

Regarding gossip, we adopt a view similar to that provided by Rosnow and Fine [55] where gossip is "small talk about personal affairs and peoples' activities with or without a known basis in fact," while rumor is seen as "unsubstantiated information on any issue or subject." We view gossip as an element that is about "concrete individuals and their concrete situations" [69, p. 34], is evaluative [70], is "about people known to the persons involved in the communication" as a form of social control (directly or indirectly), involves essentially nonpublic information [71], is often "a means of reinforcing group norms, in that people are motivated to conform in order to avoid being the target of gossip" [57, p. 2], and takes place "mutually among people in networks or groups" [72, p. 38]. Furthermore, we view the use of gossip as resulting from a deliberate and calculated *choice* to behave in a particular way, in particular situation, for a particular purpose. The specific components of such a choice (i.e., what is said, to whom, and when) have individual and collective consequences, which may or may not impact individual and group behaviors. De Vos [73, p. 21] suggests that "gossip is an unverified message about some *one* while rumor is an unverified message about some *thing*." We would augment that heuristic by suggesting gossip involves some one who you know or is within your social context, while rumor (about individuals) involves those who are not in your social context [74, 75].

In this chapter, we focus on gossip; specifically, we explore the situation where *gossip involves the manipulation of the reputation and behavior of others for the purpose of enforcing a norm of honesty in social exchange* [76, 77]. Consider that in such situations gossip "does not merely disseminate reputational information but is the very process by which reputations are decided" [78, p. 135]. There is perhaps no other social exchange context where this is more salient than in simple organizations arising from anonymous information exchange networks in the Internet.

Anonymous Information Exchange Networks

Anonymous information exchange networks are organizations that occur frequently in a variety of settings. In Internet environments, such networks take on a variety of forms. One form is the formally defined organizational mechanisms explicitly configured for information related to reputation, such as those used by eBay, Amazon, or Epinions. Another Internet form of the anonymous information exchange network is the more informal and dynamic "mission-oriented" organizations such as sites, blogs, and forums that focus on specific problem-solving contexts and domain-specific discussions, such as Java or C++ programming or specific news groups. Still another form is the dynamic types of organizations such as peer-to-peer networks exchanging bootleg music and video, based on person-to-person (P2P) architectures, and ephemeral chat rooms emerging on ghost servers in order to exchange information on hacking or other malicious or illegal behaviors. However, such anonymous information exchange networks are not limited to the Internet. For example, through the use of various devices such as written messages, secret codes, dead letter drops, disguises, and the use of aliases, a variety of secret societies and covert networks have maintained themselves as effectively anonymous information exchange networks. In both the online and offline networks, there is a semblance of, if not the fact of, anonymity. In both cases, the key to reputation lays in the messages that are passed—and so the gossip. In general, five specific properties seem to hold for these groups:

1. Participants are *contextually anonymous*.
2. Participants communicate only via *messages*.
3. Participants rely on simple forms of *trust mechanisms* to maintain the integrity of the network's functioning and purpose.

4. Participants (consequently) view their online *persona reputation* as essential, for this is the ultimate barrier to exchange.
5. Participants use *gossip* about persona reputations as the primary exchange control mechanism.

By "contextually anonymous" we mean that it the participants typically do not use their own name, but rather rely on a nickname or user identifier by which they are known for that particular exchange context. The main issue is whether that nickname or identifier can carry ancillary information about that participant *outside* of that context [79–84]. Most human networks contain sufficient information about individuals to find out (or perhaps recall) more information that may be derived outside of the context of the primary exchange. In fact, the *lack* of anonymity enables a remarkably rich and connected network to emerge [85, 86]. The types of organizations we are describing afford *isolated reputational networks*. Thus, anonymity may not always be a necessary condition for these organizations to function as we describe (e.g., recent decisions by Amazon to be less anonymous in their customer review information) if there is sufficient isolation to attenuate exogenous information about the participants.

The first two properties sufficiently constrain how decisions are made under uncertainty within this type of organization. The third property argues that, lacking detailed information about each other or any form of exogenous organizational control, these organizations rely simply on trust mechanisms of the participants. By *trust mechanisms*, we mean the set of constructs and behaviors that produce and enforce social exchanges based on individual beliefs that future exchanges with specific individuals are likely to succeed or fail. Fourth, a key component of instituting these trust mechanisms is the use of individual reputations as the barometers of exchange success. Note that we describe these as reputations of *personas* and not reputations of individuals. In accordance with our first property, the participants usually do not use their actual, traceable names, but use other (often chosen) identifiers to which we collectively refer to as personas. Given the restricted nature of communication, all information (and its consequences) is based on attachments to these personas, and as these personas may be altered quickly, there must be some mechanism that accommodates the consequences of such an easy detachment between personas and reputation [87].

That mechanism in our model (and in these networks) is gossip. We suggest that the trust mechanisms are the minimal that are necessary (and possibly sufficient) to instill control to maintain the stability of the organization when used in conjunction with gossip. Thus, *individual trust (suspicion) mechanisms* account for decisions to cooperate in uncertain environments, and these decisions are mediated by collectively visible *gossip* regarding the reputation of personas.

Hypothetical Organization, Part I: It's All Bits

The observations we make in this chapter occur for all types of organizations that have these properties. However, for this chapter we focus on a particular type of anonymous information network—those online chat rooms that illegally engage in exchanging stolen information or malware. As messages are the only way to communicate, we assume that the medium and content of exchange are also digital-based. Exchanges and payments occur in restricted forms, such as open servers on (usually compromised) hosts, are held in "secret" chat rooms, identify theft data (social security numbers, credit card numbers, information lists) [88, 89], and involve root access information to routers or personal machines [90–92], malware/bots [93, 94] and engines to make them [95]. These types of markets can be pure spot or direct exchange markets in a transaction between two actors, both of whom give and receive from each other either identical or different goods, either immediately or sequentially. Such exchanges include "any system which effectively or functionally divides the group into a certain number of pair of exchange units so that, for any one pair X-Y there is a reciprocal relationship" [96, p. 146].

On the other hand, these types of markets can operate in terms of indirect reciprocity, whereby an individual can donate information to a recipient with the expectation that such cooperation may be subsequently rewarded by someone else "in the group" other than the immediate recipient [97]. As a consequence, there must be mechanisms that account for the spread of social information used to both execute the obligation (i.e., the reciprocation) and control for behaviors that either support the structure (e.g., reward donators) or undermine the structure (e.g., those who either do not donate when they can or whose information is flawed).

Though the exchange may be immediate, and possibly indirectly reciprocal, the value is actually indeterminate until the exchanged product is engaged. Are the social security or credit card numbers valid? Is the information list bogus? Is the malware (code or script) useable? Is root access available? Are the bots available? Once engaged, the result determines, in part, whether the exchange was reciprocal or exploitive. The consequence of this drives the trust in these markets. If trust is of a sufficient level, these markets can efficiently distribute information-based products that can disrupt individuals and organizations. However, the simple structure that makes them efficient (pure information-based) also makes them vulnerable and sensitive to disruption. But what are these structures and why are they dangerous?

Hypothetical Organization, Part II: Miscreant Markets

Although the primary functions of these organizations resemble markets, and we often refer to them as markets, these types of organizations also bear a resemblance to loosely configured team or "crew." The reason is that there are hurdles to participate in the market that function as membership rites, as well as organizational norms to obey. We must first make a distinction between what the nature of the exchange may be and the reasons behind the exchange. The reasons behind the exchange can be quite diverse, ranging from challenge to entertainment. However, the dominant reason emerging (and the primary nature) is unequivocally profit, on at least one side of the exchange equation. For example, cyber-warfare (as a motive) can be viewed as conducting damage whereby the "information infrastructure is the medium target and weapon of attack, with little or no real-world action accompanying the attack" [98, p. 17]. With comparatively little resources, a country can engage in acquisition of key disruptive technological capabilities that can invoke substantial cost of another nation. However, active efforts in other countries with substantial resources (intellectual, economic) suggest attacks on U.S. infrastructures can have consequential economic impacts and such consequences are "both plausible and realistic" [99]. Acquisition of funds and malware to engage such goals drive participation in these markets.

Hackers (and other underground loosely federated groups) have distinct cultures that include in-group, out-group concepts (black hats, white hats), social structures (crews, hierarchies), roles (bitches, experts, personas), language forms (eblish), trust mechanisms (trials, rites, gossip), organizational knowledge (Who knows what? Who knows who knows what?), organizational goals (anti-Microsoft, profit), organizational learning (how do they adapt?), markets (exchange of information, selling of information), and many other elements that are little understood but essential to properly assess risk and predict/explain observed events (e.g., attacks). These hacker venues are the emergent Internet neighborhoods that are based on shared avocation rather than shared space [100]. The shared avocations range from adolescent hazing rituals to digital extortion and terrorism.

As we have noted, a remarkable element of these cultures is that virtually all communication is via simple text exchanges (e.g., chat servers) without specific identification of individuals or any physical encounters. These people do not meet each other and do not know (nor often wish to know) anything beyond what is revealed in the network of communication exchanges. This loose confederation of amorphous individuals, however, can quickly be coordinated to unleash cyber attacks (floods, worms, virii) on specific targets in a most effective manner or conduct an exchange of information that results in thousands of dollars pilfered from a bank account. Prior work on such culture is informative, but even recent work on general Internet culture only begins to elucidate core elements of this subgroup (e.g., [101, 102]). These groups can be internationally located yet locally dangerous [103]. Although there is a substantial entertainment core to hackers (e.g., [104]) and an omnipresent element of warez trading (i.e., illegal software collection and dissemination [105]), without question (and often without appreciation) these dangerous cultures are becoming increasingly characterized by two facets:

1. Most events ultimately lead to the illegal acquisition of *funds* (the exchange consequences or motives may be less or more *directly* terroristic in nature) [106].
2. Elements of the culture exist to preserve facet 1.

Much of this effort often involves a complex web of events and communication (in chat rooms) centering on compromised machines and the information obtained from them. Attempts to mitigate these dangers have had success, but the challenge grows daily [107, 108]. Even tracking compromised machines of unsuspecting individuals can be difficult.

Given our discussion of gossip and the nature of these organizations, we now try to understand the weakness in the structure. This is accomplished by examining a model of exchange in these environments, called the trust, advice, and gossip (TAG) model. Through an analysis of this model, we can examine the sensitivities of the organization to changes in key parameters. The model is implemented in an agent-based simulation of a series of exchanges in chat rooms and determines how information (including gossip) alters the efficiency and effectiveness of the exchanges. This implementation is TrustMe.

TAG Model Overview

The use of simulation in understanding organizational phenomena emerged with the early use of computers [109] and recently has regained substantial status and standing in the organizational science community [110, 111]. TrustMe is a computer simulation that implements a version of the TAG model in an online chat room context [112, 113]. We have discussed gossip; we will now address trust. As we are well aware, some form of trust is essential in virtually any form of social [114, 115] interaction. Resembling the asserted importance of gossip, Good [116, p. 32] argues that "without trust, the everyday social life which we take for granted is simply not possible." Research on trust is both diverse and contradictory. For the most part, this is because there is no generally accepted definition of trust, and, therefore, it is redefined as contexts and disciplines vary. Furthermore, trust (and its related correlates) have been defined, theorized, and investigated on many levels, such as societal [117], institutional [118], group [119], and individual [120], as well as neurological [121], and neurochemical [122].

For our purposes, we can simply view *trust* as Rousseau et al. [123, p. 395] define it—"a psychological state comprising the intention to accept vulnerability based upon positive expectations of the intentions or behavior of another," and see trust "as a matter of individual determination and involves choosing between alternatives" [124, p. 16]. Thus the "vulnerability" is encased as a willingness to risk an exchange. This necessitates two components in our model of chat room exchanges: one describes the social environment and the other describes how individuals react to that environment. First, there must be *sufficient uncertainty* in the environment such that trust plays a role influencing choices in that environment—there is a continuous element of risk of information invalidity inherent in an exchange. Simply put, not all information sources can be "trusted" to be accurate or even honest, and this condition may vary but not disappear. Second, the underlying form of the construct reflecting trust within an agent must afford states that reflect variation in responses to those environmental risks. That is, agents have decision mechanisms to absorb that risk and adjust accordingly.

As noted, the model defines an initial network of agents who often gather in chat rooms (or exchange environments similar to chat rooms) to exchange information in the general form of *advice*. In some rooms this advice may be free, while in others this may involve direct or indirect reciprocation. To an agent, the nature of the advice has potential value, such as the location of information on how to acquire virus mutation engines, or may involve more complex exchanges yielding PayPal account transfers or root access to a set of comprised machines. There is, consequently, a lag between the acquisition of the advice and the actual valuation of that advice for the recipient. We refer to these lags as *search* as a generic parameter identifying effort required to realize valuation. If an agent has a goal find out in what chat room one can acquire access to a Paypal account, then he or she may successfully engage in an exchange that yields advice (e.g., the location and form of this particular chat room) whose value is determined quickly. On the other hand, the initial exchange may fail and the search for valuation may involve several attempted exchanges to find the correct type of chat room. Each agent has a search limit, defined as the effort beyond which the agent will cease to satisfy a particular goal and either turn to another goal (if one is available) or drop out of the network entirely.

In TrustMe, there is no explicit concept of well-defined subgroups such as coalitions, friendship groups, or alliances, and, in fact, these are not required for the type of interactions addressed [125, 126]. The subgroups defined by TrustMe are solely those associations defined by trust based on communication events. TrustMe agents know each other solely by their behaviors (advice) and judgments are made based on those behaviors, directly or indirectly. In a sense, these agents are forming an "information structure" or "advice network" that may grow, shrink, or change its internal structure or density depending on the parameters of the task, the agents, and the environment. Organizations such as these are rarely without norms of behavior, and one of the primary norms is to control the quality of the advice available. As the quality of the advice is maintained, the sustained value of the network allows for mutual benefits for all (e.g., cash, access, knowledge) to be continued. The primary mechanism to control the quality of advice takes the form of gossip. In the TAG model, gossip is defined as follows:

> *Gossip* is an assertion posted to a chat room concerning the evaluation of another agent's behavior.

An example of this gossip in TrustMe looks like this:

[4.02.06 08:23] *** agent21 has joined #33
[4.02.06 08:24] <agent21> does anybody know where I can find i12?
[4.02.06 08:26] <agent04> do not trust agent21. It lies!!! :<
[4.02.06 08:27] <agent08> do not trust agent21. It lies!!! :<
[4.02.06 08:23] *** agent21 has quit the chat

In this exchange, an agent (agent21) enters a chat room and asks where it can find some information item named "i12" (an arbitrary identifier, representing some unique information such as a location for a mutation engine or an answer to some other question). In response, two agents react and generate gossip about this particular agent to the group as a warning, and agent21 leaves the chat to search elsewhere.

What is interesting about this situation is what is *not* there. Recall the prior discussion on gossip and the roles it can play in social contexts, the type of contexts that dominate our lives. This was based on research that has demonstrated the value of social groups and how that value is maintained by exchanges of social information. Recent research has also noted that there seems to be fundamental neurological apparatus to attend to socially relevant information in the environment based on perceptual cues, such as social gazes [127, 128], nonverbal cues for deception [129], emotional cues from body language [130], recognizing others through body language [131], and judging trustworthiness and threat of others through their facial cues [132, 133].

The online chat rooms as we have been describing essentially negate much of this "built in" human apparatus. Also lacking is the ability to apply explicit verbal and nonverbal strategic methods (sans technology) that are taught and employed in attempts to detect deception [134–142]. Thus, the viability of social exchange is maintained *entirely* by direct experience and gossip based on the behavior of other agents. That experience and gossip has an impact on trust, and trust, consequently, mediates the behavior [143, 144]. The implications of this are both obvious and direct:

> The viability of any TAG organization is fundamentally sensitive to how gossip and trust mechanisms interact in maintaining the collective integrity of social exchange.

The Mechanisms of Trust

Yet, the story is somewhat more complicated when the actual mechanisms underlying the decisions that accommodate trust are addressed in the TrustMe implementation [145, 146]. In general, individuals in such organizations are known by their nicknames and protect these, often with institutional methods such as a NickServ service or through more anonymous, though perhaps aggressive, methods. In these organizations, reputation is everything. Even in this simple model, consider the decisions involving messages received and messages sent. Table 7.1 summarizes some of the basic decisions. *How* these decisions are made is determined by three types of underlying additional models that relate decisions to internal states or events: trust, honesty, and gossip.

Trust Model

The trust models used to define the states of trust take the form of (predefined) state-transition diagrams where an agent's *state of trust* can take several forms, bracketed by two opposing end states (Trusting, Distrusting) and (optional) inter-

Table 7.1 Decisions Based on Advice and Gossip Contexts

Message Context	Decisions
Advice about a goal	Does the agent take advice? How does the agent handle advice from multiple, possibly conflicting, sources?
Agent receives	
Gossip about an agent	Does the agent believe the gossip? How does the agent handle gossip from multiple, possibly conflicting, sources?
Advice about a goal	Does the agent respond to a request for advice? Does the agent provide honest or deceptive advice?
Agent provides	
Gossip about an agent	When does an agent provide gossip? Does the agent provide both positive and negative gossip about agents? Does an agent provide dishonest gossip (deception)?

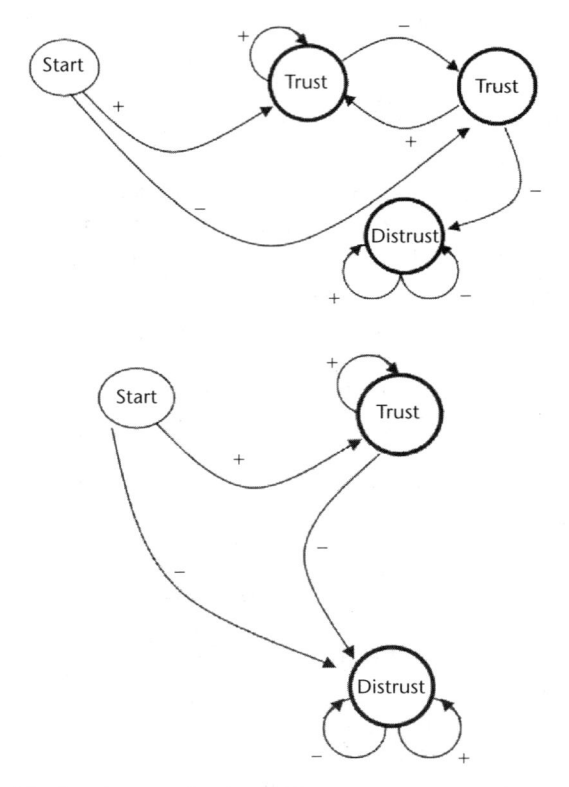

Figure 7.1 Trust models showing two levels of tolerance.

mediate states that function as event counters for positive and negative events (advice, gossip) that may impact movement toward or away from the end states. In addition, either end state can be defined as absorbing or not. In an absorbing end state, no subsequent events can dislodge an agent from that state. In a nonabsorbing end state, subsequent events can result in shifts of an agent's trust. For the examples in this chapter, and as will be explained, all agents will employ one of two unforgiving trust models. This is depicted in Figure 7.1, and will be explained later. The models differ in their tolerance for bad advice from a source. In one, agents allow bad advice only if it is followed by good advice. Two pieces of bad advice in a row from a particular agent results in an absorbing state of distrust toward that agent. In the other, a single piece of bad advice will immediately results in an absorbing state of distrust toward that agent. However, all models begin in an initially optimistic "trusting" state and realize the important (though sometimes ignored) element that social history matters and is integrated into subsequent choice (e.g., [147–153]).

Individuals in chat rooms often receive messages from multiple sources regarding a particular question or issue, which can be characterized as a threshold model [154]. In TrustMe, there are heuristic decision procedures to address these multiple advice resolution problems by one of the four advice resolution strategies. This presumes that there are $N > 1$ sources of advice to resolve. These are based on strategies to either use the trust model or not, and to randomly select from the set or base the decision on a maximum vote, as shown in Table 7.2.

Table 7.2 Conflict Resolution Options

	Use Trust Model: No	Use Trust Model: Yes
Random selection	Select advice randomly	Select advice randomly from only trusted agents
Maximum votes	Select advice with most votes	Select advice from trusted agents with most votes

Honesty Models

Research seems to demonstrate the wide variation of conditions under which people, given the absence of significant threat cues, tend to cooperate in general [155] and on the Internet in particular [153]. However, there is an element of deception in life in general [157], and this is seen as an important element of online society [158]. Here deception takes the form of choices to provide *wrong* advice to an agent. This behavior may reflect individualism in order to achieve asymmetric gains in an exchange with insensitivities to group norms, but it can also be engaged as punishment for the same behavior by those who attempt to enforce group norms and coherence [114, Ch. 11; 159, 160]. The honesty models define an agent's decisions when it notices a request for advice in a chat room. There are four models available in TrustMe: (1) always answer (honestly); (2) answer only if the requesting agent is trusted; (3) answer only if the requesting agent is trusted, but deceive the agent if it is distrusted; (4) answer every post by deception.

Gossip Model

The gossip model has three components: when to post gossip, when to believe gossip, and what part of the model gossip affects. Apart from reactions from posts defined in the honesty model, agents must consider what to do when directly experiencing bad (and good) advice from another agent. For bad advice, there are three model types for posting gossip: never post gossip; post gossip when an agent experiences bad advice; post gossip when an agent's trust model state switches to distrust.

These different models can be ambiguous in terms of their underlying causality. An agent that never posts gossip can be viewed as either a "highly moral" agent, obeying a higher norm of behavior (reminiscent of the initial chapter quotes), or can be viewed as a "highly individualistic" agent who secures private information and shirks a social norm of enforcement because of, for example, excessive (perhaps social) costs [161]. Posting gossip in response to direct experience (either immediately or lagged based on the trust state) can be viewed as an agent engaging in a socially responsible act (altruistically) or assumes that it will benefit in the future (indirect reciprocity).

On the other hand, under what conditions will an agent *believe* gossip? Similarly, there are three model types for determining the impact of gossip viewed: never believe gossip; believe gossip about some agent if the number of gossip posts exceeds some threshold of gossip tolerance, G_{tol}, which defines the "general norm of tolerance" for the group; and believe gossip about some agent only if the posting agent is trusted (may need a conflict-resolution mechanism).

Finally, what might gossip affect in the decision model for the agents? Gossip may impact whether or not an agent takes advice or provides advice to another agent, independent of any trust component, or it may directly impact a trust component. These are situations when an agent does not directly experience deception, but does indirectly through the assertions made by others. Is it meaningful to say that indirect experiences can impact trust of an agent never encountered? Should the trustworthiness of the sources of gossip figure in? Or should there be some independent "risk count" that does not directly impact trust formation [162]?

Gossip and Disruption

We explore some of the implications of this perspective by running a series of virtual experiments, where agents have goals in which they attempt to acquire information that is simply characterized as integers but symbolically stands for any information object that is transferable electronically. Accordingly, they seek out chat rooms that may have participants who possess knowledge of, or access to, those information objects. Each agent is also boundedly rational in the following three ways [163–165].

- First, agents have a limited memory of chat room contexts. Each time an agent visits/considers a chat room, it stores the context of that location in memory. This <chat_room: context> pairing is called a *chunk* [166–168]. However, there is a limit to how many of these chunks an agent can recall. Agents generate experientially based chunks as a first in, first out (FIFO) aged knowledge queue.
- Second, not only do agents have a limited number of chunks they can recall, but they also may encounter errors in recall of those chunks. The likelihood of error is based on a time-based trace degradation parameter. The decay is a simple function of the relative length of time (age) a particular memory chunk has been resident in memory, where the decay is expressed as a likelihood of accurate recall defined by $age^{log\beta}$.
- Finally, there is search limit tolerance, S_{tol}, which defines the threshold beyond which the agent will not seek to contact the network for advice for that goal and move on to another goal (if available). The reasoning behind these limits is, in part, to define a behavioral baseline level for error in the simulation. That is, it is assumed that the particular network has adjusted to the "white noise" of occasional error and accuracy that recurring (and accepted) members of the group afford, and that equilibrium is defined by the behavioral responses of the active set of members.

Setting the Baselines

All runs involve 20 agents, each with seven randomly assigned information goals to achieve. For these simulations, each information goal is independent and self-contained [169]. All agents have a primary chat room where they exchange messages. Each agent has a memory limit of 50 items of the discussed form, <chat_room: context>. The decay parameter was set $\beta = 0.9$, which results in a slight tendency toward cognitive error as time increases without reference to the chunks. Agents have an associated search tolerance of 50. The norm for the network is set to be *slightly* tolerant but unforgiving. Specifically, an agent will tolerate one piece of bad advice but not two pieces of bad advice *in a row* from a given agent; otherwise, the state shifts to an absorbing distrusting state for the source agent (as shown in Figure 7.1). Thus, each agent has a social memory of interactions with all other agents (i.e., the network is anonymous, but well defined).

An agent can work on only one information goal at a time, and requests a goal when it has either solved a previous goal or it has exceeded its search limit. When an agent receives an information goal, it first checks it own memory. If that fails, then the agent will post a message for help to the primary chat room and wait for a randomly assigned period. During such a waiting period, agents will also be able to respond to posts by other agents (such as providing advice or gossip) as they are actively monitoring the messages. If no agent responds during within the waiting period, it will engage in a default search and not check the chat room again until the task is completed (or its search limit is exhausted). On the other hand, if responses to the agent are posted, it will assess them accordingly (advice, gossip) and make a decision. If an agent follows advice, it will proceed directly to the chat room that is suggested. The results (success or failure) will drive subsequent postings to the main chat room. If failure occurs, the agent will note it and proceed to search on its own until it finds the item or exceeds the search limit.

The first thing to appreciate is the sensitivity of the equilibrium to variances in information value and how it is controlled in these environments. As noted, the equilibrium condition is based on a trust structure that accommodates a particular (in this case, relatively low) error rate. To illustrate this, we systematically degrade the information value of the advice by lowering the β value. This essentially causes more bad advice to be passed, albeit unintentionally, over the network, resulting in additional effort for agents to achieve their goals and even an increase in goals not achieved. Furthermore, we will illustrate how gossip engages structural insulation against bad advice.

The key to stabilization of these networks is an equilibrium condition of the norms of trust models and the use of gossip as a control mechanism, given the environment of advice flowing in the network. The trust ratio (T_Ratio) is the number of dyadic links between agents in the network that are judged as trustworthy as a fraction of the total possible trustworthy links. The simulated networks in this chapter are seen as starting not from a random "initialization" state, but from an equilibrium state developed over time such that all agents do trust each other (T_Ratio = 1.0). By slightly reducing the β value, the error rate in memory, and therefore incorrect advice, is increased, but is accommodated by the trust model and gossip norms.

A Virtual Experiment in Disruption

Given the sensitivities of the network to levels of bad advice, we explore the extent to which such networks can be disrupted from an equilibrium condition by inserting various numbers of deceptive agents. In this virtual experiment, a fixed number of active agents (20) interact in a primary chat room and (collectively) seek information on 140 tasks (7 tasks each). We compare two trust models reflecting differing norms of trust behavior in two networks. Both trust models are essentially the same (unforgiving) but varied in the tolerance for bad advice. The first is the model shown in Figure 7.1 upper (UnF.2), while the second one, Figure 7.1 lower (UnF.1), will switch to an absorbing distrusting state after a single piece of bad advice. This is used to depict a somewhat more established network (UnF.2) and a substantially more cautious one (UnF.1). We then cross this manipulation with two types of gossip reflecting differing norms of this type of behavior. In the first type (low level), gossip was *posted* about an agent when, and only when, an agent directly caused a distrusting state of trust to be achieved (via specific, bad advice). Thus, an agent posted gossip about another agent only once—when it reached a distrusting state. In the second type (high level), gossip was posted as in the low level, but after that it was reposted *every* time a distrusted agent posted a request for advice to the chat room, thus reaffirming, reminding, and broadcasting the credibility of source. The high level of gossip reflects a more socially intense communication pattern of grooming and maintenance. In fact, Dunbar [170, p. 681] notes that "Analysis of a sample of human conversations shows that about 60% of time is spent gossiping about relationships and personal experiences." In both types, gossip was *believed* about another agent only if the source of the gossip was trusted and the number of trusted posts exceeded the tolerance level defined by the particular trust model (i.e., two for UnF.2 and one for UnF.1).

Within each manipulation, the number of deceptive agents was systematically varied from none (0%) to 20 (100%), holding the number of agents constant (at 20). A deceptive agent is a normal agent that systematically provides *wrong* advice about sources of information. For each of these conditions, 25 replications were run. The primary dependent variables are task completions (percent of total problems solved), trust ratio (T_Ratio), and total effort expended (search steps) by the network.

Results of TAG Experiments

The results of the analyses are shown in Table 7.3. First, there were overall main effects of the trust model, gossip model, and deception on all three dependent variables. The less tolerant trust models (UnF.1) resulted in lower task completion rates, lower T_Ratios, and higher levels of effort by the network. The high level of gossip had substantial impacts on reducing the task completion rates, reducing the T_Ratio, and increasing the net effort of the group. Finally, the use of deception also lowered task completion rates, lowered the T_Ratio, and increased group effort required to find solutions.

Second, there were also significant two-level interaction effects of trust model × gossip, trust model × deception, and gossip × deception. There was no interaction between trust model and gossip on either task completion or effort. However, there was a significant interaction effect for T_Ratio: high gossip conditions lowered the T_Ratio significantly more in less forgiving networks. Trust model × deception yielded significant interactions for all dependent variables: deception had a slightly (but significantly) stronger effect on the less forgiving trust model (UnF.1) on lowering the task completion percentage, lowering the T_Ratio, and increasing the level of effort expended on the tasks. Gossip interacted significantly with deception for all dependent variables: deception occurrences under high gossip conditions resulted in

Gossip and Disruption

Table 7.3 Analysis of Variance Results

Dependent Var	Source	df	F	p
Task Completion				
	Trust Model (T)	1	99.99 ***	.00
	Gossip Model (G)	1	35515.83 ***	.00
	Deception (D)	20	1901.80 ***	.00
	T × G	1	.21	.64
	T × D	20	17.20 ***	.00
	G × D	20	840.57 ***	.00
	T × G × D	20	21.21 ***	.00
T_Ratio				
	Trust Model (T)	1	16148.5***	.00
	Gossip Model (G)	1	5991192.8***	.00
	Deception (D)	20	241181.0***	.00
	T × G	1	15992.6***	.00
	T × D	20	1160.3***	.00
	G × D	20	20351.7***	.00
	T × G × D	20	115.8***	.00
Level of Effort				
	Trust Model (T)	1	150.75***	.00
	Gossip Model (G)	1	76420.02***	.00
	Deception (D)	20	4701.91***	.00
	T × G	1	1.20	.27
	T × D	20	36.04***	.00
	G × D	20	1951.64***	.00
	T × G × D	20	41.51***	.00

***$p < .001$

substantially lower percentage of tasks completed, more effort expended, and lower T_Ratios.

Finally, all trust model × gossip × deception three-way interactions were significant: high gossip conditions had significantly more impact when the norms of trust were less forgiving and with low levels of deception.

Interpretation

Although there were many significant effects, it is important to discern the most interesting and substantial ones, relative to the issues at hand. Consider the two-way interactions that examine the (related) impact that the norms of gossip had in the presence of deceptive agents on task completion and level of effort, as shown in Figures 7.2 and 7.3. Regardless of whether the trust norms were more or less tolerant, high gossip conditions almost immediately reduced task completion percentages and raised the effort level to that of a fully disrupted network, with virtually no useful information flowing.

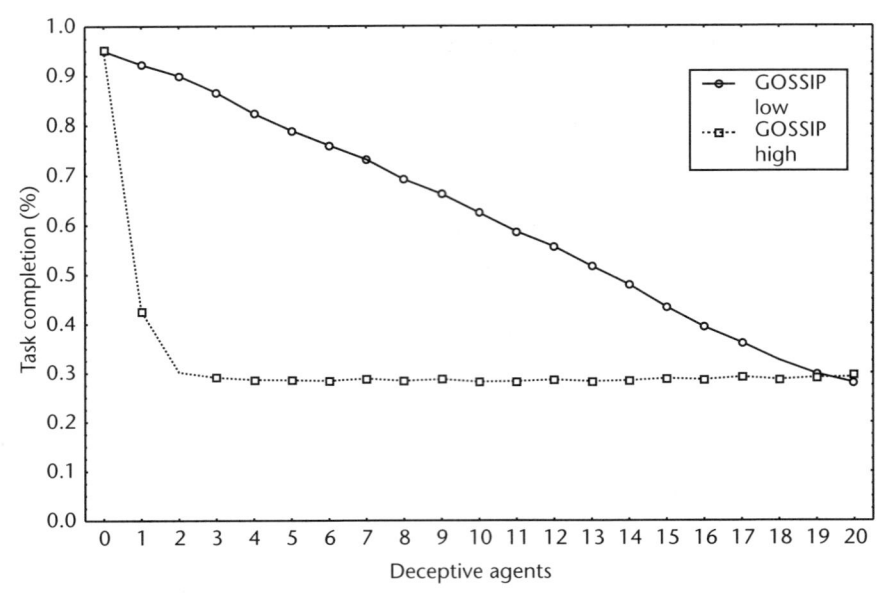

Figure 7.2 Influence of gossip norms and deception on task completion.

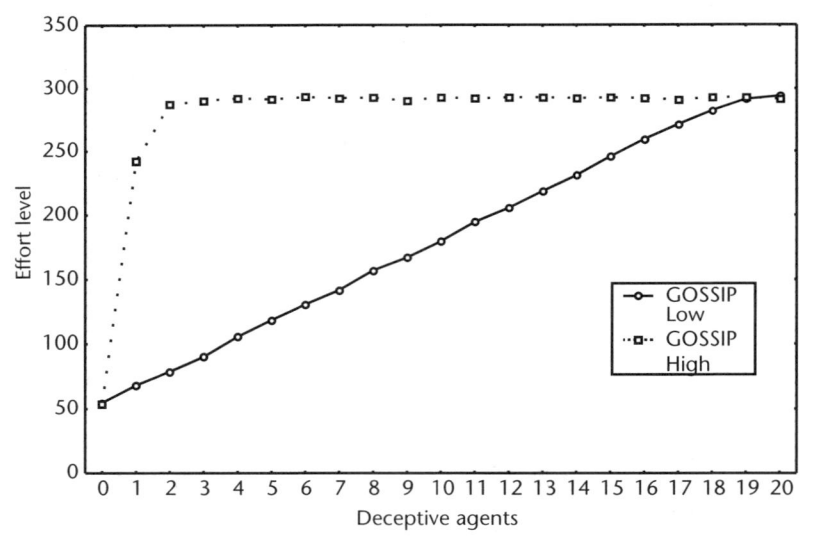

Figure 7.3 Influence of gossip norms and deception on level of effort.

Insight to this can be gained by examining the how the underlying trust network declined. In Figure 7.4, the three-way interaction of T_Ratio is graphed in response to deceptive agent presence under low and high gossip norms for both trust models. As can be seen, altering the type (and amount) of gossip had substantial impact on trust reduction, with somewhat differing nonlinear results depending on the underlying trust model. The reasons for the latter effects are based, in part, also on the timing of the gossip (based on task completion and random timings) with respect to the "stages" of trust changes for the particular model.

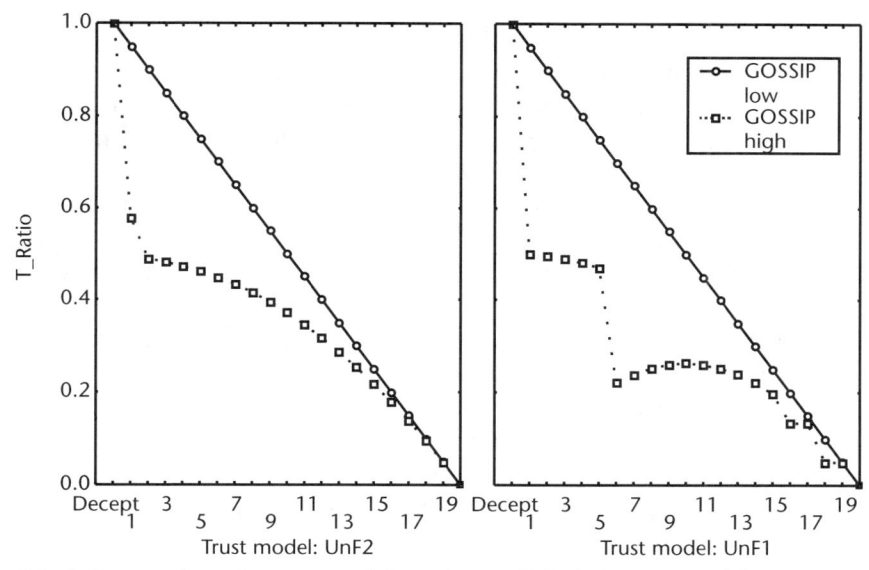

Figure 7.4 Influence of gossip norms and deception on T_Ratio by trust model.

Giving and Taking

There is a rich and growing body of social and biological evidence that humans are geared to interact in groups. A key component of that interaction involves gossip, and that gossip (and indirectly, trust) is ubiquitous in social groups. As Spacks [14, p. 263] concludes, "gossip will not be suppressed." Yet little theory or research has directly examined the value of gossip for group cohesion. Further there is little work on the role of gossip in anonymous information exchange networks, such as those we see on the Internet [171–173] or in secret societies. Still, we can find some guidance in some of the earliest work on communication networks, where Bavelas [174, p. 726] asserted that in "groups that are free of outside direction and control, it is clear that the interaction patterns that emerge and stabilize are a product of social processes within the group." This assertion holds today, where in our groups, gossip is the primary social process. The importance of this is that much of the social and biological apparatus is absent from simple chat rooms, dead letter drops, or other media used by these anonymous information exchange networks. In such groups gossip remains the lynchpin for functional (or dysfunctional) organizations of the type we referred to as TAG groups [175, 176].

In general, reputation and feedback mechanism are important information control devices, particularly on the Internet (e.g., [177–179]). Regarding our example, understanding threats from a social science perspective is providing interesting insights into the emerging problems of malware (e.g., [180]) and covert networks [181, 182], and this project in particular is derived from earlier work on simulating trust and gossip [183, 184]. Our perspective is that a common structure exists for the huge number of anonymous groups interacting to exchange information in various settings, including the Internet-based environment of chat rooms and secret societies that may or may not use the Internet. We suggest that these are essentially simple structures involving core elements of trust, advice, and gossip. It is these core elements that permit thousands of these structures to exist and to thrive. These core elements also make them susceptible to disruption.

The simulation in this chapter has demonstrated, albeit theoretically, fundamental elements of these groups: *They are extremely susceptible to high levels of socially focused gossip questioning the credibility of the key sources of information.* Recall that the high levels of gossip were achieved simply by altering the norm of when gossip was asserted. In these situations, gossip about agents was asserted whenever an agent detected a request for advice by an agent who was *not* trusted. These sensitive networks were extremely easy to disrupt, as their reliance on rapid dissemination of judgments regarding source credibility quickly "shut down" the network. Regardless of their initial levels of trust (i.e., their trust models), if the network permitted and was attuned to claims against a source, then that network quickly became dysfunctional, trust eroded, performance declined, and effort levels escalated.

Gossip is *dysfunctional* not because the agents are wasting time gossiping, but because the existence of gossip can reduce listening to *any* information, including that which is valuable. However, the underlying lesson for maintaining, rather than disrupting, such organizations is that gossip is *functional*, as it can "isolate, prune, and tune" an organization through spreading information on the viability of its less reliable components. Gossip is a mechanism of cultural learning [185], but we argue that is also a form of *organizational* learning.

These networks are susceptible to deception. It is not surprising that deception reduces performance. What is more interesting is that it increases effort, in part, because of the reduction of valuable information and, in part, because of second-order effects such as decreasing listening to any information by making the information in the gossip more prone to error. A key issue that should be further explored is when effort is increased to the extent that the group is no longer viable. Since individuals tend to leave groups when the effort or workload becomes extreme, group viability becomes a key issue.

Blind trust can mitigate, to an extent, the impacts of gossip and deception. However, this research suggests that the trust in and of itself is not enough. Research, although not well developed, suggests that certain types of deception are difficult to detect in simple computer-mediated interaction [186]. This raises the issue of sanctions. Can imposing sanctions on overt gossipers or deceivers increase the benefits of trust and decrease the difficulties raised by deception and gossip? This work suggests that such sanctions may need to be very severe and go beyond simply mistrusting individuals to denying them access to the network. Also significant are the complexities involved in crafting the decision and response apparatus that handles how messages, gossip, trust models, and social networks interact. Further work needs to be done on determining how sensitive these networks are to variations and different parameter values and events. The overall results, however, are consistent with evolutionary simulations examining defector discrimination strategies and gossip [187].

Another issue this raises is credentialing and membership. As noted, a key feature of these groups is that the members are contextually anonymous; that is, visual cues, face recognition, voice identification, and so forth are not available. However, most secret societies have developed both membership rituals and codes. Such devices are critical in that they permit trust by creating the illusion among communication partners that they "know" each other. What we have not examined here, and what needs to be considered, is the impact these devices have on the formation of the networks, their tendency to engage in gossip, and their tendency to respond to gossip through sanctions or network disruption.

In this chapter, the groups were quite small and functionally stable of the type that Ellickson [188] calls "close-knit" in that they are repeat players and can identify (in our case, via personas) individuals. Thus, individual behaviors can be detected and attributable back to a unique persona. This model is different from larger, more loose-knit networks realized by peer-to-peer protocols or in large secret societies. In these larger groups, repeat encounters are rare, identification of shirkers is difficult or impossible, and even low numbers of cooperators (relative to the millions of users) generate sufficient sources of content. Without punishment for norm enforcement, the game is to engage clever mechanisms to generate *sufficient* cooperation [189]. Nevertheless, our results do suggest that, even in these larger groups, were they to be structured in a cellular fashion or as a set of loosely connected cliques, gossip could be used to disrupt the group.

The results of our simulations demonstrate that TAG models, without exogenous contact and reference to other information about group members, require the use of some level of gossip (with an associated level of trust) to maintain the integrity of the group. However, these organizations can be extremely susceptible to infiltration of deceptive messages unless extraordinary coordinated (or sympathetic) responses occur that rapidly "shut down" deceptive agents, if detected. Chapter 5 explored an interesting example of the "cascading collapse" of a terrorist organization when a key member of that organization was revealed as a traitor. In terms of our model, the cascade occurred because trust levels and tolerance in one group were invaded by a deceptive agent that induced distrust among the other agents. As one can imagine, gossip (as accusation) was more than likely a strong source of information (real or deceptive) during those periods, leading eventually to the collapse of communication and trust, and eventually the organization itself. In real groups, the impact of gossip, although negative, may be less precipitous when generated by a truly anonymous member if the group also has membership rituals and codes that are used to screen participants.

In this chapter, the social networks were basically flat. All actors could communicate with all others. In many social networks, however, there are explicit or implicit gatekeepers who limit the flow of information. The import of such gatekeepers depends on the size, density, and structure of the networks. It is possible that certain network structures would inhibit or abet the impact of gossip. Enabling these gatekeepers to sanction those who engage frequently in gossip or deception might mitigate the impact of gossip. If this were the case, then the problem would become how one can infiltrate and gain the initial trust of such networks and which gatekeeper should be worked with. Only then can that trust be exploited and breach the insular mechanisms inherent in more amorphous and unstructured networks.

We also saw that one of the impacts of gossip was to effectively move the information exchange network from an all-to-all network to, effectively, a "star" network, as the number of trusted others dwindled. This suggests that, when gossip is to be expected, groups might organize in clusters, with only the hubs of the stars being connected. Such a structure might inhibit the flow of gossip. The strength of such a structure in inhibiting the deleterious effects of gossip is a point for further research.

We conclude with a speculative observation that bears more than passing interest in examining implications for all organizations. In this chapter, we focused on gossip rather than rumor (as we have defined them) in anonymous information exchange networks. This work was guided by our understanding of a particular type of Internet-based organization. But how might gossip and rumor be related, and how might these views generalize to other types of organizations? We suggest at least two important ways. First, both gossip and rumor can impact elements of *trust*, and that trust is an "essential component of all enduring social relationships" [190, p. 13]. Rumor can influence trust in individuals, a product or company, or an institution [57, 191, 192]. Second, it is quite feasible that under the correct conditions, gossip can potentially *mutate* into rumor. The latter point can be explained by the growth of what we call *intersecting connectivity* of communication technologies underlying the infrastructure for social relationships. Thus, "normal" organizations that rely on these technologies for their employees actually are conduits to other networks through their technology or through their employees, or both. As gossip is remarkably prevalent in organizations [51, 193], and given the available intersecting technologies available, it is plausible that many individuals are boundary spanners to their online associates. Thus, more and more individuals will have more and more of what Granovetter calls "weak ties" [154, 194], opening the gate for distal information to pass quickly. As the information (e.g., gossip) loses its context and purpose (i.e., that which has made it "gossip"), some forms can hop from group to group and mutate (given the new contexts) into a rumor, now online and subject to informational distortions and spread [195, 196].

The Internet, online games, and other activities where individuals participate in anonymous information exchange networks are becoming part of most individuals' lives. For example, instant messaging software can be acquired for free, and one estimate by the Radicati Group [197] suggests that by 2009, there will be 1.2 billion accounts worldwide. In the United States:

- 70% of adults access the Internet [198], 87% of teens 12–17 access the Internet, and 76% of these use it to access news [199];
- 53 million people (42% of U.S. Internet users) engage instant messaging [200];
- 53% of home Internet users have broadband [201], and 71% of these users get news daily from the Internet [198];
- 27% of Internet users say they read blogs [202];
- 60 million users access search engines on any given day [203];
- 35% of cell phone users engage in text messaging [204].

This rapid growth in telecommunication technology and access to the Internet means that information spreads more broadly and faster than ever before. It also means that anonymous information exchange networks can more quickly spread rumors that can have negative consequences. In 1750 riots erupted in Paris when gossip about certain police generated a rumor spread throughout the city that "the police" were abducting children to reduce vagrancy [205]. In 2006, gossip about the behavior of an individual who had burned the pages of the Qur'an, quickly spread into a rumor that resulted into the burning of a Catholic church in Pakistan [206–209]. The former event occurred (as have many others) without the current speed and spread of today's communication technologies. Today, the spread of rumors through online media have the potential to increase the number of such events and the level of participation; however, they also have the potential to reduce such activity. Studies such as this one, and others that explore some of the issues raised here, are critical for gaining insight into what is likely to happen in these events.

The Internet, online games, and other activities where individuals participate in anonymous information exchange networks are becoming part of most individuals' lives. In these contexts, whether for good or bad, gossip matters and cannot be ignored. Our results suggest that gossip can be utilized to affect the survivability of these groups, the way in which they are organized, and, of course, the reputations of the members. If such groups become more prevalent, come to control wealth and resources, and engage in activities that affect the health and welfare of millions, then it is important that we understand how to counter gossip and how to structure these groups so that they are gossip-proof.

Acknowledgments

This research was funded in part through a grant from the National Science Foundation, Computer and Information Sciences, Information and Intelligent Systems Award IIS-0084508. We thank Dr. Alexander Kott for his helpful comments on improving this chapter.

Endnotes

[1] In their interesting essay on culture, White and Dillingham [2] stress the importance and uniqueness of this capability by referring to it as the "ability to symbol"—the ability to freely and arbitrarily originate, determine, and bestow meaning upon things and events in the external world, and the ability to comprehend such meanings (p. 1). This ability underlies the emergence of culture [3]. In fact, evidences points to the ability of cultural symbols (e.g., objects) to activate reward circuitry in the brain [4].

[2] White, L., and B. Dillingham, *The Concept of Culture*, Minneapolis, MN: Burgess, 1973.

[3] White, L. *The Evolution of Culture: The Development of Civilization to the Fall of Rome*, New York: McGraw-Hill, 1959.

[4] Erk, S., et al., "Cultural Objects Modulate Reward Circuitry," *NeuroReport*, Vol. 13, No. 18, 2002, pp. 2499–2503.

[5] Stirling, R., "Some Psychological Mechanisms Operative in Gossip," *Social Forces*, Vol. 34, No. 3, 1956, pp. 262–267.

[6] Dunbar, R., N. Duncan, and A. Marriott, "Human Conversational Behavior," *Human Nature*, Vol. 8, 1997, pp. 231–246.

[7] Fine, G., "Social Components of Children's Gossip," *Journal of Communication*, Vol. 27, No. 1, 1977, pp. 181–185.

[8] Fine noted that one interesting difference between children's gossip of others and adult's gossip of others was that children did not seem to mind gossiping with the target of the gossip present. Those with children are probably aware of children's books regarding the impropriety of "negative gossip" (e.g., [9–11]).

[9] Berry, J., *A Children's Book About Gossiping*, New York: Grolier, 1988.

[10] Harrison, A., *Easy to Tell Stories for Young*, Jonesborough, TN: National Storytelling Press, 1992.

[11] Hodgson, M., *Smelly Tales*, St. Louis, MI: Concordia, 1998.
[12] Shermer, M., *The Science of Good and Evil*, New York: Times Books, 2004.
[13] Lane, M., *Spinning Tales, Weaving Hope: Stories of Peace, Justice & the Environment*, Philadelphia, PA: New Society Publishers, 1992.
[14] Spacks, P. M., *Gossip*, New York: Knopf, 1985.
[15] Stewart, P., and A. Strathern, *Witchcraft, Sorcery, Rumors, and Gossip*, Cambridge, U.K.: Cambridge University Press, 2004.
[16] In fact, the term "gossip" has referred to a kind of person and not a type of talk since prior to the nineteenth century [14, p. 27].
[17] Ben-Ze'ev, A., "The Vindication of Gossip," in Goodman, R., and A. Ben-Ze'ev, (eds.), *Good Gossip*, Lawrence, KS: University Press of Kansas, 1994, pp. 11–24.
[18] Tannen, D., *You Just Don't Understand*, New York: Ballantine, 1990.
[19] Dunbar, R., *Grooming, Gossip and the Evolution of Language*, Cambridge, MA: Harvard University, 1996.
[20] Dunbar, R., "The Social Brain: Mind, Language and Society in Evolutionary Perspectives," *Annual Review of Anthropology*, Vol. 32, 2003, pp. 163–181.
[21] Whiten, A., and R. Byrne, (eds.), *Machiavellian Intelligence II: Extensions and Evaluation*, New York: Cambridge University Press, 1997.
[22] Dunbar, R., "Gossip in Evolutionary Perspective," *Review of General Psychology*, Vol. 8, No. 2, 2004, pp. 100–110.
[23] Enquist, M., and O. Leimar, "The Evolution of Cooperation in Mobile Organisms," *Animal Behavior*, Vol. 45, 1993, pp. 747–757.
[24] Cosmides, L., "The Logic of Social Exchange: Has Natural Selection Shaped How Humans Reason? Studies with the Wason Selection Task," *Cognition*, Vol. 31, 1989, pp. 187–276.
[25] Cosmides, L., and J. Tooby, "Cognitive Adaptation for Social Exchange," in Barkow, J., L. Cosmides, and J. Tooby, (eds.), *The Adapted Mind*, New York: Oxford, 1992, pp. 163–228.
[26] Gigerenzer, G., and K. Hug, "Domain-Specific Reasoning: Social Contracts, Cheating, and Perspective Change," *Cognition*, Vol. 43, 1992, pp. 127–171.
[27] Sugiyama, L., J. Tooby, and L. Cosmides, "Cross-Cultural Evidence of Cognitive Adaptations for Social Exchange Among the Shiwar of Ecuadorian Amazonia," *Proc. of the National Academy of Sciences*, Vol. 99, 2002, pp. 11537–11542.
[28] Chiappe, D., A. Brown, and B. Dow, "Cheaters Are Looked at Longer and Remembered Better Than Cooperators in Social Exchange Situations," *Evolutionary Psychology*, Vol. 3, 2004, pp. 108–120.
[29] Also, cooperators are remembered better than those who were irrelevant to the social contract context, thus supporting the general influence of the social ecology.
[30] Grèzes, J., C. Frith, and R. Passingham, "Brain Mechanisms for Inferring Deceit in the Actions of Others," *The Journal of Neuroscience*, Vol. 24, No. 24, 2004, pp. 5500–5505.
[31] Stone, V., et al., "Selective Impairment of Reasoning About Social Exchange in a Patient with Bilateral Limbic System Damage," *Proc. of the National Academy of Sciences*, Vol. 99, 2002, pp. 11531–11535.
[32] Fiddick, L., "Domains of Deontic Reasoning: Resolving the Discrepancy Between the Cognitive and Moral Reasoning Literatures," *The Quarterly Journal of Experimental Psychology*, Vol. 57A, No. 4, 2004, pp. 447–474.
[33] Fiddick, L., M. Spampinato, and J. Grafman, "Social Contracts and Precautions Activate Different Neurological Systems: An fMRI Investigation of Deontic Reasoning," *NeuroImage*, Vol. 28, 2005, pp. 778–786.
[34] Kurzban, R., and M. Leary, "Evolutionary Origins of Stigmatization: The Functions of Social Exclusion," *Psychological Bulletin*, Vol. 127, No. 2, 2001, pp. 187–208.
[35] MacDonald, G., and M. Leary, "Why Does Social Exclusion Hurt? The Relationship Between Social and Physical Pain," *Psychological Bulletin*, Vol. 131, No. 2, 2005, pp. 202–223.
[36] Casebeer, W., and P. Churchland, "The Neural Mechanisms of Moral Cognition: A Multiple-Aspect Approach to Moral Judgment and Decision-Making," *Biology and Philosophy*, Vol. 18, 2003, pp. 169–194.

[37] Haidt, J., "The Moral Emotions," in Davidson, R., K. Scherer, and H. Goldsmith, (eds.), *Handbook of the Affective Sciences*, Oxford, U.K.: Oxford University Press, 2003, pp. 852–870.

[38] Moll, J., et al., "The Neural Basis of Human Moral Cognition," *Nature Reviews: Neuroscience*, Vol. 6, 2005, pp. 799–809.

[39] Takahashi, H., et al., "Brain Activation Associated with Evaluative Processes of Guilt and Embarrassment: An fMRI Study," *NeuroImage*, Vol. 23, 2004, pp. 967–974.

[40] Gluckman, M., "Gossip and Scandal," *Current Anthropology*, Vol. 4, No. 3, 1963, pp. 307–316.

[41] Paine, R., "What Is Gossip About? An Alternative Hypothesis," *Man, New Series*, Vol. 2, No. 2, 1967, pp, 278–285.

[42] Cox, B., "What Is Hopi Gossip About? Information Management and Hopi Factions," *Man*, Vol. 5, 1970, pp. 88–98.

[43] Goodman, E., and A. Ben-Ze'ev, (eds.), *Good Gossip*, Lawrence, KS: University Press of Kansas, 1994.

[44] Haviland, J., *Gossip, Reputation, and Knowledge in Zinacantan*, Chicago, IL: University of Chicago, 1977.

[45] McAndrew, F., and M. Milenkovic, "Of Tabloids and Family Secrets: The Evolutionary Psychology of Gossip," *Journal of Applied Social Psychology*, Vol. 32, 2002, pp. 1064–1082.

[46] Rysman, A., "Gossip and Occupational Ideology," *Journal of Communication*, Vol. 26, No. 3, 1976, pp. 64–68.

[47] Axelrod, R., *The Evolution of Cooperation*, New York: Basic Books, 1984.

[48] Gürerk, O., Z. Irlenbusch, and B. Rockenbach, "The Competitive Advantage of Sanctioning Institutions," *Science*, Vol. 312, April 7, 2006, pp. 108–111.

[49] Prietula, M., and D. Conway, "Evolution of Metanorms," Working Paper, Goizueta Business School, Emory University, Atlanta, GA, 2006.

[50] Kniffin, K., and D. S. Wilson, "Utilities of Gossip Across Organizational Levels: Multi-Level Selection, Free-Riders, and Teams," *Human Nature*, Vol. 16, No. 3, 2005, pp. 278–292.

[51] Noon, M., and R. Delbridge, "News from Behind My Hand: Gossip in Organizations," *Organizational Studies*, Vol. 14, No. 1, 1993, pp. 23–36.

[52] Allport, G., and L. Postman, *The Psychology of Rumor*, New York: Holt, Rinehart & Winston, 1947.

[53] They define what they call the "basic law of rumor," in which they define the intensity of a rumor as varying with the product of its importance and ambiguity [54, p. 33].

[54] Allport, G., and L. Postman, *The Psychology of Rumor*, New York: Holt, Rinehart & Winston, 1947.

[55] Rosnow, R., and G. Fine, *Rumor and Gossip: The Social Psychology of Hearsay*, New York: Elsevier, 1976.

[56] Knopf, T., *Rumors, Race, and Riots*, New Brunswick, NJ: Transaction Books, 1975.

[57] Koenig, F., *Rumor in the Marketplace: The Social Psychology of Hearsay*, Dover, MA: Auburn House, 1985.

[58] Rosnow, A., "Inside Rumor: A Personal Journey," *American Psychologist*, Vol. 46, No. 5, 1991, pp. 484–496.

[59] Rosnow, R., "On Rumor," *Journal of Communication*, Vol. 24, No. 3, 1974, pp. 26–38.

[60] Shibutani, T., *Improvised News: A Sociological Study of Rumor*, New York: Bobbs-Merrill, 1966.

[61] Kapferer, J., "A Mass Poisoning Rumor in Europe," *Public Opinion Quarterly*, Vol. 53, 1989, pp. 467–481.

[62] Reeve, A., *Turn Me On, Dead Man: The Beatles and the "Paul-Is-Dead" Hoax*, Bloomington, IN: Author House, 2004.

[63] Turner, P., *I Heard It Through the Grapevine: Rumor in African-American Culture*, Berkeley: University of California Press, 1993.

[64] Rose, A., "Rumor in the Stock Market," *Public Opinion Quarterly*, Vol. 15, 1951, pp. 461–486.

[65] Bartholomew, R., "The Martian Panic: Sixty Years Later," *Skeptical Inquirer Magazine*, November/December 1998, available May 15, 2006, http://www.csicop.org/si/9811/martian.html.

[66] Ponting, J., "Rumor Control Centers: Their Emergence and Operations," *American Behavioral Scientist*, Vol. 16, No. 3, 1973, pp. 391–401.

[67] Rude, G., *The Crowd in the French Revolution*, Oxford, U.K.: Oxford University Press, 1967.

[68] Fay, S., *The Origins of the World War, Volume I*, Toronto: Free Press, 1968.

[69] Taylor, G., "Gossip As Moral Talk," in Goodman, R., and A. Ben-Ze'ev, (eds.), *Good Gossip*, Lawrence, KS: University Press of Kansas, 1994, pp. 34–46.

[70] Wert, S., and P. Salovey, "A Social Comparison Account of Gossip," *Review of General Psychology*, Vol. 8, No. 2, 2004, pp. 122–137.

[71] Szvetelszky, Z., "Ways and Transmissions of Gossip," *Journal of Cultural and Evolutionary Psychology*, Vol. 1, No. 2, 2003, pp. 109–122.

[72] Stewart, P., and A. Strathern, *Witchcraft, Sorcery, Rumors, and Gossip*, Cambridge, U.K.: Cambridge University Press, 2004.

[73] de Vos, G., *Tales, Rumors, and Gossip*, Englewood, CO: Libraries Unlimited, 1996.

[74] Note that rumors can involve events that are relevant in your social context (such as purchased products or firms) or even individuals if they are unknown or ill-specified, such as Satanic cults [75].

[75] Victor, J., "Satanic Cult Rumors As Contemporary Legend," *Western Folklore*, Vol. 49, No. 1, 1990, pp. 51–81.

[76] Gossip, as it is communication, can obviously serve many functions. For example, Rosnow and Fine [54] suggest three categories of *gossip* based on its function: informative (news exchanges about the social environment), moralizing (manipulative toward norm specification and control), and entertainment (including maintenance of communication patterns). Foster [77] provides a 2×2 taxonomy of *gossipers* based on two axes: social activity (high, low) and network access (high, low).

[77] Foster, E., "Research on Gossip: Taxonomy, Methods, and Future Directions," *Review of General Psychology*, Vol. 8, No. 2, 2004, pp. 78–79.

[78] Emler, N., "Gossip, Reputation, and Social Adaptation," in Goodman, R., and A. Ben-Ze'ev, (eds.), *Good Gossip*, Lawrence, KS: University Press of Kansas, 1994, pp. 117–138.

[79] Marx [80] identifies several forms and rationales for anonymity, and the general roles and policies for Internet anonymity are under open discussion from a variety of perspectives (e.g., [81–84]).

[80] Marx, G., "What's in a Name? Some Reflections on the Sociology of Anonymity," *The Information Society*, Vol. 15, 1999, pp. 99–112.

[81] Froomkin, A., "Legal Issues in Anonymity and Pseudonymity," *The Information Society*, Vol. 15, 1999, pp. 113–127.

[82] Hoffman, D., T. Novak, and M. Peralta, "Information Privacy in the Marketspace: Implications for the Commercial Uses of Anonymity on the Web," *The Information Society*, Vol. 15, 1999, pp. 129–139.

[83] Nissenbaum, H., "The Meaning of Anonymity in an Information Age," *The Information Society*, Vol. 15, 1999, pp. 141–144.

[84] Teich, A., et al., "Anonymous Communication Policies for the Internet: Results and Recommendations of the AAAS Conference," *The Information Society*, Vol. 15, 1999, pp. 71–77.

[85] Milgram, S., "The Small World Problem," *Psychology Today*, Vol. 2, 1967, pp. 60–67.

[86] Watts, D., *Small Worlds: The Dynamics of Networks Between Order and Randomness*, Princeton, NJ: Princeton University Press, 1999.

[87] Friedman, E., and P. Resnick, "The Social Cost of Cheap Pseudonyms," *Journal of Economics & Management Strategy*, Vol. 10, No. 2, 2001, pp. 173–199.

[88] Consider a recent example, where a single laptop computer was stolen. This laptop contained personal information, such as social security numbers and Internal account numbers, of 230,000 customers and advisors [89].

[89] Dash, E., "Ameriprise Says Stolen Laptop Had Data on 230,000 People," *New York Times Online*, January 26, 2006, http://www.nytimes.com.

[90] A rootkit is a general term referring to software (e.g., Hacker Defender, HE4HOOK) that allows undetected access to a machine and grants root (administrative) privileges [91]. Such software can now exist in the BIOS flash memory [92].

[91] Hoglund, G., and J. Butler, *Rootkits: Subserting the Windows Kernel*, New York: Addison-Wesley, 2005.

[92] Lemos, R., "Researchers: Rootkits Headed for BIOS," *SecurityFocus*, January 26, 2006, http://www.securityfocus.com.

[93] Malware is a general term for exogenous software that functions "with bad intentions" such as virus, worms, and spyware. Not only is the software is distributed, but often exchanged are methods to hide them, even in voice over Internet protocol (VOIP) traffic [94].

[94] Judge, P., "How Bad Is the Skype botnet Threat?" *Techworld*, January 25, 2006, http://www.techworld.com.

[95] These are easy-to-use applications that generate malware or mutate signatures of existing malware to facilitate intrusion. Thus, these "mutation engines" are exchanged (in simple form) or sold (in more exotic forms), resulting in a substantial transfer of functional "expertise."

[96] Lévi-Strauss, C., *The Elementary Structures of Kinship* (J. H. Bell, J. R. von Sturmer, and R. Needham, Trans.), Boston, MA: Beacon Press, 1969.

[97] Alexander, R., *The Biology of Moral Systems*, New York: de Gruyter, 1987.

[98] Shimeall, T., P. Williams, and C. Dunlevy, "Countering Cyber War," *NATO Review*, Winter 2001/2002, pp. 16–18.

[99] Billo, C., and W. Chang, *Cyber Warfare: An Analysis of the Means and Motivations of Selected Nation-States*, Institute for Security Technology Studies, Dartmouth College, Hanover, NH, 2004.

[100] Dertouzos, M., *What Will Be: How the New World of Information Will Change Our Lives*, San Francisco, CA: HaperEdge, 1997.

[101] DiMaggio, P., et al., "Social Implications of the Internet," *Annual Review of Sociology*, Vol. 27, 2001, pp. 307–336.

[102] Wilson, S., and L. Peterson, "The Anthropology of Online Communities," *Annual Review of Anthropology*, Vol. 31, 2002, pp. 449–467.

[103] On December 17, 2003, 13 NASA sites were attacked by the Brazilian hacking group DRWXR. The attacks by the hackers were politically motivated and the message left behind protested the war in Iraq. A video clip showing U.S. soldiers was also attached.

[104] Turgeman-Goldschmidt, O., "Hacker's Accounts: Hacking As Social Entertainment," *Social Science Computing Reviews*, Vol. 23, No. 1, 2004, pp. 8–23.

[105] Goldman, E., "A Road to No Warez: The No Electronic Theft Act and Criminal Copyright Infringement," *Oregon Law Review*, Vol. 82, 2003, pp. 369–432.

[106] Recent reports identify the threats of nation-states engaging in this activity [99].

[107] Phishing refers to a variety of fraudulent methods to get responders to divulge personal financial information, such as credit card numbers, account numbers, and passwords, often by hijacking the look of established brands (e.g., Citibank). For example, the Anti-Phishing Working Group noted that in February 2005, there were 13,141 new and unique phishing e-mails reported, including 64 hijacked brands [108].

[108] APWG, "Phishing Activity Trends Report," Anti-Phishing Working Group, February 2005, http://www.antiphishing.org.

[109] Cyert, R., and J. March, *A Behavioral Theory of the Firm*, Englewood Cliffs, NJ: Prentice Hall, 1963.

[110] Carley, K., and M. Prietula, (eds.), *Computational Organization Theory*, Hillsdale, NJ: Erlbaum, 1994.

[111] Prietula, M., K. Carley, and L. Gasser, (eds.), *Simulating Organizations: Computational Models of Institutions and Groups*, Menlo Park, CA: AAAI/MIT Press, 1998.

[112] The difficulty, legal and ethical, of studying chat room behavior in situ is also complicated by the culture. Hudson and Bruckman [113] reported that they were "kicked out" of 63.3% of chat rooms they entered when they announced the purpose of their visit—and also kicked out of 29% of the chat rooms where they did not!

[113] Hudson, J., and A. Bruckman, "Go Away: Participant Objections to Being Studied and the Ethics of Chatroom Research," *The Information Society*, Vol. 20, 2004, pp. 127–139.

[114] Coleman, J., *Foundations of Social Theory*, Cambridge, MA: Harvard University Press, 1990.

[115] Luhmann, N., *Trust and Power*, New York: John Wiley, 1979.

[116] Good, D., "Individuals, Interpersonal Relations, and Trust," in Gambetta, D., (ed.), *Trust: Making and Breaking Cooperative Relations*, Oxford, U.K.: Basil Blackwell, 1988, pp. 31–48.

[117] Giddens, A., *The Consequences of Modernity*, Stanford, CA: Stanford University Press, 1990.

[118] La Porta, R., et al., "Trust in Large Organizations," *American Economic Review: Papers and Proceedings*, Vol. 87, No. 2, 1997, pp. 333–338.

[119] Williams, M., "In Whom We Trust: Group Membership as an Affective Context for Trust Development," *Academy of Management Review*, Vol. 6, No. 3, 2001, pp. 377–396.

[120] Rotter, J., "Generalized Expectances for Interpersonal Trust," *American Psychologist*, Vol. 26, 1971, pp. 443–452.

[121] McCabe, K., "Goodwill Accounting in Economic Exchange," in Gigerenzer, G., and R. Selten, (eds.), *Bounded Rationality: The Adaptive Toolbox*, Cambridge, MA: MIT Press, 2001.

[122] Kosfeld, M., et al., "Oxytocin Increases Trust in Humans," *Nature*, Vol. 435, June 2, 2005, pp. 673–676.

[123] Rousseau, D., et al., "Not So Different After All: A Cross-Discipline View of Trust," *Academy of Management Review*, Vol. 23, No. 3, 1998, pp. 393–404.

[124] Misztal, B., *Trust in Modern Societies*, Oxford, U.K.: Blackwell, 1996.

[125] This is not an arbitrary distinction; it is based on a theoretical model called the *social agent matrix* [126]. The model social agent matrix provides a two-dimensional categorization scheme that specifies the kind of knowledge required by the agent(s) (in terms of increasingly complex social situations) and the kind of information-processing capabilities required by the agent(s) to operate with that knowledge in order to exhibit various kinds of individual and collective phenomena.

[126] Carley, K., and A. Newell, "The Nature of the Social Agent," *Journal of Mathematical Sociology*, Vol. 19, No. 4, 1994, pp. 221–262.

[127] Holmes, A., A. Richards, and S. Green, "Anxiety and Sensitivity to Eye Gaze in Emotional Faces," *Brain and Cognition*, Vol. 60, 2006, pp. 282–294.

[128] Pelphrey, K., R. Viola, and G. McCarthy, "When Strangers Pass: Processing of Mutual and Averted Social Gaze in the Superior Temporal Sulcus," *Psychological Science*, Vol. 15, No. 9, 2004, pp. 598–603.

[129] Grèzes, J., S. Berthoz, and R. Passingham, "Amygdala Activation When One Is the Target of Deceit: Did He Lie to You or to Someone Else?" *NeuroImage*, Vol. 30, 2006, pp. 601–608.

[130] de Gelder, B., "Towards the Neurobiology of Emotional Body Language," *Nature Reviews: Neuroscience*, Vol. 7, March 2006, pp. 242–249.

[131] Loula, F., et al., "Recognizing People from Their Movement," *Journal of Experimental Psychology: Human Performance and Perception*, Vol. 31, No. 1, 2005, pp. 210–220.

[132] Suslow, T., et al., "Amygdala Activation During Masked Presentation of Emotional Faces Predicts Conscious Detection of Threat-Related Faces," *Brain and Cognition*, Vol. 61, No. 3, August 2006, pp. 243–248.

[133] Winston, J., et al., "Automatic and Intentional Brain Responses During Evaluation of Trustworthiness of Faces," *Nature Neuroscience*, Vol. 5, No. 3, 2002, pp. 277–283.

[134] Inbau, F., et al., *Criminal Interrogations and Confessions*, 4th ed., Gaithersburgh, MD: Aspen, 2001.

[135] The research supporting the viability and reliability of the latter methods is mixed and somewhat complicated methodologically, though generally favoring lower actual abilities and success rates approaching chance or uninformed base rates [136–142].

[136] DePaulo, B., et al., "Cues to Deception," *Psychological Review*, Vol. 129, No. 1, 2003, pp. 74–118.

[137] Ekman, P., and M. O'Sullivan, "Who Can Catch a Liar?" *American Psychologist*, Vol. 46, No. 9, 1991, pp. 913–920.

[138] Kassin, S., and C. Fong, "'I'm Innocent!: Effects of Training on Judgments of Truth and Deception in the Interrogation Room," *Law and Human Behavior*, Vol. 23, No. 5, 1999, pp. 499–516.

[139] Kraut, R., and D. Poe, "On the Line: The Deception Judgments of Customs Inspectors and Laymen," *Journal of Personality and Social Psychology*, Vol. 39, 1980, pp. 784–798.

[140] Mann, S., A. Vrij, and R. Bull, "Detecting True Lies: Police Officers' Ability to Detect Suspects' Lies," *Journal of Applied Psychology*, Vol. 89, No. 1, 2004, pp. 137–149.

[141] Vrij, A., "Why Professionals Fail to Catch Liars and How They Can Improve," *Legal and Criminological Psychology*, Vol. 9, 2004, pp. 159–181.

[142] Vrij, A., and R. Taylor, "Police Officers' and Students' Beliefs about Telling and Detecting Trivial and Serious Lies," *International Journal of Police Science & Management*, Vol. 5, No. 1, 2003, pp. 41–49.

[143] In a study of online bulletin board gossip, Harrington and Bielby [144] demonstrated the central role that online reputation and honesty play in the absence of other types of social bonds.

[144] Harrington, C., and D. Bielby, "Where Did You Hear That? Technology and the Social Organization of Gossip," *The Sociological Quarterly*, Vol. 36, No. 3, 1995, pp. 607–628.

[145] In crafting computational models, any hypothesized construct or component must be articulated in code, for in the code resides the theory at hand. Details of TrustMe implementation are found in [146].

[146] Prietula, M., "TrustMe Reference Manual," Working Paper, Goizueta School of Business, Emory University, Atlanta, GA, 2006.

[147] Berg, J., J. Dickhaut, and K. McCabe, "Trust, Reciprocity, and Social History," *Games and Economic Behavior*, Vol. 10, 1995, pp. 122–142.

[148] King-Casas, B., et al., "Getting to Know You: Reputation and Trust in a Two-Person Economic Exchange," *Science*, Vol. 308, April 1, 2005, pp. 78–83.

[149] Again, evidence for the central role of social contexts is suggested by underlying neurological responses to social cues and events such as positive trust signals [150], unreciprocated altruism [151], and social reward frustration [152]. Lacking specific information on risk, these could be considered ambiguous choices that seem to also be associated with specific neural circuitry [153].

[150] Zak, P., R. Kurzban, and W. Matzner, "The Neurobiology of Trust," *Annals of the New York Academy of Science*, Vol. 1032, 2004, pp. 224–227.

[151] Rilling, J., et al., "Opposing BOLD Responses to Reciprocated and Unreciprocated Altruism in Putative Reward Pathways," *NeuroReport*, Vol. 15, No. 16, 2004, pp. 2539–2543.

[152] Siegrist, J., et al., "Differential Brain Activation According to Chronic Social Reward Frustration," *NeuroReport*, Vol. 15, No. 16, 2005, pp. 1899–1903.

[153] Hsu, M., et al., "Neural Systems Responding to Degrees of Uncertainty in Human Decision-Making," *Science*, Vol. 310, December 9, 2005, pp. 1680–1683.

[154] Granovetter, M., "Threshold Models of Collective Behavior," *American Journal of Sociology*, Vol. 83, No. 6, 1978, pp. 1420–1443.

[155] Brown, D., *When Strangers Cooperate: Using Social Conventions to Govern Ourselves*, New York: The Free Press, 1995.

[156] Constant, D., L. Sproull, and S. Kiesler, "The Kindness of Strangers: On the Usefulness of Electronic Weak Ties for Technical Advice," in Kiesler, S., (ed.), *Culture of the Internet*, Mahwah, NJ: Lawrence Erlbaum Associates, 1997, pp. 303–322.

[157] Bailey, F., *The Prevalence of Deceit*, Ithaca, NY: Cornell University Press, 1981.

[158] Castelfranchi, C., and Y. Tan, (eds.), *Trust and Deception in Virtual Societies*, Norwell, MA: Kluwer Academic, 2001.

[159] Gibbs, J., "Sanctions," *Social Problems*, Vol. 14, 1966, pp. 147–159.

[160] Horne, C., "Sociological Perspectives on the Emergence of Social Norms," in Hechter, M., and K.-D. Opp, (eds.), *Social Norms*, New York: Russell Sage Foundation, 2001, pp. 3–34.

[161] Kandori, M., "Social Norms and Community Enforcement," *Review of Economic Studies*, Vol. 59, 1992, pp. 63–80.
[162] Reactions to deceit vary depending on whether you were deceived or you witness (or in our case, witness via gossip) deceit by others [129].
[163] The strong form of bounded rationality was originally conceived to characterize the effects of a restricted rational agent on the assumptions and conclusions of economic and administrative theory [164] and was highly influenced by research from sociology, social-psychology, and cognitive psychology. Bounded rationality ranges from strong forms that hypothesize the underlying cognitive (e.g., [165]) or institutional [109] apparatus to less restrictive derivatives that address various organizational and economic issues at micro and macro levels.
[164] Simon, H., "From Substantive to Procedural Rationality," in Latsis, S., (ed.), *Method and Appraisal in Economics*, Cambridge, U.K.: Cambridge University Press, 1976.
[165] Newell, A., *Unified Theories of Cognition*, Cambridge, MA: Harvard University, 1990.
[166] A *chunk* in this sense is a bond established between two symbolic knowledge references—chat room and context. The concept of a chunk in a full architecture can be found, for example, in Soar [165]. This refers not to the typical isolated seven chunks in semantically free working memory [166], but to task specific–related chunks accessible through the long-term working memory [167, 168].
[167] Miller, G., "The Magical Number Seven Plus or Minus Two: Some Limits on Our Capacity for Processing Information," *Psychological Review*, Vol. 63, 1956, pp. 81–97.
[168] Ericsson, K. A., "Protocol Analysis and Expert Thought: Concurrent Verbalizations of Thinking During Experts' Performance on Representative Tasks," in Ericsson, K. A., et al., (eds.), *Cambridge Handbook of Expertise and Expert Performance*, Cambridge, U.K.: Cambridge University Press, 2006, pp. 223–242.
[169] What we mean is that there are no complex decomposable goals where $N > 1$ sources must be sought. An example of a complex goal would be the following: If a hacker wants to launch a denial of service (DoS) attack against some target, she may have to find (1) zombie software components, (2) delivery and execution components, and (3) a set of compromised hosts. Thus, *three* goals must be achieved before the task can be completed and may, in fact, involve three different transactions.
[170] Dunbar, R., "Coevolution of Neocortical Size, Group Size and Language in Humans," *Behavioral and Brain Sciences*, Vol. 16, No. 4, 1993, pp. 681–735.
[171] Recall that we explicitly distinguish Internet gossip in a group from Internet rumor (e.g., [172, 173]).
[172] Ericsson, K. A., and W. Kintsch, "Long-Term Working Memory," *Psychological Review*, Vol. 102, 1995, pp. 211–245.
[173] Borden, D., and K. Harvey, (eds.), *The Electronic Grapevine: Rumor, Reputation, and Reporting in the New On-Line Environment*, Mahwah, NJ: Erlbaum, 1998.
[174] Bavelas, A., "Communication Patterns in Task-Oriented Groups," *The Journal of the Acoustical Society of America*, Vol. 22, No. 6, 1950, pp. 725–730.
[175] Some work is beginning on attempting to discover deception by linguistic and paralinguistic cues (e.g., [176]).
[176] Zhou, L., "An Empirical Investigation of Deception Behavior in Instant Messaging," *IEEE Trans. on Professional Communication*, Vol. 48, No. 2, 2005, pp. 127–160.
[177] Dellarocas, C., "The Digitization of Word of Mouth: Promise and Challenges of Online Feedback," *Management Science*, Vol. 49, No. 10, 2003, pp. 1407–1424.
[178] Resnick, P., and R. Zeckhauser, "Trust Among Strangers in Internet Transactions: Empirical Analysis of eBay's Reputation System," in Baye, M., (ed.), *The Economics of the Internet and E-Commerce: Advances in Applied Microeconomics*, Greenwich, CT: JAI Press, 2002.
[179] Resnick, P., et al., "Reputation Systems," *Communication of the ACM*, Vol. 43, No. 12, 2000, pp. 45–48.
[180] Chen, L.-C., and K. Carley, "The Impact of Countermeasure Propagation on the Prevalence of Computer Viruses," *IEEE Trans. on Systems, Man and Cybernetics—Part B: Cybernetics*, Vol. 34, No. 2, 2004, pp. 823–833.
[181] Carley, K., J. Lee, and D. Krackhardt, "Destabilizing Networks," *Connections*, Vol. 24, No. 3, 2001, pp. 31–34.

[182] Tsvetovat, M., and K. Carley, "Structural Knowledge and Success of Anti-Terrorist Activity: The Downside of Structural Equivalence," *Journal of Social Structure*, Vol. 6, No. 2, April 2005, http://www.joss.org.

[183] Prietula, M., "Advice, Trust, and Gossip Among Artificial Agents," in Lomi, A., and E. Larsen, (eds.), *Dynamics of Organizations: Computational Modeling and Organization Theories*, Cambridge, MA: MIT Press, 2001.

[184] Prietula, M., and K. Carley, "Boundedly Rational and Emotional Agents: Cooperation, Trust, and Rumor," in Castelfranchi, C., and Y.-H. Tan, (eds.), *Trust and Deception in Virtual Societies*, Norwell, MA: Kluwer Academic, 2001.

[185] Baumeister, R., L. Zhang, and K. Vohs, "Gossip As Cultural Learning," *Review of General Psychology*, Vol. 8, No. 2, 2004, pp. 111–121.

[186] Hollingshead, A., "Truth and Lying in Computer-Mediated Groups," *Research on Managing Groups and Teams*, Vol. 3, 2000, pp. 157–173.

[187] Nakamaru, M., and M. Kawata, "Evolution of Rumours that Discriminate Lying Defectors," *Evolutionary Ecology Research*, Vol. 6, No. 2, 2004, pp. 261–283.

[188] Ellickson, R., *Order Without Law: How Neighbors Settle Disputes*, Cambridge, MA: Harvard University, 1991.

[189] Strahilevitz, L., "Charismatic Code, Social Norms, and the Emergence of Cooperation on the File Swapping Networks," *Virginia Law Review*, Vol. 89, May 2003, pp. 505–595.

[190] Seligman, A., *The Problem of Trust*, Princeton, NJ: Princeton University Press, 1997.

[191] Fine, G., and P. Turner, *Whispers on the Color Line: Rumor and Rage in America*, Berkeley, CA: University of California Press, 2001.

[192] Robertson, R., "Rumours: Constructive or Corrosive," *Journal of Medical Ethics*, Vol. 31, 2005, pp. 540–541.

[193] Kurland, N., and L. Pelled, "Passing the Word: Toward a Model of Gossip and Power in the Workplace," *Academy of Management Review*, Vol. 25, No. 2, 2000, pp. 428–438.

[194] Granovetter, M., "The Strength of Weak Ties: A Network Theory Revisited," *Sociological Theory*, Vol. 1, 1983, pp. 201–233.

[195] Bordia, P., and R. Rosnow, "Rumor Rest Stops on the Information Highway: Transmission Patterns in a Computer-Mediated Rumor Chain," *Human Communication Research*, Vol. 24, No. 2, 1998, pp. 163–179.

[196] Dresner, D., and H. Farrell, "Web of Influence," *Foreign Policy*, November/December, accessed May 1, 2006, http://www.foreignpolicy.com.

[197] The Radicati Group, *Instant Messaging Market, 2005–2009*, Palo Alto, CA, 2005.

[198] Horrigan, J., "For Many Home Broadband Users, the Internet Is the Primary News Source," *Online News: Pew Internet & American Life Project*, March 22, 2006, http://www.pewinternet.org.

[199] Lenhart, A., M. Madden, and P. Hitlin, "Youth Are Leading the Transition to a Fully Wired and Mobile Nation," *Teens and Technology: Pew Internet & American Life Project*, July 27, 2005, http://www.pewinternet.org.

[200] Shiu, E., and A. Lenhart, "How Americans Use Instant Messaging," *Pew Internet & American Life Project*, September 1, 2004, http://www.pewinternet.org.

[201] Horrigan, J., "Broadband Adoption at Home in the United States: Growing but Slowing," *33rd Annual Telecommunications Policy Research Conference, Pew Internet & American Life Project*, September 24, 2005, http://www.pewinternet.org.

[202] Rainie, L., "The State of Blogging," *Data Memo: Pew Internet & American Life Project*, January 2005, http://www.pewinternet.org.

[203] Rainie, L., and J. Shermak, "Search Engine Use," *Data Memo: Pew Internet & American Life Project*, November 2005, http://www.pewinternet.org.

[204] Rainie, L., and S. Keeter, "Cell Phone Use," *Data Memo: Pew Internet & American Life Project*, April 2006, http://www.pewinternet.org.

[205] Farge, A., and J. Revel, *The Vanishing Children of Paris: Rumor and Politics Before the French Revolution*, Cambridge, MA: Harvard University Press, 1999.

[206] Catholic Online, "Catholic Church Burned After Desecration Rumor," February 22, 2006, http://www.catholic.org.

[207] With regard to politics and religion, some of the largest networks of "local" associations, it is estimated that 84 million Internet users engaged in the 2004 political campaign [208] and that 64% of Americans have used the Internet for religious purposes [209].

[208] Rainie, L., M. Cornfield, and J. Horrigan, "The Internet and Campaign 2004," *Pew Internet & American Life Project*, March 6, 2005, http://www.pewinternet.org.

[209] Hoover, S., L. Clark, and L. Rainie, "Faith Online," *Pew Internet & American Life Project*, April 7, 2004, http://www.pewinternet.org. Fisher, D., "Rumoring Theory and the Internet," *Social Science Computer Review*, Vol. 16, No. 2, 1998, pp. 158–168.

CHAPTER 8
Crystal Ball: Quantitatively Estimating Impacts of Probes and Interventions on an Enemy Organization

Katya Drozdova and John Kunz

While previous chapters discussed qualitative ways to acquire knowledge about enemy organizations, assess their structures and behaviors, and understand their potential malfunctions, this and the following chapters build upon such qualitative insights with quantitative approaches. Here, our objective is not only to determine organizational behavior, but also its quantitative characteristics. One way to learn more about an organization in this manner is to use probes—actions that produce detectable and measurable responses—to elucidate or confirm the organization's structure, behavior, and risk profile. To design an effective probe, however, one needs a way of estimating how the organization's structure and processes relate to the quantitative characteristics of such detectable responses. Similarly, in planning disruptive interventions against enemy operations, one may wish to insure that the disruption achieves a certain quantitatively assessable degree (e.g., one needs a way to quantify the likely intervention impact in order to develop optimal strategies against the enemy). This chapter discusses analytical and computational methods for addressing these needs towards the broader goal of helping strengthen U.S. national and homeland security.

To make our discussion more concrete, we introduce a specific enemy test-case scenario and discuss its broader implications for countering enemy organizations. To analyze this scenario, we use organization and network theories as well as risk analysis, optimization, and other techniques integrated into the *virtual design team* (VDT) research framework—a validated approach for computational organization modeling and analysis applications. The scenario describes a hypothetical enemy mission to plan and execute a suicide-bombing attack by a human-guided torpedo on a U.S. military vessel in a foreign port. The feasibility, considerable damage potential, ongoing enemy intent, and lasting threats to U.S. interests and assets in such situations make this attack scenario a realistic danger.

While the human-guided torpedo is a simple weapon, the enemy will need an organization to secretly prepare and successfully conduct this mission. The case analysis illustrates fundamental pressures and choices that such an organization must deal with in order to survive and succeed in its mission. In turn, to prevent this attack and dismantle this enemy organization, it is necessary to understand and have the tools for analyzing how it functions and where it might fail or be vulnerable to counteraction.

Organizational Analysis Approach

Our approach treats an organization as a complex system of individuals connected through a combination of social networks (individuals linked through authority and informal relations) and task networks (individual tasks linked through communication and coordination requirements). These networks constitute and support organizational structures shaping the complexity that an organization must deal with to succeed in a mission [1]. The total *mission complexity*—that is, the complexity of the system of people working on tasks in a given environment toward achieving specific mission objectives—thus consists of organization complexity and task complexity (8.1). *Organization complexity* reflects social-network interdependencies, and *task complexity* reflects task-network interdependencies.

Total Mission Complexity = Organization Complexity + Task Complexity   (8.1)

Our results suggest that understanding the relationship between organization and task complexity offers insights into organizational performance and vulnerability to intervention.

This chapter discusses how to plan probes and interventions in order to exploit such vulnerabilities and explores their sensitivity to different assumptions about enemy networks. Specifically, we explore two alternative types of possible enemy network structures: scaled and scale-free [2, 3]. Scaled refers to normal distribution, and scale-free refers to exponential distribution, of network links density used as a proxy measure of complexity in network structures [4]. Studies of networks point to systematic differences between scaled and scale-free networks, including their differential responses to internal failures and external interventions. Scale-free networks are characterized by having a few highly connected nodes, or hubs, and tend to be more robust against random node failures but fragile to attacks on their hubs. Alternatively, scaled networks are less sensitive to targeted attacks, but they tend to collapse when a critical fraction of nodes fails (see, e.g., [4, 5]). These network properties affect mission complexity and shape the options that determine not only organizational performance but also its survival—especially in hostile environments, where enemy organizations often attempt to conceal their activities.

Organization's Options for Dealing with Mission Complexity

For a given (constant) mission complexity (limited by the necessary tasks and available people and resources), an organization's resource allocation options range from using fewer people allocated to more tasks to using more people with each individual conducting fewer tasks on average; see (8.1). When an organization optimizes for fewer individuals, their social network—that is, organization complexity—will be lower and thus easier to manage and conceal, but on average each individual will have to conduct more work with increasing task complexity. This interdependency makes the mission vulnerable to task failure. Alternatively, when an organization optimizes for lower task complexity (more streamlined, independent, and self-contained tasks), the mission will be resilient against task failures but will require more people. This results in a resilient task structure but also higher organization complexity—which is more difficult to control and conceal—making the organization more vulnerable to intervention (Figure 8.1).

These complementary alternatives define a spectrum of organizing possibilities. Along this spectrum, different organizations will choose different ways for dealing with the opposing pressures of organization and task complexity. Their design of constitutive task and social networks will reflect these choices and determine points where the organization will be resilient and where it will be vulnerable to intervention. To optimize interventions, one aims to identify and target these enemy vulnerabilities. Operationally, one confirms potential vulnerabilities through probes by measuring enemy network responses to subtle actions. Quantitative analysis of how alternative candidate networks perform under different conditions then predicts their respective risks and vulnerabilities, suggesting disruption opportunities.

Specifically, using computational tools such as the VDT organization modeling and simulation system developed by researchers at Stanford University's Center for Integrated Facility Engineering, one may quantitatively model these differences by adjusting computational parameters that represent observed or assumed organization attributes and preferences. For example, some groups prefer and optimize for secrecy (e.g., espionage and terrorist networks), others favor overt impact (e.g., a traditional state military and diplomatic corps). When missions require tradeoffs, some organizations will value individual lives or the integrity of their social network, even at the risk of occasional task failure, whereas others will focus on completing certain tasks at all costs. Organizations may further attempt to optimize for both social and task spaces, making them more difficult to distinguish. Nevertheless, these distinctions will persist because of the opposing pressures generated by mission complexity, as discussed earlier, providing the opportunity to identify and neutralize enemy organizations.

Enemy Organization Dynamics and Counteraction Strategies

To recognize how these opposing pressures play out in a specific enemy organization, consider a concrete example of a terrorist organization similar to al Qaeda.

Organization complexity
Reflecting social-network structure

Task complexity
Reflecting task-network structure

Mission complexity

Figure 8.1 Opposing pressures of mission complexity, as in (8.1). When organization complexity decreases, task complexity must increase, and vice versa, provided an organization optimizes its resources for a mission of given complexity.

Such organizations structure major missions around fewer people with limited interactions (i.e., for low organization complexity of their mission-critical social networks). Thereby they often avoid detection as well as limit the knowledge that authorities may gain about the organization from informants or agent capture. For example, al Qaeda's continued operations and attacks despite U.S. and allied counteractions may suggest that this type of organization is highly resilient, with its fault-tolerant behaviors reflecting these strategies for protecting individual nodes from detection and the network from propagation of damage due to node failures [6].

However, despite these strategies, an organization cannot disguise its total mission complexity. Hence, during a mission, reduced organization complexity results in increased task complexity (Figure 8.1). As complexity shifts from social network to task network, task interdependency requirements, delays, and failure risks increase. (Thus, for example, observed delays between major al Qaeda attacks may not only be due to external prevention efforts, but also internal corrections necessary to account for failed tasks in undetected missions.) Task complexity also generates detection and disruption targets. The critical distinction—that our approach makes explicit—is that these targets are in the task networks, not social networks, and thus interventions against task networks are likely to be more effective. The case analysis considers how to disambiguate the predicted types of structure and resilience properties of these networks and estimate their behaviors.

Actionable Implications

The ability to disambiguate and estimate the resilience and behavioral properties of organizational networks has actionable implications. For example, enemy behavior that is consistent with optimizing for social network resiliency, while tolerating task complexity and potential performance degradation, suggests that the enemy is willing to abandon or reschedule certain tasks for the sake of protecting its mission-critical social networks. Actionable implications are to probe and track these tasks, delaying intervention until the broader network and potentially more valuable targets (e.g., command and control structures) become exposed. Alternatively, enemy behavior that is consistent with optimizing for task resiliency, even at the expense of sacrificing some individuals and exposing their organizational links, suggests high perceived importance of these tasks to the enemy—and likely high security risk of their completion to the U.S. or allied interests and assets (e.g., such tasks may be part of a mission to acquire weapons of mass destruction capability). Actionable implication in this case is to intervene as soon as possible.

In this chapter, we describe methods explicitly designed for generating such actionable insights through the use of analytical and quantitative modeling tools enabled by the general VDT framework. This framework integrates empirical knowledge about organizations with underlying organization theory and computational techniques. In turn, this approach enables improved planning capabilities toward the goals of detection, prevention, and elimination of enemy organizations.

Main Methods

Our approach relies on three basic methods. First, when analyzing a specific organization, we identify key real-world aspects of its mission, structure, and environment in order to accurately map the relevant operational properties onto their computational representations. Second, our analysis models the organization's social and task networks to identify how it operates toward completing a given mission. Third, probabilistic simulations of organizational models quantitatively estimate and describe the likely properties of organizational structure and behavior as well as risks and problems that can trigger its malfunction and disruption. The estimated parameters and behavior patterns provide benchmarks for assessing probe and intervention options toward strategy development and optimization.

The analytical foundations behind VDT-based probe and intervention analysis largely derive from information-processing organization theory as well as risk and network analysis techniques. Specifically, the VDT research program has developed new micro-organization theory and embedded it in software tools that enable the creation of virtual computer models of a given organization; probabilistic simulation of its capabilities, limits, and risks; as well as quantitative and graphical analyses of the effects of different possible probes and interventions. The quantitative model feedback supports decision-makers in choosing an optimal plan of action to reliably produce desired impacts upon their target organization [7]. These VDT methods have been used and validated in many domains, including academic research, as well as practical applications in fields such as defense, aerospace, construction, electronics, and software engineering, among others [8].

The quantitative analysis here uses a commercial implementation of the VDT approach, SimVision software platform (for the educational version, see epm.cc). Whereas VDT and SimVision were designed for optimizing and improving organizations, here we use it with reverse objectives of deoptimizing and destroying enemy organizations. In combination, these methods provide the framework for linking the operation-analytical and computational aspects necessary to generate meaningful, measurable, and actionable results.

We now describe the joint capability of this general framework by applying it to the specific enemy scenario analysis, which allows us to highlight the broader implications of our approach for developing reliable strategies toward detecting and defeating enemy organizations.

Developing Strategies Against Enemy Organizations

Probe and intervention options are building blocks for strategy development. Probes are subtle, narrowly focused, and economical actions that are performed to understand the structure and properties of an enemy organization by assessing its responses to small perturbations. Interventions involve broadly based actions large enough to produce significant disruptive effects on enemy networks performance. The strategic objective is to find optimal combination of probes for inferring the enemy organization sufficiently accurately to plan interventions that have high probability of producing desired effects with predictable reliability and low friendly cost. The ability to estimate the quantitative impacts of such different actions enables strategy optimization.

At the operational level, probe and intervention approaches may vary depending on their objectives, constraints, and the available knowledge about enemy capabilities and intensions. For example, attempts to prevent an impending high-consequence attack may focus on aggressive disruption. Alternatively, actions against frequent lower scale attacks may initially focus on subtle probes and response tracking to identify the broader organizational network and its resilience properties and use that information to plan a more comprehensive response. Finally, the state of the enemy system or its behavior may not be clear to an operational planner at particular points in time (e.g., when intelligence about the enemy organization and operation is incomplete). Thus, the planner may need additional analyses and clarifying probes to estimate the state of the enemy. For planning and evaluating such action options, the computational VDT-based approach involves two stages: probe identification and intervention planning.

Probe Identification

We use heuristic search strategies to identify probes that best discriminate among candidate enemy organization model states. Toward this goal, a planner develops relevant heuristics based on practitioner inputs, operational knowledge, and computational organization theory or prior research. The planner tests different combinations of potential actions and compares simulation results. The planner then chooses to operationalize probes that test hypothesized enemy organization properties to confirm them or identify the needed model adjustments. In this case, one wants to implement the probes such that they elicit useful information without alerting the enemy, so that the estimated models reflect the current enemy state as accurately as possible. The enemy state may nevertheless evolve (e.g., due to internal or environmental developments), and the planner will need to keep abreast of potential changes. To fill this need, the VDT-based probe analysis capability allows fast and easy updating of enemy models and generating additional feedback to assess the likely changes and their impacts.

Based on the operational results of a probe and analysis of the associated VDT case, the planner then updates the organization models and analyzes them to identify additional potential weaknesses and likely targets for disruption. The planner implements this analysis by conducting additional probes, such as impeding effectiveness of most centralized actors or slightly perturbing the interdependent tasks whose detection would point to organizational vulnerabilities. The computational analysis then attempts to find optimal combinations of probes to confirm enemy structures and properties. Based on lessons learned, the planner may then design targeted intervention actions to induce desired enemy responses.

Intervention Planning

To support disruptive intervention planning, we also use the VDT-based parametric models of an organization and probabilistic simulations of its behavior to estimate organizational performance and risk profile under different conditions. Additional work within the VDT research framework has recently demonstrated a genetic program (GP) that selects a family of possible interventions to optimize predicted organization performance [9]. The GP model defines a generic set of candidate interventions, such as adding or removing assigned actors—modeled as full-time equivalents of work (FTEs) assigned to a task for a scheduled duration—from the social network of participants. Another candidate intervention, among other possibilities, is to change properties of the organization, such as coordination strategies for conducting interdependent tasks. These GP-based methods aim to minimize optimization functions—such as (8.2)—where the factor weights vary depending on the values of the respective factors allowing nonlinear weighting of different predicted values:

$$\text{Function} = \sum \begin{pmatrix} \text{Schedule Weight} * \text{Schedule Duration} + \text{FTE Weight} * \\ \text{Added FTEs} + \text{Coordination Risk Weight} * \\ \text{Coordination Risk} \end{pmatrix} \quad (8.2)$$

When used in reverse mode—that is, for maximizing rather than minimizing risks—this method also applies to the goals of disrupting and destroying enemy organizations.

Analysis of any particular organization will require some baseline facts and operational context. Then, by assessing hypothesized enemy models and quantifying the effects of potential counteractions, the computational tools such as VDT can improve decision-makers' understanding of their target organization and help them plan appropriate responses using the methods outlined here. The following scenario analysis describes this process in more explicit detail, applying the VDT framework and tools to the general planning of enemy counteraction strategies in the context of U.S. national and homeland security needs. (We refer to VDT as the underlying analytical and computational approach behind specific software implementations, such as SimVision used in this research under academic license from ePM.)

Test-Case Scenario: A Human-Guided Torpedo Attack on a U.S. Military Vessel in a Foreign Port

Our test case considers a suicide-bombing attack mission by a terrorist organization utilizing a homemade human-guided torpedo weapon to target a U.S. military vessel in or near a foreign port. For example, the attack's location could be the port of Aden, Yemen, where U.S. military ships refuel, or other such strategically important areas at the crossroads of major military and commercial activity.

Making this low-tech weapon does not require much engineering knowledge nor imagination—one could get instructions off the Internet and background from history books as these weapons date back to at least World War II (see, e.g., [10, 11]). A simple yet robust torpedo needs a metal cylinder packed with explosives and capped with a detonator designed to explode on impact. For greater impact, the suicide bombers riding the torpedo could accelerate by simply releasing compressed oxygen. This weapon is easy to deliver undetected in busy ports and tourist areas. It could be made to look like common watercraft for underwater exploration powered by two riders in scuba diving gear (one steering and targeting and the other navigating and accelerating to cause detonation). These materials and techniques are accessible to terrorist organizations.

Being easy to make and deliver undetected, the human-guided torpedo could cause much damage if an attack succeeded. However, a successful attack would require organizing individuals with pertinent skills to gather intelligence on a potential target; provide resources; plan, command, and coordinate the mission; administer logistics; and execute the attack. This organization would also need to work in secrecy to avoid detection. While operationally difficult, these requirements are not prohibitive. There are many enemy organizations with the requisite experience, technical expertise, resources, and organizational reach—as well as lasting hostile intent against the United States and its allies—to attempt or facilitate this attack. Examples range from terrorist groups such as al Qaeda, Hamas, and Hezbollah to state-supported elements such as the Iranian Revolutionary Guards and various paramilitary, insurgent, and subversive groups worldwide. During execution, this weapon is easy to maneuver and conceal in a busy seaport, as well as difficult to identify—generally, among many objects moving in water and, specifically, as having hostile intent—and is even harder to destroy without advanced knowledge of the operation.

Thus, if the mission reaches execution phase, the attack could be catastrophic. The likely consequences could include not only casualties, property damage, and operational disruption, but also propaganda value for the enemy as well as economic setbacks (e.g., disruption of transshipment routes) and exposed national security vulnerability for the United States. Likely political complications may also stifle U.S. response (e.g., due to U.S. loss of face as well as destabilization of friendly regimes in the area whose cooperation would be needed). For instance, if the attack took place in Aqaba, the pro-U.S. Yemeni government would be hard-pressed to balance its support for U.S. counterterrorist strategies and its neglect, or inability to curb, the ongoing extremist activities (e.g., [12]). These realities will affect the success and costs of any preventive or defensive strategy against such attacks and enemy organizations.

To prevent the attack, one would need to detect and disrupt the enemy's organizational activity in advance. In particular, one would need to know how the organization works, its weak points, and potential failure modes. However, learning these details may be difficult. A sophisticated enemy organization would compartmentalize mission-critical information and disguise its overt activity. In this scenario, a scuba-diving shop would make credible cover in an area where there are many such shops and also potential targets. Assembling the suspicious items (e.g., explosive payload) shortly before the attack would reduce detection and prevention risk. Al Qaeda used similar strategies to succeed in attacking the USS Cole with a simple boat laden with explosives and piloted by suicide bombers. Though countermeasures have been taken, U.S. and allied interests in this area remain vulnerable—as demonstrated by continued terrorist attacks. Examples include the 2002 attack on an oil tanker in the Persian Gulf; the 2005 attack on U.S. Navy ships USS Ashland and USS Kearsarge docked in the port of Aqaba, Jordan; and the 2005 attack on a cruise ship off the coast of Somalia [13].

Enemy Organization Model

To provide an accurate and useful computational analysis, we apply the operational context to generate modeling assumptions and interpret results. This includes identifying the necessary agents, tasks, and resources for organizing and executing this attack, as well as factors shaping enemy decisions, structures, and behaviors. The attack's feasibility requirements and likely consequences discussed earlier are among such key factors. From the operational context, we derived the enemy model for the VDT-based analysis.

The VDT-based computational analysis typically involves five steps. First, one outlines the actors, tasks, mission, and information-processing requirements in a given operational context. This involves representing the social and task networks for the mission plan from start to finish. Second, from this outline one creates a graphical baseline VDT model. Third, one assigns actor, task, and organization properties that govern their behaviors. Behavioral parameters reflect skill and experience requirements, task duration and interdependencies, and known preferences that reflect ways in which the organization tends to work. Fourth, one runs probabilistic simulations to estimate organization behavior, confirm or adjust the baseline model, and generate performance feedback and risk assessments. Fifth, based on feedback, one analyzes the sensitivity of organizational performance, fault tolerance, and resilience properties to various changes by altering model parameters and rerunning the simulations. Simulation outputs suggest probe and intervention options to cause desired organizational change.

Following these steps, Table 8.1 outlines the baseline model elements and context. Figure 8.2 graphically represents the resulting organization model. The social network for this mission consists of a headquarters team (HQ); advanced intelligence team; operational commander; administrator for local logistics; two engineers for developing, testing, and launching the torpedo as well as training the suicide bombers to use it; two bombers for powering, navigating, and targeting the weapon; a financial courier for delivering funds; and a communications courier for videotaping the attack for subsequent propaganda use by the headquarters.

In the model these actors' behavioral properties reflect the mission requirements. The operational commander, administrator, and both engineer actors in the baseline model are skilled in their respective tasks and experienced in terrorist operations. The lead bomber is also highly skilled (in handling explosives). The HQ role is to provide ideological and strategic vision and to channel funds, without directly participating in the mission, and hence they do not need to be skilled in the mission tasks. Because of organizational secrecy and information compartmentalization strategies, the mission-critical team will not have prior experience working together.

Once the organization model has been developed, planners need to know how it would perform over time and under uncertainty—in order to develop optimal counteraction strategies. We illustrate this analysis by considering a probe and two interventions against this enemy mission. We use the mission-complexity framework, as summarized by (8.1), to elucidate the different effects of these counteractions upon alternative possible network designs—scaled and scale-free—that the enemy might employ to structure its mission activities. Because these two network types have very different resilience properties and failure modes [2], it is important that inter-

Table 8.1 Scenario Outline for the Human-Torpedo Attack Used for Deriving Modeling Assumptions

Task Precedence and Duration	Tasks	Actors*
0 (6–12 months)	Gather initial intelligence: scout potential target areas; place inconspicuous markings to assist attack (e.g., a chain underwater for torpedo guidance)	Advanced intelligence team (intel)
0 (3–6 months)	Torpedo R&D: design and build the weapon away from the intended target	Engineers 1 and 2 (E1*, E2*)
0 (1–2 weeks)	Torpedo proof of concept (POC): test prototypes, optimize design, assure feasibility	E1*
1 (1–2 weeks)	Decide strategy: choose target, method, and timing; appoint operational commander	Headquarters (HQ)
1 (1–3 months)	Recruit bombers: the operational commander (OC) identifies and develops agents that can be appointed as bombers	OC*
1 (1–3 months)	Gather intelligence: obtain target information (e.g., docking schedules) for attack execution planning	Intel team
2 (3 months)	Plan and command the operation: develop plan, activate agents, coordinate tasks, arrange funding	OC*
3 (5 days)	Deliver funds: deliver money from HQ or intermediaries to administrator or designated agents	Financial courier
4 (3–5 months)	Administer logistics: maintain a local safe house and cover, procure materials	Administrator*
4 (3–5 days)	Assemble torpedo on site: assemble the weapon at safe house, test and balance on location underwater	E1*
4 (2–4 days)	Train bombers: E2 trains Bomber 1 (B1) and Bomber 2 (B2) in diving, how to handle the torpedo, navigate undetected, and so on away from target location	E2*, B1*, B2*
5 (2 days)	Conduct "go" meeting: About 4–7 days before the attack, OC meets with mission-critical agents and gives final attack instructions	OC*, administrator*, E1*, B1*, B2*
6 (1 hour)	Deliver torpedo to predetermined launch place	B1*
7 (1 hour)	Execute attack: administrator and E1 help launch the torpedo and go into hiding; bombers execute	Administrator*, E1*, B1*, B2*
7 (1 hour)	Record attack: a communications courier records the attack upon launch signal	Communications courier
8 (1 day)	Claim responsibility: the courier delivers the recording to Al Jazeera (AJ) or HQ (months after attack)	Communications courier

* indicates mission-critical role.
The Task Precedence and Duration column indicates parallel tasks by the same task precedence number and lists approximate task duration in parenthesis.

vention planners know what type of enemy network they are targeting. We show how VDT tools enable probe analysis to distinguish these alternative network types based on their estimated behavior patters. Then, simulations show the different effects of interventions on these alternative networks distinguished by their different structural complexity characteristics.

Enemy Mission Complexity Characteristics

Recall that total mission complexity consists of the complexity of social and task networks within a mission (Figure 8.1). Because for a terrorist organization, any social

Test-Case Scenario: A Human-Guided Torpedo Attack on a U.S. Military Vessel in a Foreign Port 201

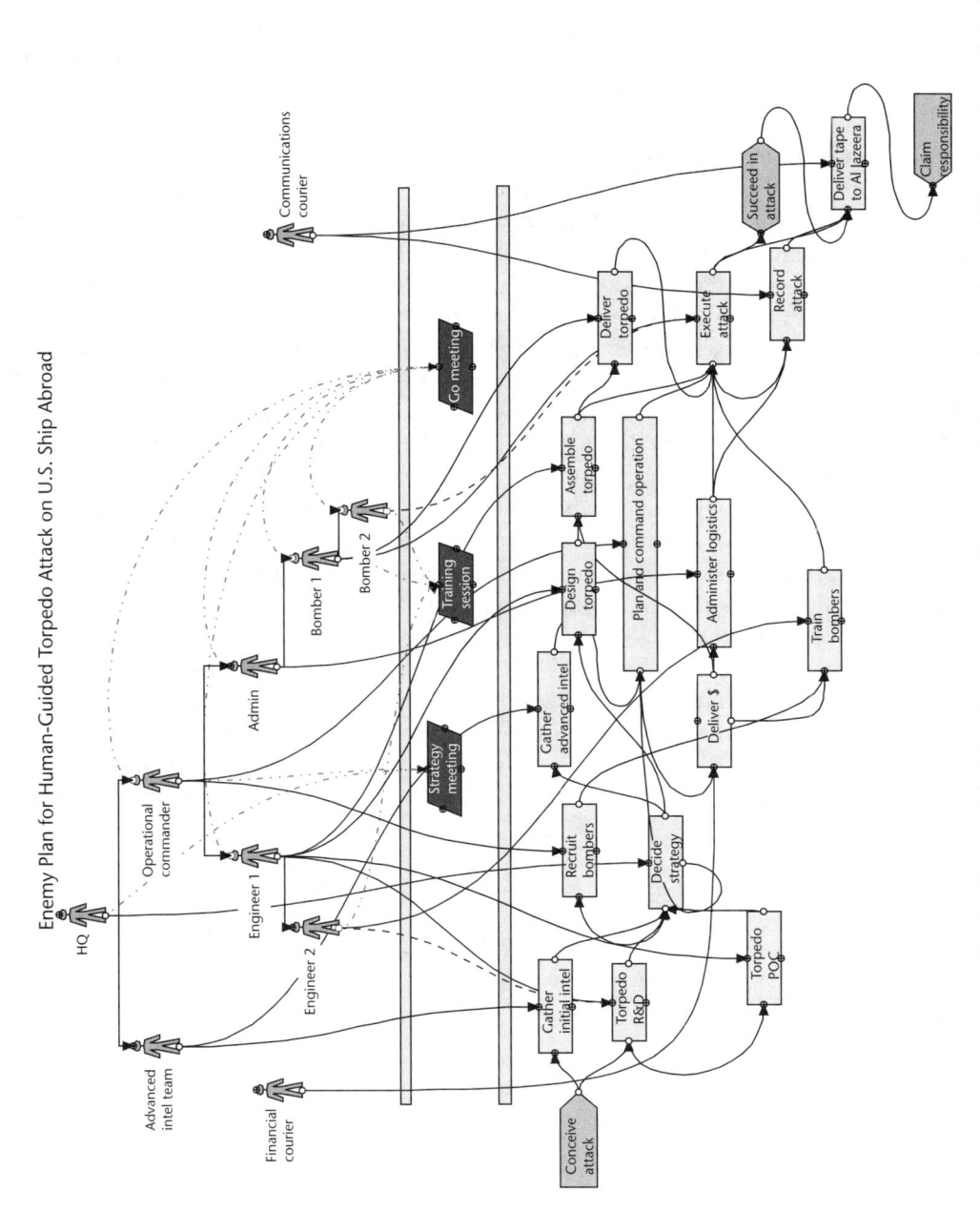

Figure 8.2 A VDT graphical representation of the baseline model of the plan to execute a human-guided torpedo attack on a U.S. ship (outlined in Table 8.1). Each task, represented as a rectangle, can coordinate or initiate corrective work, or rework, in dependent tasks—for example, in case of task or information exchange failure (coordination and rework links not shown for simplicity).

network node or link risks exposing the organization to counterterrorist detection and intervention, sophisticated clandestine terrorist groups limit this risk by organizing missions around as few individuals and connections as possible. That is, they aim to reduce organization complexity. The social network structure discussed in the scenario reflects this organizing principle. Each agent has zero to three evenly distributed authority links to other actors (subordinate or superior). The advanced intelligence team and operational commander (OC) report to headquarters at different points in time. The administrator and head engineer 1 report to OC. Engineer 2 reports to engineer 1. Bomber 1 reports to the administrator, and bomber 2 reports to bomber 1. The couriers are loosely associated with HQ and have no direct role in implementing the attack (see Figure 8.2). This structure represents a scaled social network (with a mean of 0.7 and variance of + −0.7 links per actor-node).

Keeping this social network structure—and thus this level of organization complexity—constant, we present analyses of three cases of task network topology: baseline, scaled, and scale-free (Table 8.2). All cases derive from the baseline case that has 16 tasks with a combination of sequential and parallel precedence relationships (shown in Figure 8.2). All cases have the same actors, tasks, task precedence, and work volumes, but they differ in rework link characteristics. Rework links indicate task dependencies and additional information exchanges necessary to correct for task failure. The baseline case has a fixed number of two rework links per task. The scaled case has a random distribution (mean 2, variance + −2) in the number of rework links. The scale-free case has uneven distribution that is an increasing exponential distribution with a minimum of zero and a maximum of five links. All the cases in Table 8.2 have the same tasks, actors, and task precedence (as shown in Figure 8.2), but the task work volumes vary in the number of rework links—that is, links to other tasks that indicate task dependencies and additional necessary information exchange activities in case of task failure. In VDT, these alternatives are modeled by creating and simulating additional model cases derived from the same baseline design. The three resulting cases are: (1) baseline case with constant number

Table 8.2 Three Test Cases of the Enemy Organization's Task Network

	Case 1: Baseline	Case 2: Random Number of Rework Links (Scaled)	Case 3: Baseline plus Increased Rework Link Frequency Distribution (Scale-Free)
Number of tasks	16	Same	Same
Work volume	1 day–6 months	Same	Same
Number of coordination links per task	0–2	Same	Same
Task complexity	Medium	Same	Same
Error and information exchange probability	0.1	Same	Same
Rework: measured by number of rework links per task	2	0(3 tasks) 1(0) 2(9) 3(2) 4(2)	0(2 tasks) 1(1) 2(2) 3(3) 4(4) 5(4)

of rework links per task, (2) scaled case with a random number of rework links per task, and (3) scale-free case with an increasing rework link frequency up to a peak that more than doubles the mean of the other cases.

This setup implies that, to disrupt the enemy organization, there is value in using quick, low-cost probes to determine whether the network is scaled or scale-free, and if it is scale-free, to identify the nodes that have greatest centrality, as those will be most vulnerable to disruption.

Estimating Impacts of an Example Probe

A successful probe will elicit desired information (e.g., provide evidence to distinguish whether the enemy task-network structure is scaled or scale-free in this example) without disrupting the enemy's work. The planner needs to choose the probe carefully to generate the needed quantifiable effects without alerting the enemy. VDT simulations enable the testing of different action options to determine what sort of actions and at what intensity level would likely constitute a successful probe.

An example probe considers a disinformation action such as introducing false rumors about changed U.S. ship movement schedules at a port in the suspected area of attack. This probe would in effect increase environment uncertainty for the enemy organization without actually changing facts on the ground. One way to model increased environmental uncertainty is by changing project noise probability, which is the probability of actors being distracted from their assigned tasks. This VDT model parameter reflects organizational reality that actors do not always sufficiently complete their tasks on the first attempt. Many tasks require some rework (i.e., corrective work in case of task failures or information-processing problems), and related or interdependent tasks require communication between the workers performing them. In addition, every project suffers some level of noise and distraction, whether it is due to errors, impromptu meetings, or new information. The noise probability specifies the likelihood of these delays and distractions. Comparing case simulation results with different probability settings gives one an idea of the consequences of making changes to the model [14].

Simulations results show that a change from baseline 0.1 to 0.7 noise probability generates the desired effect. The probe allows the mission to complete on time in both scaled and scale-free cases (Figure 8.3) but generates substantially different patterns in actor behaviors (Figure 8.4). In particular, the graph of OC's volume of additional communications and coordination activity over time (estimated backlog of several extra FTE work days to complete his work) shows considerably different patterns in the scaled versus scale-free task networks (Figure 8.4). The OC having the most centralized network position and having to update others as a result of new information explains this prediction.

This approach suggests a way to distinguish the most centralized actor in the task network by estimating the differential impact of a probe on actor backlog patterns in the scaled versus scale-free network cases. In the scale-free task network, the actor performing the most centralized task would represent an information hub. This actor would likely need to communicate with others to coordinate rework in light of the new information (i.e., in this case, probe-induced disinformation). This activity in the scale-free case triggers larger delays than in the scaled structure case.

Estimated disinformation probe effects on mission completion
- The probe allows on-time mission completion in both scaled and scale-free task network cases
- Each network model case is compared to baseline (solid bars indicate task duration with probes)

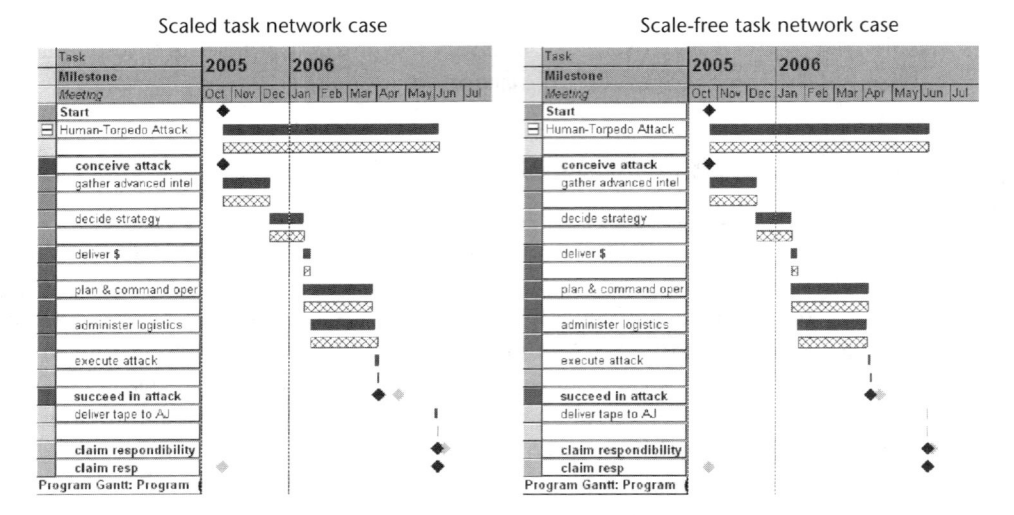

Figure 8.3 Results of simulating the human-guided torpedo attack mission with a disinformation probe. The probe affects enemy organization uncertainty (modeled by increasing probability of actor distraction from tasks resulting in additional communications and rework), but allows the mission to complete on time, as shown in the nearly identical project schedule Gantt charts. The charts compare scaled and scale-free task network cases to a baseline case, respectively.

Estimated disinformation probe effects on actor backlog patterns (backlog refers to increased communications and coordination activity over time compared to schedule).
Effects in scale-free task network case:
- Differentiation between operational commander and administrator backlogs
- Increased operational commander backlog, decreased administrator backlog

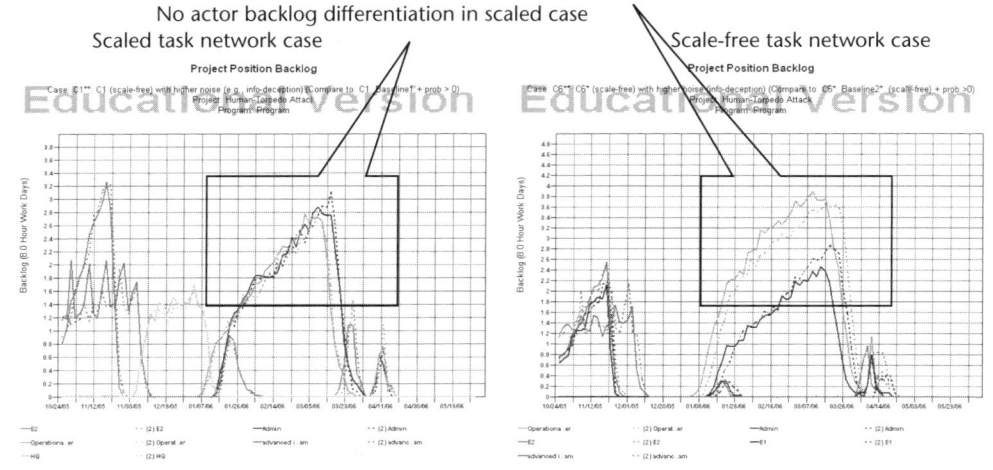

No actor backlog differentiation in scaled case

Figure 8.4 A disinformation probe elucidates different actor backlog patterns, allowing the identification of the OC actor as having responsibility for the most centralized task measured by highest backlog due to extra communications and coordination activities.

The probe results also illustrate the opposing pressures of mission complexity. Note that in the social network, the OC is not necessarily the most network-central actor—no actor is, since the social network links have random distribution per actor. This social network structure causes the complexity to shift to the task network, which becomes increasingly complex with the associated vulnerability to hub-node disruption.

Thus, inferring the enemy task-network structure will enable predicting its resilience properties as well as its vulnerability points, such as the tasks assigned to the commander. Next, we use these probe results to estimate the impacts of potential interventions.

Estimating Impacts of Example Interventions

To estimate the conditions that will likely induce enemy mission failure, the analysis example explores the impacts of two interventions: increasing the complexity of the most network-central task and decreasing the skill of the actor responsible for such a task. We use VDT to model these interventions against three potential enemy task-network types: baseline, scaled, and scale-free (Table 8.2)—keeping the given scaled social-network structure constant (which reflects an organization optimized for evading detection).

For each case (in Table 8.2), we considered three situations:

1. Baseline, without intervention.
2. Increasing the complexity of the most centralized task so that, in comparison with the baseline, dramatically (five times) more rework is required by dependent tasks when the independent task encounters detected problems. In the baseline case, each identified failure initiated dependent task rework of 10 percent of the dependent task work volume. In the increased rework case, each identified failure initiated dependent task rework of 50 percent of the dependent task work volume.
3. Replacing the actor responsible for the most centralized task with a new one that has lower skill (e.g., if the highly skilled actor is captured or incapacitated and has to be replaced by a lower skilled substitute). The general predicted impact is that tasks performed by this actor will take longer. It is necessary to run the simulation to predict the quantitative impact on task duration and to estimate the quantitative measures of that impact.

For comparison, the preintervention predicted schedule durations of the baseline, scaled, and scale-free cases (case 1 column of Table 8.3) are within 4 percent of each other, which is a normal statistical variation for VDT model simulations. However, simulations analysis elucidates and quantifies the significantly different responses of the scaled versus scale-free network structures to the example interventions.

The results of our analysis (Table 8.3) show that the scaled task network is indeed more resilient to interventions than the scale-free task network in this scenario. Two different interventions show the same pattern—namely, that the scaled network has less extreme response to disruptive intervention than the scale-free case.

Table 8.3 Predicted Durations to Complete the Tasks of Figure 8.2 Under Different Conditions

Case (rework link frequency distribution)	*Percent Change in Predicted Mission Duration*		
	1. Baseline: prior to intervention	2. Increased rework volume of the longest network-central task	3. Decreased skill of actor responsible for longest network-central task
Baseline (constant)	0 (B-1)	+43 (B-2)	+16 (B-3)
Scaled (normal)	+4 (S-1)	+29 (S-2)	+16 (S-3)
Scale-free (rising exponential)	+4 (SF-1)	+Infinite (SF-2)	+20 (SF-3)

The predicted baseline duration was 244 days. The numbers in the table show the percent change in predicted duration relative to the baseline prior to intervention. (Parentheses indicate the name of the individual case.)

The scaled case with large increase in rework load (S-2)—where rework again refers to corrective actions necessary to redo or complete the tasks disrupted the by intervention—has shorter predicted duration than the baseline case with the same intervention (B-2). In this case, the savings in rework from those tasks that have fewer than average number of rework links exceed the growth in rework from those tasks that have greater than average number of rework links. Alternatively, task duration in the scale-free cases is the same or greater in all three network structure cases.

One case (SF-2 in Table 8.3) had so much predicted rework that the simulation did not finish, indicating that the intervention drove the task complexity so high that the task could never be completed. This again exemplifies the opposing pressures of mission complexity, where the organization with a social network of low complexity (e.g., loosely structured to evade detection and limit networkwide damage due to potential detection, as is the case in our scenario) must compensate by a task network of high complexity. This complex task network is vulnerable to performance degradation and ultimately inability to complete the mission.

Because the scale-free task network case showed greater response to intervention of incapacitating the actor assigned to the most centralized task—and in the probe analysis we found that the OC is more likely to be responsible for the most centralized tasks—one can conclude that interventions in the scale-free case will be more effective when targeted against the OC's activity. His incapacitation or disruption is more likely to result in the organization's inability to complete the mission, resulting in attack prevention.

In practical counterterrorist or other security planning applications of this analysis, the question now arises: having computationally identified a good target for disruption, how would one actually detect this person or activity? To address this question in our example analysis, we extrapolate from the probe results. The result that the mission completes on time (without intervention), even though there is an increase in the OC's backlog, suggests that the OC will have to compensate for his backlog by changing behavior. For example, he may choose to save time by calling mission participants instead of conducting the scheduled project meeting in person. This communications medium choice would facilitate on-time completion of the necessary information exchange, but the use of communications technology would potentially leave electronically detectable traces. One would need to know what to look for in the communications patterns. The computational analysis using tools such as VDT can help distinguish the relevant patterns.

Analytical and Numerical Methods Underlying VDT and Related Approaches

The VDT research program and tools enable the creation and analysis of virtual computer models of a given organization or project, including the actors, tasks, processes, and linkages among them. Probabilistic simulations of these models produce quantitative estimates of organizational characteristics, suggesting performance-optimizing options based on predicted organizational dynamics and feedback measures [8, 15]. Overall, the VDT approach combines engineering design principles with insights from real-world organizations and organization theory.

The analytical foundation of VDT builds upon information-processing theories of Galbraith [16, 17], March and Simon [18], and the underlying ideas about the impacts of uncertainty and environment on organizational developments [1, 19, 20]. Galbraith first made the observation about contingent behavior of organizations [16, 17], and the VDT research builds on and confirms the importance of contingency [21].

The VDT computational framework represents organization as a dynamic system of interrelated actors that exchange and process information to achieve a specific set of tasks. The modeler specifies relevant organization properties (e.g., actor skills and authority relationships, task sequence and interdependency, levels of uncertainty, time and resource limitations, and so on). The system then simulates information-processing activities and uses quantitative criteria (e.g., for measuring risk of delay, cost overrun, workflow breakdown, and the like) to assess the organization's overall information-processing capacity and its performance implications. This capacity depends on information channel limitations as well as humans' bounded rationality, time, and attention [15].

Other approaches related to VDT in both theory and implementation include Burton and Obel's macrocontingency theory–based model of organizations, Masuch and Lapotin's system using nonnumerical computing paradigms derived from artificial intelligence to model organizational decision-making and to predict the impact of structure on performance, and Carley et al.'s extensions to this model to include learning and communication between actors (see [8]). This work has influenced VDT development, and we discuss its technical aspects in the context of SimVision software implementation.

Computationally different from VDT approaches that share some underlying theories include Burton and Obel's [22] knowledge-based expert system, OrgCon, and a variety of systems developed by the Computational Analysis of Social and Organizational Systems (CASOS) project at Carnegie Mellon University. OrgCon seeks to assess organization efficiency and effectiveness based on multiple-contingency theory of organizations. This theory accounts for contingencies associated with organization size, technology, environment, strategy, leadership style, and organizational culture. The system maps these contingencies onto design parameters such as decision authority, information processing, coordination, control, and incentives [22]. For a particular organization, the system determines parameter values based on practitioner responses to a series of questions designed to elicit the practitioner's view of where his or her organization fits into the OrgCon framework. OrgCon's rule-based engine then identifies possible misfits for the user to consider toward improving the organization's performance [23]. In contrast, CASOS efforts focus not only on improving but also on degrading and disrupting organizations, modeled as systems of intelligent adaptive agents (human and artificial). Computational implementations include NetWatch simulation tools for studying covert networks and DyNet, a computational model for network destabilization, among others (see www.casos.cs.cmu.edu).

Computationally similar to VDT approaches that come from different theories include *critical chain* (CC) and *critical path method* (CPM) for helping managers improve project performance (see [22] for detailed overview and comparison). Both CPM and CC rely on systems and graph theory, and CC also uses theory of constraints. Both approaches aim to minimize project duration subject to resource constraints while meeting the particular constraints of time, cost, and scope. CPM focuses on single projects adopting a local perspective for dealing with uncertainty, while CC takes a global perspective with respect to both single and multiple projects. In managing the available resources, CPM aims to maximize utilization of all resources, whereas CC uses buffers in baseline scheduling to help insure on-time completion while maximizing utilization of bottleneck resources [24, 25]. Similarly to VDT, these methods model organizations as projects and use risk analysis and Monte Carlo simulations to estimate uncertainty effects on project performance. Major differences lie in the scope of analysis and feedback. VDT explicitly represents not only direct work assigned to project participants, but also the indirect work such as coordination and the need for corrective actions that may arise due to failure propagation among concurrent activities. Incorporating a wider range of parameters enables VDT, and its implementations such as SimVision, to assess organizational structures, processes, and risks more comprehensively.

Organization Modeling and Strategy Development with SimVision

SimVision models and simulates actual (if known) or hypothesized candidate organization social networks (Figure 8.5). It enables an analyst to predict or confirm organization structure and estimate its performance in a given project (an instance of organizational mission). The project model typically consists of a finite set of tasks, their sequence and interdependencies, the assignment of actors to particular tasks, and information exchange links necessary to complete the project (Figure 8.6).

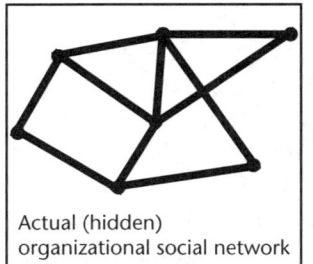

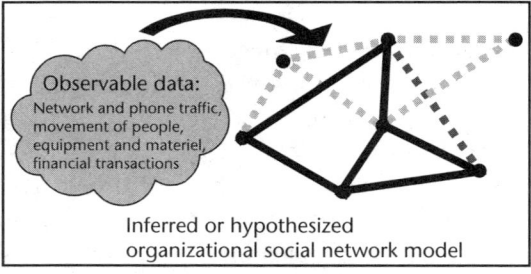

Figure 8.5 A generic organizational social network model. The panel on the left shows an actual social network. The panel on the right shows inferred or hypothesized model. Nodes represent actors and links represent coordination between actors.

Organization Modeling and Strategy Development with SimVision

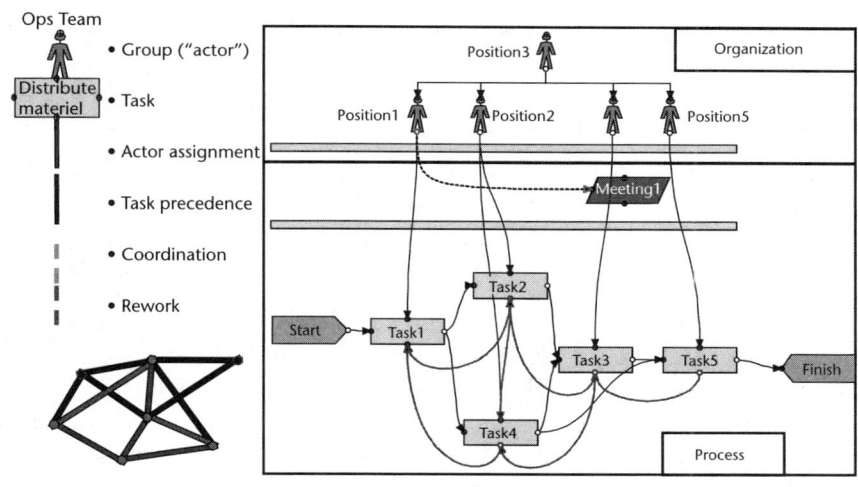

Figure 8.6 Example generic organization mission model. The upper part represents the social network where positions indicate persons, teams, or departments. The lower part represents the task network in the temporal order of work process from left to right. Vertical alignment indicates concurrent tasks. Links among tasks represent task interdependencies such as information inputs, communications, and rework links. Top-down links represent assignment of positions to tasks. The middle part of the figure shows key project meetings.

SimVision uses a Monte Carlo discrete event simulation of actors processing information to predict organization behavior, including the planned task sequence and duration specified by the baseline model as well as the volume, distribution, and impacts of nondirect (hidden) work such as waiting, exception handling, and corrections to failed tasks. The models have two parts: the generic simulation system and the instance model that describes specific actors, tasks, activities, and attributes. Both parts involve behavior rules and default variable calibration values that affect the Monte Carlo set points in the simulation [8].

A generic behavior is modeled using the following rules. The simulator breaks up the model tasks into multiple (about 100) subtasks, any of which can fail probabilistically. Verification failure probability (VFP) is the likelihood that an individual activity subtask will fail and initiate an attempt to create corrective rework. The modeler sets the initial VFP (proj.VFPexternal) for an instance project, normally in the range 0.01–0.1. The VFP affects the volume of coordination and rework that a task will require to complete. If the VFP grows during a simulation, the task will take dramatically more effort and time (hidden work) to complete than expected, or it may never complete because of the large volume of coordination and rework. During a simulation run, this failure probability changes according to the following example formula representing a behavior rule (8.3):

$$\text{task.VFPexternal} = \text{proj.VFPexternal} * \text{Solution Complexity Effect} * \text{Actor Skill Effect} \tag{8.3}$$

where behavior rule adjustment coefficients that reflect task complexity (solution complexity effect) or actor skills (actor skill effect) are determined by values in calibration matrices (Table 8.4).

Table 8.4 Sample Factor Matrix for Skill Mismatch

		Actor Required Skill Level		
		High	*Medium*	*Low*
Actor skill	High	0.5	0.7	0.9
	Medium	0.7	1	1.2
	Low	0.9	1.2	1.5

Generic project attributes are set as follows. An instance project model has a number of tasks (e.g., "deliver supplies"). A modeler creates values for the instance task such as: work volume (FTE days), successor tasks (names), coordination-dependent tasks (names of other tasks with which this task must coordinate), and rework-dependent tasks (names of other tasks that must do rework if a subtask of this task fails), skills required (selected from a predefined skills matrix such as Table 8.4), requirement complexity (high, medium, low), solution complexity (high, medium, low), uncertainty (high, medium, low) and fixed cost ($). Similarly, the modeler defines generic actor attributes such as their work capacity (in terms of FTEs), skills, experience, task, and supervision assignments. The simulator uses these inputs to execute the simulation of a project.

Simulation results include estimates of time and total effort to complete a project, as well as measures of process quality and risk (Figure 8.7). Predicted organizational performance estimates include volume and distribution (by task and actor) of direct and hidden work and impacts of hidden work such as project risks. For example, SimVision predicts risks of task and project schedule delay by dynamically simulating participant workloads and identifying backlogs (i.e., extra time or work requirement to correct or complete one's tasks) associated with coordination and other problems.

By altering SimVision inputs and rerunning simulations, a modeler considers and compares effects of different action options toward developing desired effects upon an organization. Based on model analyses and, when available, observations of the real world, the modeler determines which options best meet the planning objectives, such as in our case developing strategies against enemy organizations.

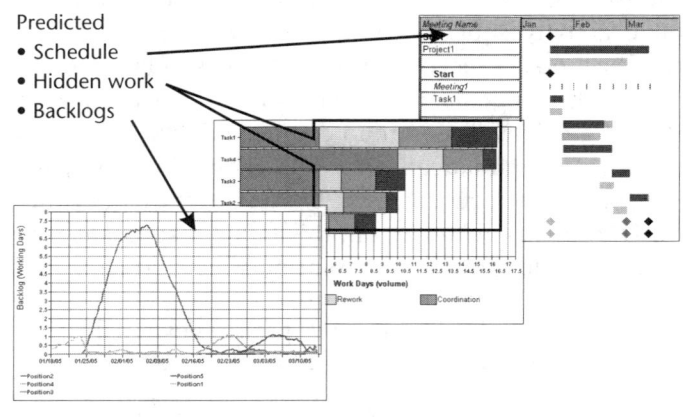

Figure 8.7 Example predicted organizational performance metrics. Upper right-hand panel shows predicted work schedule by task (Gantt chart). Middle panel shows duration of hidden work such as rework, coordination, and decision wait in addition to originally scheduled work (input to the simulation). Lower left-hand panel shows predicted actor backlog for a given mission schedule.

Implications for Detecting and Defeating Enemy Organizations

Based on established organization theory and extensive empirical validation, the analytical framework and computational tools discussed here support the analysis of a broad range of organizations. Complex and evolving real-world organizations, however, do not always fit the precise categories of computer models. Furthermore, the nature of enemy organizations and hostile environments in which they operate often force them to use clandestine and deceptive tactics, thereby concealing information that could help authorities design effective counteractions. To obtain information, one needs to know what to look for and how to measure effects—the needs that our approach aims to help address. The modeler also needs to work within the operational context—or alongside practitioners if possible—in order to translate accurately and faithfully the operational realities onto the computational parameters. Thus, linking the computational and operational issues is crucial for obtaining actionable results.

Towards this goal, informed by an operational perspective, the VDT-based framework discussed here enables systematic assessments of enemy organizations and the ability to use probabilistic estimates to compensate for potential lack of data, which is likely when dealing with hostile and difficult-to-study situations. In particular, focusing on the enemy's mission complexity—composed of organization and task complexity; see (8.1)—and analyzing its relationship to the resilient properties of the underlying social and task networks enables one to infer organizational vulnerabilities. These vulnerabilities can then be confirmed through probes and exploited through interventions as described here. Returning to our initial example, consider a terrorist organization such as al Qaeda. In the presence of hostile opponents with incomplete information, such organizations often make choices in favor of their social network survivability at the expense of task performance. These choices lead to structuring important missions for low organization complexity with fewer agents and limited information exchanges. However, missions structured for fewer agents require greater task interdependency, and the resulting increased task complexity makes them vulnerable to task failure (see Figure 8.1). Corrective actions will require additional information exchanges, thereby generating targets for detection and destruction of the enemy task networks. Then, by using tools like VDT to estimate how potential probe and intervention options may affect such networks, modelers may assist practitioners in the development of more effective strategies against enemy organizations.

References

[1] Carroll, T., and R. M. Burton, "Organizations and Complexity: Searching for the Edge of Chaos," *Computational and Mathematical Organization Theory*, Vol. 6, No. 4, 2002, pp. 319–337.

[2] Barabasi, A. L., *Linked: The New Science of Networks,* Cambridge, MA: Perseus Publishing, 2002.

[3] Barabasi, A. L., and R. Albert, "Emergence of Scaling in Random Networks," *Science*, Vol. 286, 1999, pp. 509–512.

[4] Barabasi, A. L., "Taming Complexity," *Nature: Physics*, Vol. 1, November 2005, pp. 68–70.

[5] Dodds, P., D. J. Watts, and C. F. Sabel, "Information Exchange and the Robustness of Organizational Networks," *PNAS*, Vol. 100, No. 21, 2003, pp. 12516–12521.

[6] Drozdova, K., "Organizatoinal Fault-Tolerance and Technology Risks: How Organizations Use Technology to Counter or Cloak Their Vulnerabilities," Ph.D. Dissertation, (in preparation), New York University, 2006.

[7] Levitt, R. E., et al., "Simulating Project Work Processes and Organizations: Toward a Micro-Contingency Theory of Organizational Design," *Management Science*, Vol. 45, No. 11, 1999, pp. 1479–1495.

[8] Kunz, J. C., et al., "The Virtual Design Team: A Computational Simulation Model of Project Organizations," *Communications of the Association for Computing Machinery*, Vol. 41, No. 11, 1998, pp. 84–92.

[9] KHosraviani, B., and R. E. Levitt, "Organization Design Using Genetic Programming," *Proceedings of 2004 North American Association for Computational Social and Organizational Science (NAACSOS) Conference*, Pittsburgh, PA, June 27–29, 2004.

[10] Borghese, I. V., *Sea Devils: Italian Navy Commandos in World War II*, Annapolis, MD: Naval Institute Press, 1995.

[11] Hobson, R., *Chariots of War*, Shropshire, U.K.: Ulric Publishing, 2004.

[12] AFP, "Qaeda Prisoner Escape Strains US-Yemen Ties," April 29, 2005.

[13] MIPT Terrorism Knowledge Base, May 2006, http://www.tkb.org/Home.jsp.

[14] ePM SimVision User Guide 2005.

[15] Levitt, R. E., et al., "The 'Virtual Design Team': Simulating How Organization Structure and Information Processing Tools Affect Team Performance," in Carley, K. M., and M. J. Prietula, (eds.), *Computational Organization Theory*, Hillsdale, NJ: Lawrence Erlbaum Associates, 1994.

[16] Galbraith, J., *Organization Design*, Reading, MA: Addison-Wesley, 1977.

[17] Galbraith, J. "Organization Design: An Information Processing View," *Interfaces*, Vol. 31, 1974, pp. 28–36.

[18] March, J., and H. Simon, *Organizations*, Cambridge, MA: Blackwell Publishers, 1993.

[19] Lawrence, P., and J. Lorsch, *Organization and Environment: Managing Differentiation and Integration*, Cambridge, MA: Harvard University Press, 1967.

[20] Burns, T., and G. Stalker, *The Management of Innovation,* New York: Oxford University Press, 1994.

[21] Jin, Y., et al., "The Virtual Design Team: A Computer Simulation Framework for Studying Organizational Aspects of Concurrent Design," *Simulation*, Vol. 64, No. 3, 1995, pp.160–174.

[22] Burton, R., and B. Obel, *Strategic Organizational Diagnosis and Design: The Dynamics of Fit*, Boston, MA: Kluwer, 2004.

[23] Samuelson, D. A., "Designing Organizations: CMOT Launches Success on a Solid, Scientific Foundation," *OR/MS Today*, December 2000.

[24] Lechler, T. J., B. Ronen, and E. A. Stohr, "Critical Chain: A New Project Management Paradigm or Old Wine in New Bottles?" *Engineering Management Journal*, Vol. 17, No. 4, 2005, pp. 45–58.

[25] Galloway, P., "CPM Scheduling—How Industry Views Its Use," *Cost Engineering*, Vol. 48, No. 1, 2006, pp. 24–29.

CHAPTER 9
Organizational Armor: Design of Attack-Resistant Organizations

David L. Kleinman, Georgiy M. Levchuk, Yuri N. Levchuk, Candra Meirina, Krishna R. Pattipati, and Sui Ruan

Let us reverse our focus of attention. Instead of finding ways to attack vulnerabilities of organizations, we now explore how to design organizations that can successfully resist attacks. The ability to engineer organizations that can withstand hostile interventions or rapidly adapt to them is of paramount importance in highly competitive or hostile environments. To every sword, there is (or there could be) a shield. Some organizations are less susceptible to the kinds of attacks discussed so far in this book. The questions are, can we determine what features of these organizations make them less susceptible to specific attacks, and can we design for those features? This chapter addresses these important questions, as we present a general methodology for designing resilient organizations.

While the previous chapter utilizes a model-based methodology for the organizational performance predictions to analyze enemy organizations and to devise interventions that disrupt enemy's operations, this chapter focuses on engineering organizational features for coping with disruptions. We use the methodology to design organizations that are robust to hostile interventions, to detect the need to adapt, and to determine superior adaptation options.

Our approach draws upon research in team and organizational performance that has spanned industrial and organizational psychology, business management, operations research, and decision-making in command and control. Some of the earlier work (e.g., see [1] for the literature review) examined the empirical relationships between the organizational structure and performance to analyze dimensions of structure and to draw distinctions between "hard" and "soft" performance criteria. While studying the entrepreneurship in small- and medium-size high-technology manufacturing firms, Naman and Slevin [2] showed that a measure of "fit" between the organizational structure and mission strategy can be used to predict the ensuing organizational performance. In Ryan and Schmit [3], a climate-based measure of person-environment (P-E) fit was developed for use in organizational and individual assessment. They described studies with a Q-sort measure of climate and fit (the organizational fit instrument) to discover ways in which P-E fit information can be used in organizational development. Many researchers (e.g., [4–9]), focused on employing analytic methods to manage and improve organizational performance.

Organizational adaptation is at the focus of portfolio theory [10]. It uses risk-based organizational misfit and other performance drivers to define conditions under which declining performance warrants adaptive organizational change. A model for evaluating the effectiveness of different organizational decision-making adaptation strategies was proposed in [11], using an object-oriented design approach implemented as a Colored Petri net. Carly and Svoboda [12] model organizational adaptation as a simulated annealing process. Burton et al. [13] develop and apply a rule-based contingency misfit model to empirically test multicontingency theory of organizational strategies and designs. The model in [13] is a set of if-then misfit rules, in which misfits lead to a loss in performance. Hashemian [14] provides an in-depth discussion of systemic approaches and design principles (e.g., segmentation, modularity, and so on) for engineering adaptive mechanical systems.

A large body of research addresses the security of computers and computer networks and the ability to resist attacks over the Internet. For example, a discussion about the specific types of attack techniques can be found in [15]. A case study in survivable network system analysis is presented in [16], while [17] gives an overview of trends that affect the ability of organizations (and individuals) to use the Internet safely. Arbaugh et al. [18] propose a life-cycle model for system vulnerabilities and then apply it to three case studies to reveal how systems often remain vulnerable long after security fixes are available. Kendall [19] presents a database of computer attacks for the evaluation of intrusion detection systems. The Government Accounting Office discusses risks to information security from computer attacks on the Department of Defense systems [20]. Findings from the eCrime Watch Survey [21] indicate that the number and sophistication of Internet attacks continues to rise, along with the sophistication of defense mechanisms.

This chapter augments a multiobjective structural and process optimization for designing an organization to execute a specific mission [22, 23]. The organizational design methodology from [22] and [23] applies specific optimization techniques at different phases of the design, efficiently matching the structure of a mission (in particular, the one defined by the courses of action obtained from mission planning) to that of an organization. It allows an analyst to obtain an acceptable tradeoff among multiple mission and design objectives, as well as between computational complexity and solution efficiency (acceptable degree of suboptimality).

How Organizations Cope with Disruptions

We first examine different methods organizations can use to cope with hostile interventions, such as deliberate external attacks or accidental harmful events leading to malfunctions. We also look at design features that mitigate the adverse effects of hostile events.

Different organizations serve very different purposes. Some organizations are formed to plan and execute a specific mission (and are dispersed afterward); their success is measured in terms of how well they perform on that mission. Other organizations face a large number of diverse missions throughout their existence and may shape or even choose their own missions. One aspect of success for these latter types of organizations is how well they are able to survive as a whole and preserve their constituent parts (as measured, for example, in terms of the scope and severity of losses incurred). In general, the organization's success can be measured along two dimensions: (1) performance (gains, costs), and (2) attrition (severity and scope of losses, survival rate). Superior organizational design for hostile environments aims at addressing both of these aspects.

As shown in Figure 9.1, there are different ways organizations can be successful in the face of hostile interventions.

Some organizations withstand disruptions by being insensitive to intervention(s). For example, an organization would be able to tolerate distortions in a communication channel, if it has alternative channels at its disposal. This fault-tolerant design feature (i.e., the ability to cope with unfavorable events and still maintain performance) is often referred to as robustness. Robust organizations are resistant to disruptions that affect processes for which the organizational design includes redundant processes that can achieve similar results [24].

Low sensitivity to disruptions can also be achieved by having its processes inaccessible to the potential enemy. The organization may limit external access to its infrastructure, hide (e.g., code) its communications, and disguise its operations. The concomitant feature of the design is known as security. Secure design limits the amount of damage that the enemy can inflict on an organization.

A second way of coping with unfavorable events is to *preempt* the potential causes of hostile disruptions. This includes impeding or disabling adverse actions of the enemy before they occur. This strategy can be effective if the organization can detect the precursors of adverse actions (e.g., detect the onset of an attack by terrorists) and if it has effective means to deny the potential enemy (e.g., by capturing potential terrorists or eliminating their resources). The key design features that facilitate preemption are intelligence gathering and analysis processes (directed at the known enemy) and the preventive planning and execution. The former relies in part on the information-gathering assets, while the latter may require various shaping resources. In general, however, the application of preemption as a strategy is limited, because organizations often have limited visibility into their enemy's plans.

A third way of coping with adverse events is to *block* the propagation of their effects, before they cause a substantial decrement in performance or material losses. The harmful effects may propagate both within (e.g., failures or errors) and outside (e.g., disinformation and propaganda) organizations. For example, a business organization may find some of its employees lacking the necessary skills and competencies. An adequate and timely training may remedy this situation before it is too late. The organization needs to first detect the undesirable events and then block the propagation of their adverse effects. For example, an organization may detect internal errors and correct them (or block their further propagation) in time to prevent catastrophic failures. This would impede the propagation of errors into catastrophic failures.

Another alternative involves *reversing* the undesired effects. For example, an organization may be able, by monitoring its own processes, to dynamically identify the internal bottlenecks, damaged or jammed communication channels, equipment malfunctions, fatigue, or incapacitation of its key operators. The organization may augment processes with additional resources, repair or replace faulty equipment,

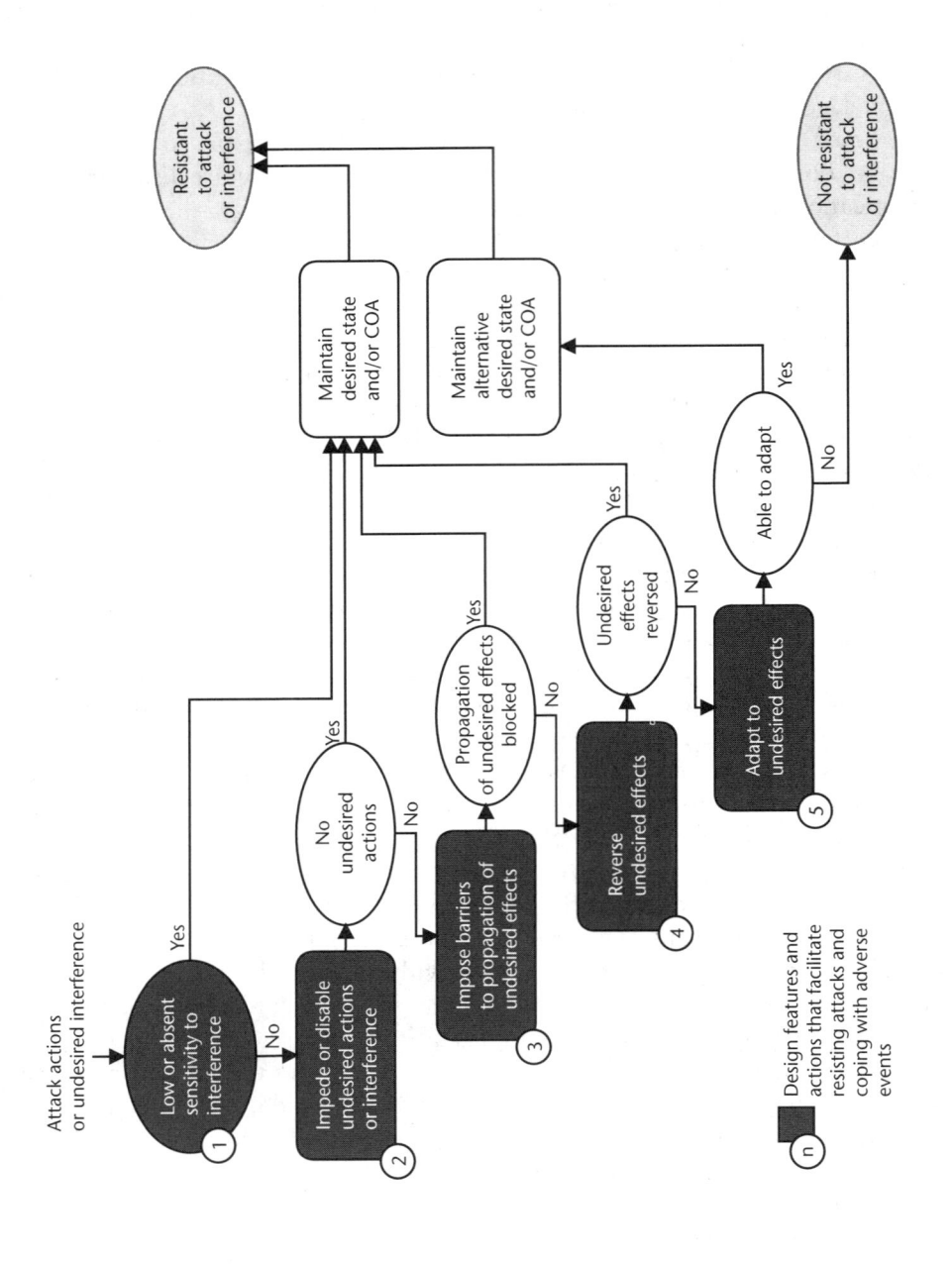

Figure 9.1 Five ways of coping with disruptions.

and replace or rearrange operators (while providing the opportunity for fatigued operators to recuperate). These actions would restore the ability of the organization to maintain its original level of performance. The key design features that facilitate the blocking of effects propagation or reversing them are the internal process monitoring and assessment, coupled with a means to block the internal flows and to repair, recharge, or replace resources (including the human resources and infrastructure). We also note that the same actions may simultaneously entail both blocking the propagation of effects and reversing them (as with the error correction example mentioned earlier).

Finally, when mitigation techniques are ineffective, and loss in performance is imminent, an organization may try to *adapt* its design (i.e., processes and structure) to cope with the effects of adverse events. For example, military organizations that face high-threat, time-critical targets may be forced to shorten their decision loops and reassign cognitive and communication resources to provide behavioral adjustments in response to changing situations in the battlespace. In the commercial sector, fierce competition is forcing companies to search for novel ways to adapt to adverse changes, such as the introduction of superior products by the competition, disruptions in the supply chain, long-term shortages in qualified workforce, and so on. In general, adaptation can be driven not only by the avoidance of failure, but also by a desire to seize the opportunity. According to McKinsey [25], corporations must make sweeping organizational changes in order to capture the opportunities of today's economy. References [26–28] point out that modern military organizations—such as the Joint Task Force Command Centers, Air Operation Centers, and Expeditionary Strike Groups—need to match their processes with new mission challenges brought about by the network-centric technologies and novel military doctrines (e.g., the effects-based operations). Whatever the motivation, adaptation is often a painful process, as it forces organizations to abandon their customary ways of doing business in favor of novel, potentially superior, but unfamiliar and untested, practices.

The adaptation process must consider not only the target (postadaptation) state of an organization, but also the transition path. The organizational design for *adaptability* should reflect both its flexibility in allowing the necessary internal changes and its ability to find and execute the effective adaptation strategies. Reference [29] describes a formalism for generating alternative adaptation options and adaptation paths (i.e., different ways in which an organization can adapt to the same state) and for choosing paths that are superior with respect to the cost, effort, or time required for adaptation.

One interesting finding of empirical research is that organizations tolerate certain adaptations better then others [30, 31]. For example, human-team-in-the-loop experiments described in [32] suggest that organizations find it easier to adapt from divisional forms to functional forms, while they find it more difficult to adapt in the opposite direction. This suggests that certain organizational designs may be superior in promoting adaptation. Based on an analysis of how several industries and disciplines (from construction to Internet applications to software development to mechanical engineering) address design for systems adaptability, it can be argued that some of the design features that facilitate organizational adaptation include function-based segmentation (modularization), plug-and-replace modularity, relative autonomy of components, and heterogeneity (Table 9.1). Other features that promote adaptation have to do with behavior indoctrination, sensitivity to external forces, and sensitivity to changes in the environment in which the organization operates. Successful organizations continuously strive to keep their operations in line with the external forces affecting their objectives, strategy, products, and people [33].

Organizational adaptation comes at a cost and brings about uncertainty, at least in the short term. On the other hand, the insensitivity of robust organizations to external forces can result in a mismatch with its environment in the long run, bringing its eventual demise. Therefore, an organizational design that combines the features of robustness with adaptability may be of interest when designing organizations to operate in hostile environments.

Next, we look at attacks directed at specific organizational design components (structural elements and processes). We also consider the issues of robustness and adaptability as they relate to the attack-resistant design of organizational components.

Attack-Resistant Design Solutions

Figure 9.2 illustrates the key components of organizational design. A specific organization may have these elements and processes represented to different degrees, possibly associated with a slightly different meaning and different implementation. However, most organizations consume information or supplies and execute actions to produce (tangible or information-related) products or effects. Most organizations bring together human operators and decision-makers, who communicate and coordinate in accomplishing their missions. All but the most trivial organizations maintain some degree of specialization for their personnel and utilize some form of management to ensure control and facilitate coordination. The organizational design links these components together to align cognitive, communication, and other resources to render specific organizational behaviors and capabilities.

Depending on the nature of the mission, certain components play a more direct role in determining organizational performance, while other components may play

Table 9.1 Design Features That Facilitate Organizational Adaptation

Design Feature	Description	Example
Function-based segmentation (modularization)	Subordination of the physical structure of a system to its functional structure	Standard modules for agile manufacturing that allow periodic factory restructuring to meet new production demands
Plug-and-replace modularity	System components designed as easily detachable modules	Modular attachment of memories and processors to a computer system to facilitate repair and upgrade
Autonomy of components	Independence of components	Independence of flight control system elements (e.g., engine, rudder) from mission control system elements (e.g., sensors)
Heterogeneity	Multifunction capabilities and multidisciplinary expertise	Using military forces for search and rescue and for disaster relief; using flight-critical sensors for vehicle health monitoring

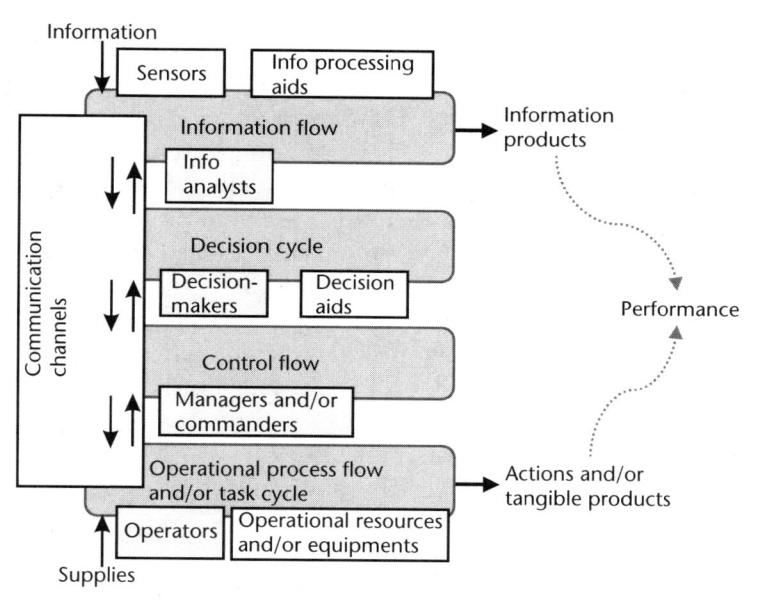

Figure 9.2 Key components of organizational design.

supporting roles. For example, in manufacturing organizations, the operators who are involved in the production of goods have a more direct and immediate impact on the quality of products than their managers who assess the production quality and report it to the upper management. In agile manufacturing plants, such as those producing pharmaceuticals, the information flow has a more immediate effect on performance, because sensed information is used frequently to adjust the process parameters. The information-based outputs produced by research organizations ultimately define their performance. The role of communication among software teams, for example, depends on how coupled their assignments are.

These examples suggest that each specific mission and the corresponding organizational design may prescribe different supporting-supported relationships among the components in Figure 9.2. Each of the design components can be attacked or may experience malfunctions. The primary effects of attacks on specific elements may propagate with different time delays into secondary, and, possibly adverse, effects on other elements. For example, overloading human decision-makers may reduce their psychomotor skills and cognitive capacity, thereby causing loss of information, incorrect decisions, or inaccurate actions. Depending on the roles that different design components play in determining organizational performance, different means (e.g., blocking the propagation of adverse effects, corrective actions, or adaptation) may be better suited to cope with adverse consequences.

In addition to facilitating internal relationships among design components, organizations also maintain external relationships with their environment (e.g., the supply and demand); these could also become targets of enemy attacks. For example, lowering prices by the competition may be used to lower the demand for a product of a given organization and consequently to reduce the cash flow. Means to resist such attacks may include actions directed both internally (e.g., restructuring to reduce the cost of offering) and externally (e.g., broadcasting the key competitive advantages to stimulate the demand).

The key element of an organization is its human capital [34]. Many business organizations declare their human capital to be their greatest asset. Humans within organizations play varied roles, from operators to managers to information analysts and decision-makers (Figure 9.2). In addition to direct attacks (e.g., high mission tempo, stress), humans may be affected indirectly by attacks on other design components (e.g., by denial of supplies needed to maintain readiness, by denial or distortion of information needed to maintain situational awareness, or by injecting tasks that are incongruent with human capabilities). The direct effects on human operators from these attacks range from fatigue (and reduced workload threshold) to incapacitation. The concomitant organizational effects include failed operational processes, decreased accuracy of actions, failed coordination, missed communications, reduced quality of (tangible and information) outputs and decisions, and deficiencies in situational awareness.

Typical organizational design solutions for resisting attacks directed at the human capital include the following:

- Periodic training to increase workload and stress tolerance thresholds;
- Keeping reserve staff or having a part of the workforce significantly underloaded, coupled with rotation of the reserve staff to maintain proficiency;
- Having multidisciplinary and multifunctional experts within the workforce;
- Rotation of positions to increase mutual awareness;
- Protection and security.

The inclusion of training processes, reserve capacity, and heterogeneous expertise allows organizations to be robust to attacks on their human assets by dynamically reallocating their human resources or by reallocating affected tasks to different operators. The increased mutual awareness and multifunctional expertise facilitate organizational adaptation.

Attacks targeting the flows of information and of tangible materials and supplies across an organization and between the organization and its environment range from denial of supplies to denial or distortion of information to attacks on communication and transportation channels. Specific examples of such attacks include disrupting the supply chains, supplying false or inaccurate information, attacking sensors and other information-gathering resources to limit access to information, and jamming the communication channels. Indirect attacks targeting organizational flows include inducing the channel overload (e.g., by increasing false alarm rates) and bottlenecks (e.g., by imposition of additional tasks, information, or supplies). The direct effects from attacks on organizational flows and channels range from transmission delays to disabled channels to flow interruptions. The concomitant organizational effects include missed communications and failed coordination, gaps in situational awareness, reduced quality of decisions, delayed (tangible and information) outputs, and failed operational processes.

Design solutions to resist attacks directed at their information and material flows and communication channels include the following:

- Increased connectivity to allow for multiple ways to transmit flows;
- Establishing redundant channels;

- Antijamming and channel security equipment;
- Maintaining alternative suppliers for similar products;
- Linking to multiple and diverse information sources.

Other more advanced solutions [35] include using modeling and simulation process assessment tools to anticipate bottlenecks and optimize network configurations and flow routing rules and processes (e.g., routing, flow schedules, and queuing policies).

Organizational Design Formalism

Measuring organizational effectiveness is difficult, in part because of the potential diversity of knowledge, skills, abilities, and other characteristics (KSAOs) among humans within the organization. Individual characteristics often play a dominant role in defining organizational behavior. At the same time, an organizational design (i.e., an organizational structure and processes) could help overcome inherent human limitations by distributing work and decentralizing the information processing and decision-making functions within the organization. A high-quality design can deliver a synergistic multiplier to the combined efforts of individual decision-makers, while an inadequate design may impede that collective effort and make it ineffective.

Developing a systematic procedure for designing organizations is an ongoing challenge to the scientific community. A key prerequisite to successful organizational engineering is the appropriate formalism in quantifying and modeling organizations and their missions. Table 9.2 summarizes the organizational design formalism based on the normative modeling approach for quantitatively modeling an organization and its mission [36].

This formalism can be used to quantitatively model missions and organizations and to evaluate the concomitant (mis)match between a mission and the organization tasked to execute the mission. In addition, the mission structure can be used to quantify and measure the state of organization's knowledge about its mission prior to the mission execution. This knowledge can then be compared with the actual mission (i.e., ground truth) in order to quantify mission uncertainty.

To quantitatively assess organizations, we need to specify the performance criteria that differentiate one organization (of the same type) from another. Then, an objective function can be defined as a weighted sum of these criteria. The criteria for successful organizational performance form a natural hierarchy. For example, an efficient task execution is useless unless it contributes to achieving a specific organizational goal; in turn, the value of achieving a goal may depend on the degree to which the desired effect is induced. Whether or not the level of communication is appropriate depends on the degree to which it contributes to achieving organizational objectives. While a posteriori recognition of organization's success can be relatively straightforward (although it still may be subjective), measuring the degree to which different organizational processes impact organizational performance could be tricky.

Table 9.2 Elements of the Normative Model for Mission-Driven Design of Organizations

Organization

Structural Dimensions and Elements	Standardized Organizational Processes That Enable Distributed Operations	Dynamic Processes
(Superior-subordinate) command hierarchy among human agents (i.e., decision-makers)	Communication standards (rules, procedures)	Developing strategies to achieve mission objectives
Resource allocation among human agents (i.e., a mapping that specifies which decision-maker controls which resources)	Coordination methods transferring information	Mission monitoring, including the detection of critical events and tracking mission progress
Communication structure and processes for transferring information		Choosing courses of actions (tasks)
		Designating responsibilities for reacting to events and performing mission tasks
		Dynamic task allocation and scheduling (including dynamic resource allocation)
		Communication

Mission

Static Mission Structure	Mission Dynamics
Goals	Distribution of event occurrence times
Enabling relationships among goals	Probabilistic description of how task outcomes propagate to spawn new events
Events	Probabilistic description of task attributes
Impeding relationships (i.e., threats) and enabling relationships (i.e., opportunities) between events and goals	
Task precedence and information prerequisites	
Task input-output relationships	
Task resource requirements	
Task expertise requirements	
Task workload impact	
Courses of action (i.e., sequencing of tasks) to achieve goals	
Event-driven courses of action	

Precursors for Superior Organizational Performance

The organizational theory (e.g., [37–39]) has identified several generic ingredients critical to an organization's success, including the following:

- Awareness of the environment;
- Expertise;
- Capability (derived from resources and human expertise);
- Self-awareness;
- Timely management;
- Coordination;
- Communications.

Awareness of the environment ensures that the organization knows the mission environment in order to react to it appropriately. Expertise enables the organization to decide what needs to be done and to know what to do when. Capabilities define what can be done, who can do what, and who will be able to do what is needed. Self-awareness enables the organization to distribute mission responsibilities and task load appropriately. Timely management enables the organization to distribute the task load dynamically and to ensure that things are accomplished in a timely fashion. Coordination allows the organizational elements to join forces to complete complex tasks, as well as to synchronize efforts and to apply synergy. Finally, communication provides the means for sharing information and knowledge, passing along warnings and orders, and enabling coordination.

Enemy attacks could inhibit one or more of the key ingredients of organization's success, causing organizations to fail. Some effects of the attacks can be instantaneous; other effects take longer to manifest themselves. As we noted earlier, different design components contribute differently to defining organizational performance under different circumstances. Fortunately, there exists a generic design parameter that can be used to predict organizational performance, as well as to predict the effects of specific attacks. This design parameter is known as congruence between an organization and its mission, defined by the so-called congruence theory [6, 40].

Based on the empirical observation that an organization operates best when its structure and processes fit, or match, the corresponding mission environment [41], congruence theory states that the better an organization is matched to the overall mission, measured using multiattribute performance measures of workload, resource allocation, expertise, and communication and coordination, the better that organization will perform (see Figure 9.3). Mismatches, if they could be measured dynamically, can be used to identify the need for an organization to adapt. In other words, we may predict that the congruent organizations will perform well, while the incongruent organizations will perform poorly. Hence, if an attack has disrupted an organization to cause it to have a significant mismatch with its mission, this organization then is likely to exhibit subpar performance, unless a proper match is restored. If we can use quantitative measurements of organizations and their missions to assess congruence and predict organizational performance, we can then assess the effects of specific attacks on performance and devise the required adjustments to restore performance. We can also strive to devise design solutions to preserve or recover congruence in the face of attack.

The model-based experimental research, conducted as part of the Adaptive Architectures for Command and Control program [42], used human-in-the-loop

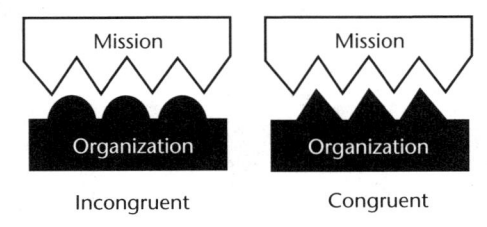

Figure 9.3 Notional informal visualization of (mis)match between a mission and an organization.

experiments to validate the congruence hypothesis. The A2C2 research used the distributed dynamic decision-making team-in-the-loop real-time war-gaming simulator to compare the performance of different organizations for the simulated mission scenario(s) [43].

Findings from team-in-the-loop experiments using the DDD simulator gave the empirical validation of congruence hypothesis (see Figures 9.4 and 9.5 for a high-level summary of A2C2 experiment 4 findings). They showed not only that the congruent organizations outperformed their less congruent counterparts [44, 45] (Figure 9.4), but also that the organizational architecture type differently affected team processes (Figure 9.5), as was predicted by models [45, 46]. The key finding from this research was that the structural (in)congruence between an organization and its mission allows researchers to predict how the organization would perform under different situations.

Figure 9.4 shows that the six-node model-based organization (whose architecture was obtained via an algorithm-based design optimization procedure described in [36]) achieves superior performance to that achieved by the conventional JTF organization (whose architecture was optimized for the same mission by the subject matter experts). Moreover, Figure 9.4 shows that the four-node model-based organization (also obtained via a design procedure from [36]) achieves similar performance to that achieved by the conventional JTF organization, despite 33 percent manning reduction.

Figure 9.5 shows that the congruent model–based architectures require less communication (this was one of the design criteria for the A2C2 experiment 4 mission), as engineered capabilities at each command node reduced wasteful internode coordination. Also, better and timelier use of communication channels by the congruent architectures supported anticipatory behavior (a performance predictor).

The experimental validation of congruence hypothesis, albeit limited, paves the way for a quantitative methodology (described later) to design superior organizations for specific missions. In many practical cases, however, the process congruence between the organization and the mission can only be assessed a posteriori (i.e., from simulation-based performance predictions or monitored data).

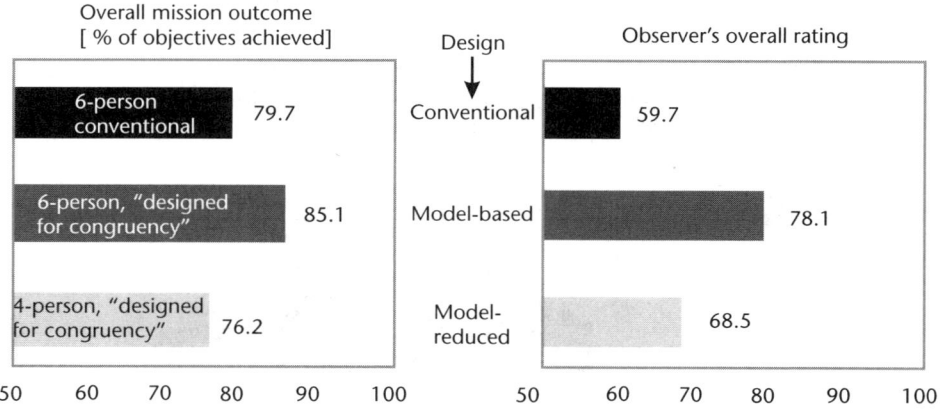

Figure 9.4 Organizational performance for congruent versus conventional organizations—A2C2 experiment 4 (summary).

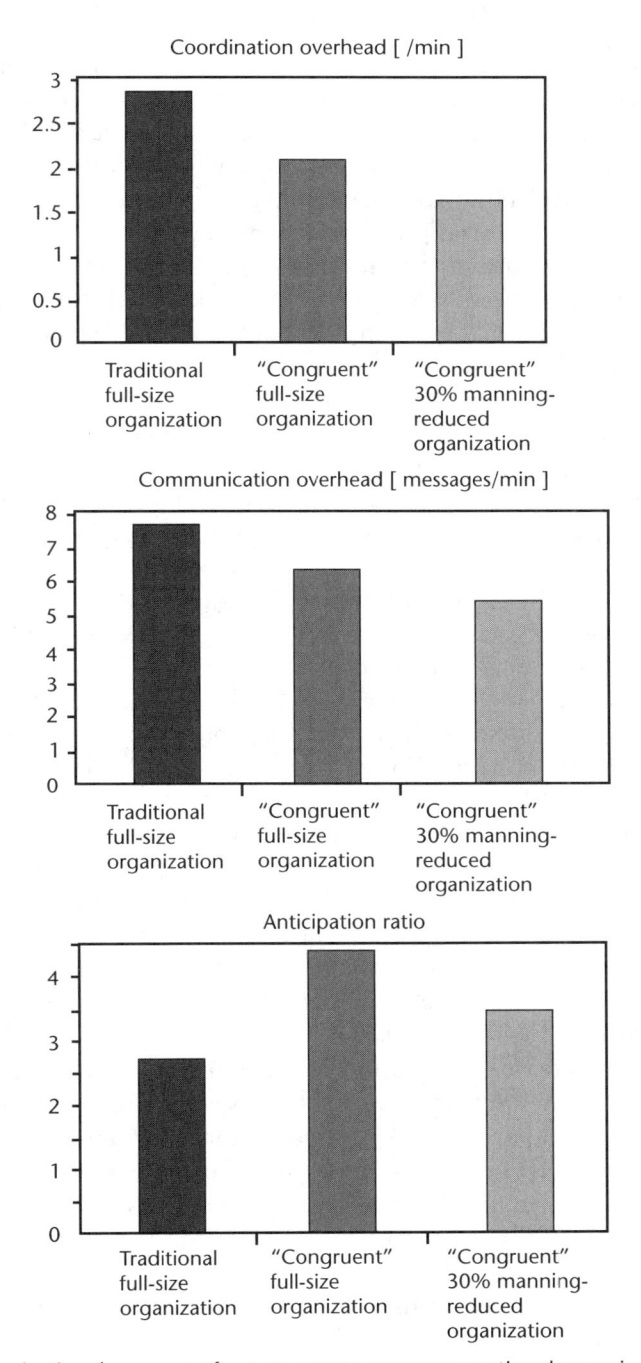

Figure 9.5 Organizational processes for congruent versus conventional organizations—A2C2 experiment 4 (summary).

A Computational Approach for Predicting Organizational Performance

To design an organization best suited for its mission, it is necessary to generate performance predictions in order to optimize its expected performance. The ability to generate performance predictions that will closely match the reality is essential to engineering superior organizations. How closely can we predict the performance of a given organization in a given operational setting? Will the potential diversity among the KSAOs of human members of an organization limit our ability to predict its performance? How does uncertainty in the mission environment propagate into uncertainty in organizational performance?

When designing organizations to operate in uncertain military environments, the specifics about many scenario events and mission parameters often are inaccessible a priori, with only estimates available to the designer. Once the mission execution starts, the actual mission scenario, coupled with particular decision and operational strategies employed by the organization, determines the ensuing organizational performance. If performance predictions generated by models used for the design agree with the actual organizational performance, this would allow one to select superior designs for the prospective mission out of several alternatives. In other words, the organizational design can then be optimized for the specific mission.

A computational approach for predicting organizational performance [47] utilized an executable synthetic model of a generic organization (illustrated in Figure 9.6) to generate performance predictions for a given organization and known mission in probabilistic terms (e.g., specified in terms of the mission event frequencies and required courses of action). The approach used the distributed computer model of an organization to synthesize a network of decision-making and communicating agents that jointly process the event-driven distributed mission tasks. The approach utilized the organizational performance model from the *team optimal design* (TOD) methodology for iteratively designing superior organizations to optimize their predicted performance and process measures [36]. The TOD model accounts for communication overhead and assesses the task delays due to instantaneous overload of a decision-maker. Monte Carlo simulations were used to generate model-based organizational performance predictions and to predict variations in performance [48].

The executable organizational model [47] was experimentally validated in a set of human-in-the-loop experiments that tested several command-and-control organizations and compared their expected performance (predicted by the corresponding synthetic models) on specific missions with the observed performance of manned organizations. Findings from these experiments [49] showed that researchers were able to successfully predict how the organizations would perform under different situations (Figure 9.7).

This approach can be used to predict organizational performance loss resulting from various hypothesized attacks. Coupled with the normative design methodology (described later) for engineering organizations that are congruent with their missions, this approach can be used to determine the best ways to adapt to specific hostile events and to reverse (or negate) the adverse effects of attacks. Next, we describe a systems engineering–based normative methodology that employs analytic methods to manage and improve organizational performance and to design superior organizations.

A Computational Approach for Predicting Organizational Performance

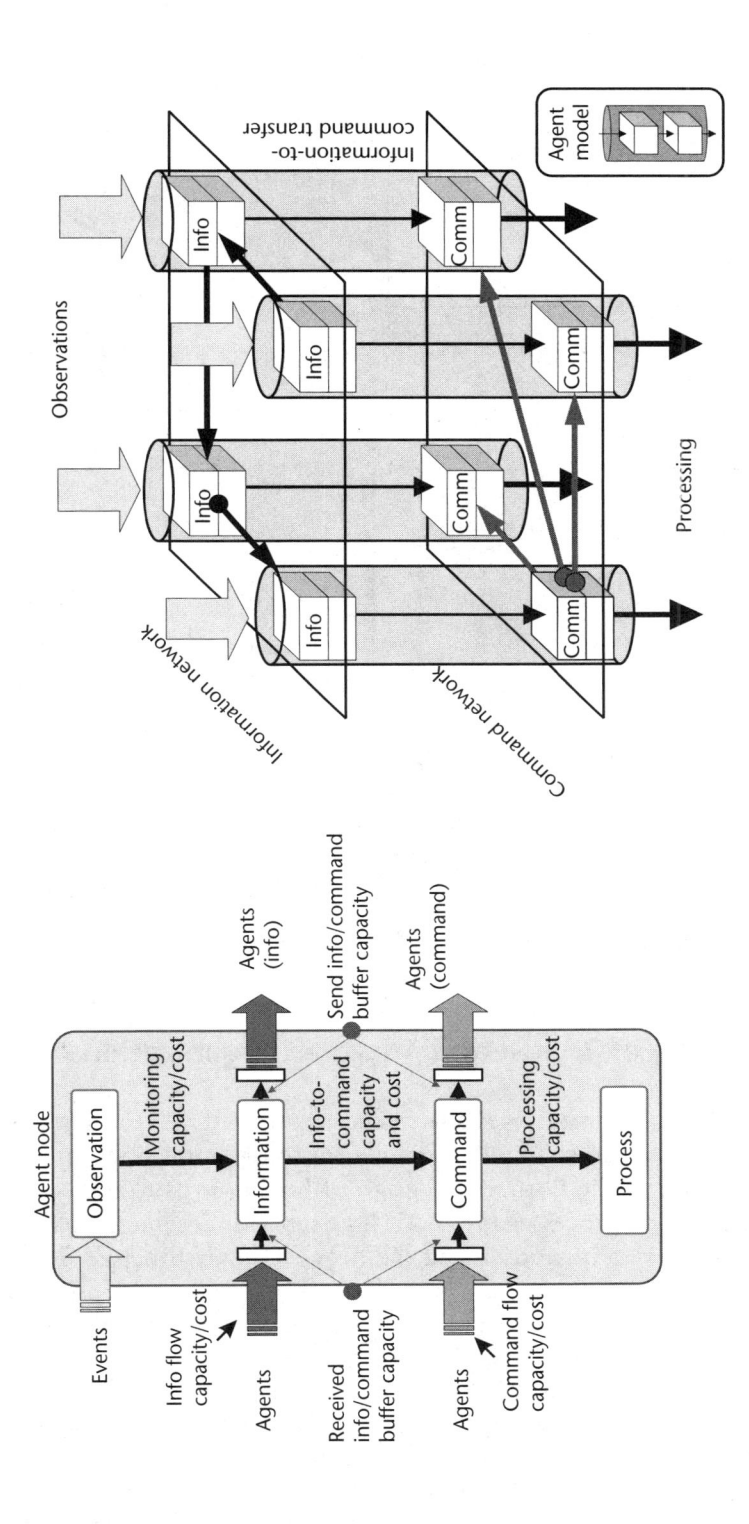

Figure 9.6 Elements of a synthetic agent-based model of a generic organization.

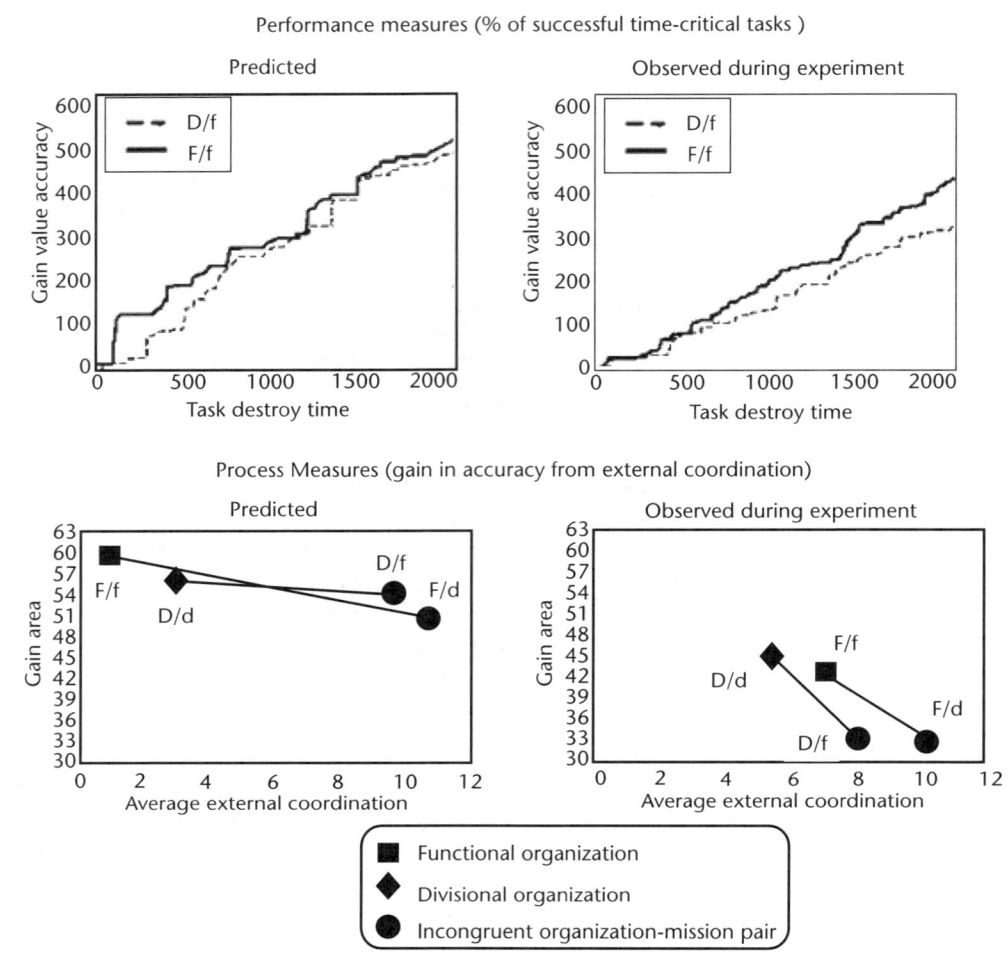

Figure 9.7 Example model-predicted versus empirical organizational performance and process measures for A2C2 experiment 8.

Normative Design of Robust and Adaptive Organizations

The concept of organizational congruence motivated the research on designing mission-based organizations and ultimately led to the application of systems engineering techniques to the normative design of human organizations to optimize their performance and processes [4, 50, 51]. Congruent design methodology provides a baseline for designing and testing the degree of match between a mission and an organization.

The systems engineering approach for designing superior, congruent organizations is as follows. First, a quantitative model that describes the mission and the organization is built. The mission description may include the dynamic events and the task environment, the goals, the desired end effects, potential actions to induce the desired effects, and so on. Next, different performance criteria, and some enabling process criteria, are combined into an objective function, and an organizational design is generated to optimize this objective function (see [36] for a detailed formal description of the organizational design methodology and algorithms). Figure 9.8 provides a high-level overview of the iterative design process. Note that other sequencing of phases than in Figure 9.8 may be preferred, depending on which factors are the most critical for a specific mission.

The approach in Figure 9.8 generates an organization that is well matched to a specific mission. However, in the face of severe external impacts and mission changes (such as time stress or fatigue, delayed and uncertain information, the emergence of unforeseen tasks, new technologies, different strategic options on the part of one's adversaries, and so on), organizations must be flexible to maintain superior performance. By flexibility, we mean the attributes of robustness (i.e., the ability to maintain short-term performance in the presence of environmental changes through process modifications) and adaptability (i.e., the ability to maintain high-quality performance in the presence of mission changes by adjusting decision processes and team structures).

The approach for designing congruent organizations has been extended in [52] (Figures 9.9 and 9.10): (1) to construct robust organizations capable of processing a range of expected missions, and (2) to construct adaptive organizations capable of online structural reconfiguration or strategy adaptation to cope with unforeseen changes in the mission or in the organization. The multimission robust organizations were shown to sustain required levels of performance in dynamic environments without having to alter their structures. Robustness in an organization introduces redundancies in task-resource allocation resulting in a stable organization with respect to environmental perturbations, and decision and processing errors. Evidently, this insensitivity results in a higher operating cost or in slightly degraded performance (compared to the performance of optimized congruent organization, finely tuned to its mission) for each specific submission, but it minimizes the organization's fragility [45].

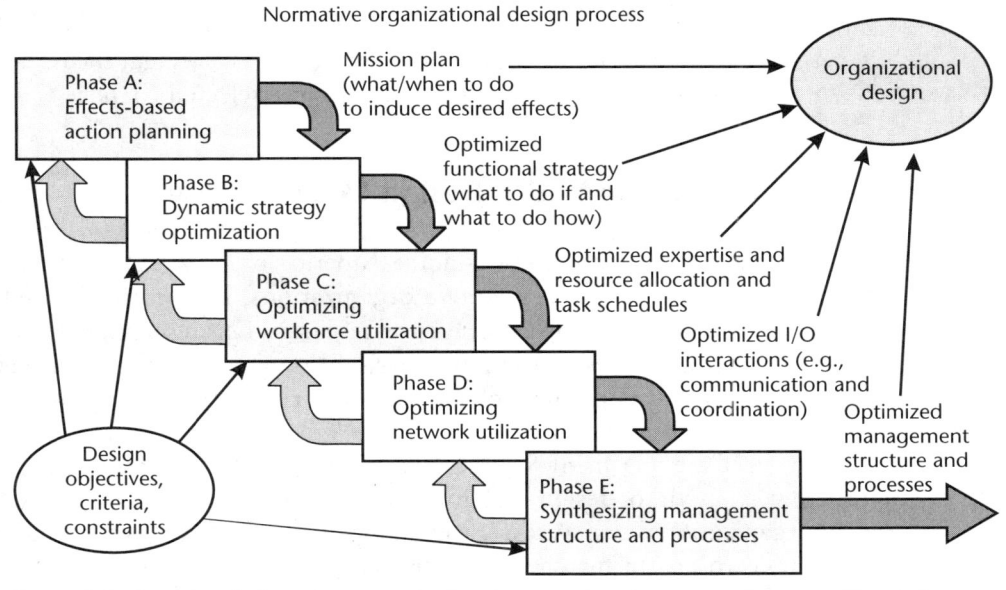

Figure 9.8 Iterative design of an organizational structure and processes for a specific mission.

Criterion for robust design:
root-mean square performance measure behavior over selected range of missions

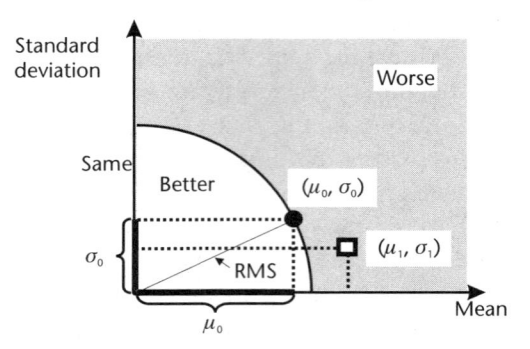

Robust design algorithm

Input:
Mission scenario, mission uncertainty, organizational parameters, robustness parameter
Output: Robust organization $O_e^{(K)}$
Step 1. Generate mission realizations $M_1, M_2, \ldots, M_K$
Step 2. Construct mission concatenation
$$M_e^{(K)} = M_1 \rightarrow M_2 \rightarrow \ldots \rightarrow M_K$$
Step 3. Find organization $O_e^{(K)}$ to execute mission $M_e^{(K)}$ using iterative organizational design process

Figure 9.9 Highlights for robust organizational design algorithms (see [47] for algorithm details).

Adaptive organizations are able to generate new strategies or reconfigure their structures to potentially achieve even higher performance. Organizational adaptation process can be significantly simplified when specific causes of adaptation, or adaptation triggers (e.g., changes in the mission environment, resource failures, and so on), are anticipated a priori. After a suitable adaptation option (e.g., strategy shift, resource reallocation, hierarchy reconfiguration) is selected, the organization needs to coordinate among its members to realize the selected change.

Empirical Validation of Normative Design Methodology

The normative methodology for designing superior organizations, and its extensions for designing robust and adaptive organizations, has been validated by the A2C2 human-team-in-the-loop experimental research. This research has shown that the formal algorithm-based, automation-assisted organizational design methodology is able to synthesize organizations that are superior to the ones optimized by the subject matter experts (see Figure 9.5).

For the A2C2 experiment 8, described in detail in [45], a reverse engineering approach was used to design mission scenarios that specifically (mis)matched selected organizational structures. Varied allocations of resources to decision-makers were used to create matches and mismatches between task-resource requirements and decision-maker–resource capabilities by manipulating the need for

Empirical Validation of Normative Design Methodology

Input-output notations for adaptive design algorithm

Input: Mission scenario N consisting of morphing stages $M_1, M_2, \ldots, M_K$; set of organizations $S_0 = \{O_j, j=1,\ldots,L\}$
Reconfiguration costs $C^R(O_j, O_j)$;
Performance costs $C^P(O_j, M_j)$;
Reconfiguration and performance cost weights W^R and W^P.
Output: Adaptation strategy $\pi = \pi(i)$, $i = 1,\ldots,K$ and its cost Θ_π where $\pi(i) \in \{1,\ldots,L\}$ and $O_{\pi(i)}$ is the structure in which to execute morphing stage M_i.

Choosing optimal adaptation strategy (path)-illustration

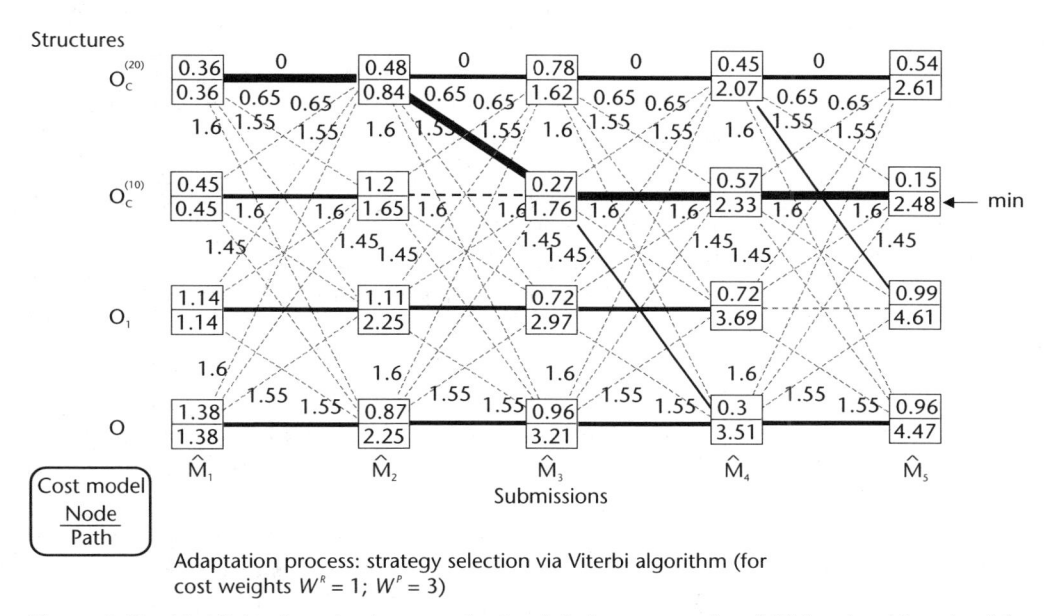

Adaptation process: strategy selection via Viterbi algorithm (for cost weights $W^R = 1$; $W^P = 3$)

Figure 9.10 Highlights for adaptive organizational design process (see [47] for algorithm details).

multidecision-maker task processing (reducing the need for multidecision-maker processing in congruent cases, and increasing it in the incongruent ones). As multidecision-maker task processing required communication and asset synchronization among the decision-makers participating in task execution, it resulted in increased task execution latency. Based on the scheduling algorithms [22, 23] that are part of the normative organizational design process, we further increased the interdecision-maker dependence in incongruent cases by specifying a precedence structure among tasks that must be executed by different decision-makers.

To effectively test our congruence concepts empirically, a distance metric between the two contrasted organizational structures (measuring the degree of asset control dispersion based on the resource capabilities of the assets controlled by organizational members) was maximized to counter the inevitable experimental variance when dealing with human teams. The experimenters preselected functional (F) and divisional (D) organizational structures (Figure 9.11). These architectures represent two extreme cases of organizational structures and therefore are suited for

Functional vs. divisional organizations

Decision-maker nodes of a divisional (D) organization

Decision-maker nodes of a functional (F) organization

	Strike	BMD	ISR	AWC	ASuW/mines	SOF/SAR
CVN	2F18S	xxx	1UAV	2F18A	1FAB 1MH53	1HH60
DDGA	8TLAM	3ABM 4TTOM	1UAV	6SM2	1FAB, 2HARP	1HH60 1SOF
DDGB	8TLAM	3ABM 4TTOM	1UAV	6SM2	1FAB, 2HARP	1HH60 1SOF
CG	8TLAM	3ABM	1UAV	6SM2	1FAB, 2HARP 1MH53	1HH60
FFG	2F18S	xxx	1UAV	2F18A 4SM2	1FAB, 2HARP 1MH53	1HH60
DDGC	8TLAM	3ABM 4TTOM	1UAV	6SM2	1FAB 2HARP	1HH60 1SOF

Mission scenarios
f – congruent to a functional organization scenario
d – congruent to a functional organization scenario

Scenario	AAW	Mines	ASuW	BMD	Strike	SAR	SOF
f	39	10	34	20	75	16	13
d	49	9	34	20	69	16	13

AAW = antiair warfare tasks
Mines = mine-clearing operations
ASuW = antisurface warfare
BMD = ballistic missile defense
Strike = strike warfare
SAR = search and rescue
SOF = special/ground operations
ISR = Intel/surveillance/recon

Figure 9.11 Functional (F) and divisional (D) organizational C2 architectures, and f and d mission scenarios for A2C2 experiment 8.

congruence analysis. Two scenarios, termed functional (f) and divisional (d), were designed to create the matched situations for Ff (functional structure and functional scenario) and Dd (divisional organization and divisional scenario) cases, and to create mismatches for Fd and Df cases. The human-in-the-loop experiment was conducted using eight teams and is described in detail in [53].

Findings from the A2C2 experiments 8 and 9 [54] empirically validated the variation of congruence hypothesis, namely, the hypothesis from the contingency theory that the proper choice of an organizational structure is contingent upon the specific task environment—that is, how the task activities are structured (see Figure 9.12 for findings from A2C2 experiment 8).

These experimental findings showed that the ability of organizations to exhibit superior performance depends on the actual task parameters (e.g., the type of tasks, the attributes of corresponding tasks such as processing times and resource requirements, uncertainty in the information on task parameters, and task tempo) and on the organizational constraints (e.g., the operational resources available, the capacity of communication channels, and the level of training and expertise of personnel). These experiments have demonstrated the following:

1. Match between the organization and mission can be understood in terms of requirements for interteam coordination that can reduce the speed of mission execution;
2. We can reverse engineer the mission—essentially to attack the weaknesses of one organization while nicely matching another type;

Reverse-Engineering Organizational Vulnerabilities 233

Figure 9.12 Findings from A2C2 experiment 8.

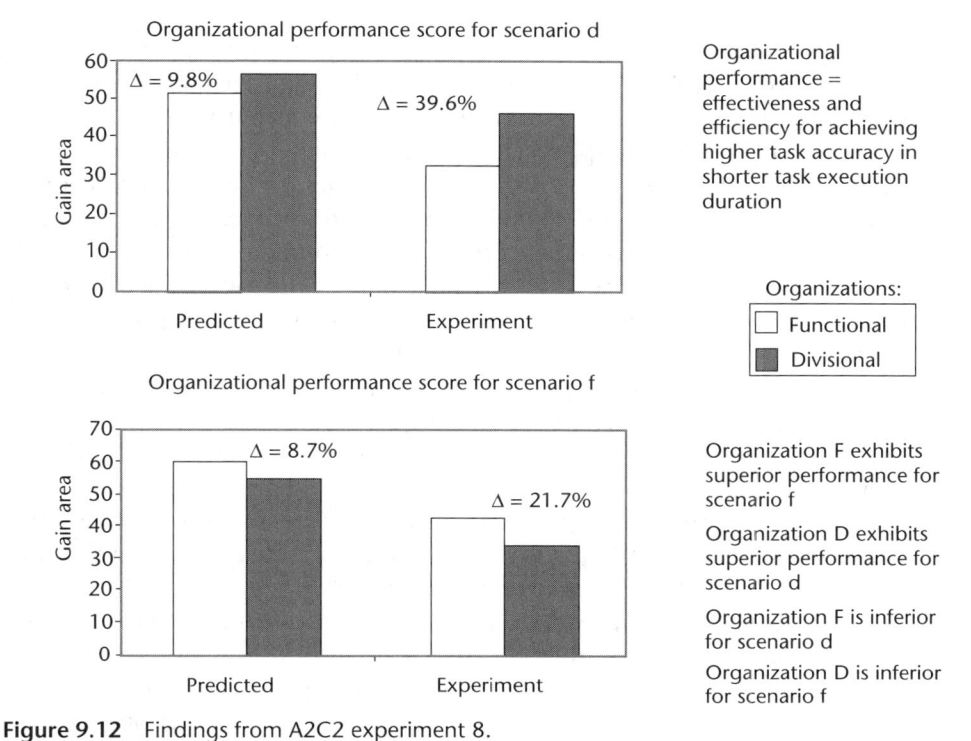

3. Resource allocation is an important determinant of organizational performance;
4. Excessive coordination results in reduced mission effectiveness.

These findings not only validated our normative organizational design methodology, but also suggested ways to reverse engineer conditions to attack the weaknesses of enemy organizations. The extensions to the normative design model [52, 54] enabled the model-driven design of conditions (e.g., imposed mission tasks) and interventions (e.g., targeted infrastructure elements) to maximize the destabilizing effects when planning to attack enemy organizations.

Reverse-Engineering Organizational Vulnerabilities

The corollary of the congruence hypothesis from contingency theory is that no organization is universally superior. As was described earlier, we can examine organizations for vulnerabilities and then design missions for which the organizations have poor fit. When organizations face these incongruent missions, they are likely to perform poorly. Once this happens, the organizational malfunctions (that are not necessarily products of human limitations, but rather the results of organizational design limitations for a given incongruent mission) will begin to accumulate and in some cases propagate into catastrophic failures.

Different organizations (e.g., see robust and adaptive designs described earlier) may show different tolerances to specific mission and organizational pressures. Certain task failures or malfunctions can force some organizations to undergo a self-reinforcing cycle of deterioration (e.g., when an anticipated destabilizing feedback loop develops and leads to instabilities in which errors get magnified rather than attenuated). We are specifically interested in identifying those failures and malfunctions that are likely to cause severe destabilizing effects on an enemy organization, from which the enemy would not be able to recover quickly or easily.

In addition to reverse engineering organizations to assess where each specific organization is most vulnerable, we are also interested in (1) identifying the bottlenecks of each design and assessing the relative contributions of the design components on organizational performance, (2) identifying those errors that a specific organization would tolerate with ease and those malfunctions that will cause the organization to fail, (3) assessing how organizational processes and structure affect shared situational awareness, (4) predicting with confidence how undermining different organizational processes and elements would affect the ability of the organization to successfully fulfill its missions, (5) assessing the degree of robustness of an organization to various failures, and (6) predicting situations for which it would be most difficult for the organization to adapt.

Congruent design methodology provides a baseline for designing and testing the degree of match between a mission and an organization. Also, as was shown earlier, this design methodology allows one to apply reverse engineering in order to design mission scenarios that match and mismatch a given organizational structure. This approach can be used to reduce the performance efficiency of a given organization. This methodology is based on forcing an organization to cope with various situations (missions, submissions, tasks, and task patterns) with which the organization is poorly matched and thus cannot perform efficiently, as well as with situations that require the organization to perform processes that might be observed (e.g., communications). The latter may be needed when certain information about the enemy organization is not known and needs to be inferred via intelligent probing.

The *process assessment methodology* in [35] allows one to identify the design bottlenecks whose overload (accidental or intentional) can cause organizations to fail. This approach can be used to reverse engineer the vulnerabilities of a given organization. Potential failure precursors range from (1) the span of control overload, to (2) mismatch between assigned responsibilities and operational capabilities, to (3) increased information turn-around cycle, to (4) overdependence of the organization on specific elements (e.g., information sensors, critical and unique resources, and knowledge hubs), to (5) the likely points of communication network overload and stovepipes.

By knowing the design of a specific organization, the mutual influence relationships among organizational parameters (Figure 9.13) can be assessed quantitatively, and parameters can be rank-ordered in terms of criticality to the organization's vulnerability. While the organizational parameters adhere to the general principles of distributed processor systems, the extent of their mutual relationships depends on the specific organizational structure and processes. This knowledge can be used to determine the best courses of applying external pressure to destabilize the organization.

As an illustration, added schedule pressures (e.g., more tasks with deadlines) can constrain or compress the duration of activities, thus increasing the cumulative load

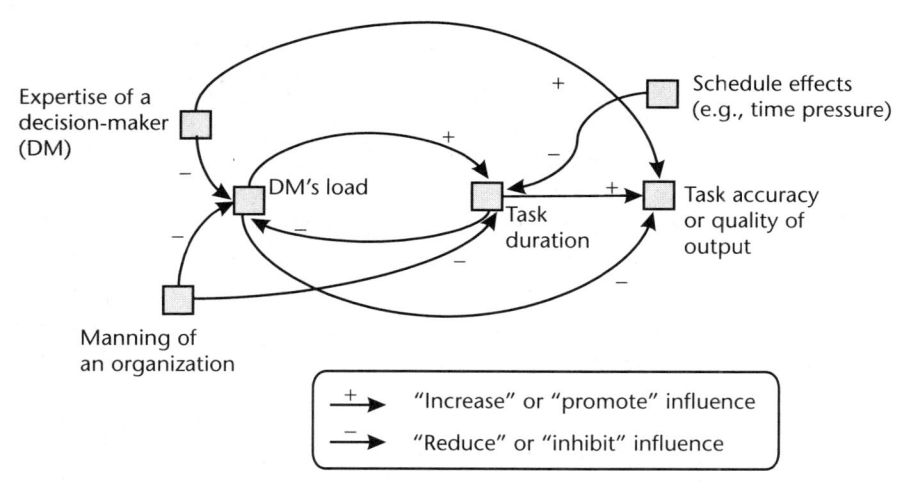

Figure 9.13 Mutual influence among (a subset of) organizational parameters.

on the organization (per unit time). On the other hand, the shortage of manning or expertise may lead to increased load for each decision-making agent, possibly extending the duration of activities due to capacity constraints. Better expertise and higher manning lead to reduced load, because the work is distributed into smaller pieces and each piece is processed more efficiently. However, the dependency of load on manning is nonlinear, because the communication overhead that accompanies higher manning would eventually outweigh the benefits of partitioning the work into smaller and smaller pieces for distributed processing. Also, manning of organizational units affects the quality of produced information outputs indirectly—that is, through other mechanisms such as through overload, expertise (mis)match, or work deficit.

The examples of organizational vulnerabilities that can be induced by external influence and that may eventually result in systemic failures include the following:

- *Overload:* When not enough time is available to complete the activities, it results in a work deficit, which can be partially (and in some cases fully) compensated for by working above the normal workload level (but at or below the workload capacity threshold). Occasional or short-term workload above the normal workload level has no or minimal detrimental effect on the quality of outputs produced. However, systematic or long-term workload above the normal workload level results in increased levels of fatigue or stress, which is likely to have a substantial detrimental effect on the quality of information processing and decisions made.

- *Work deficit (i.e., inability to complete mission task requirements due to time constraints):* When not all the work deficit can be compensated for by working above the normal workload level, the remaining work deficit (work that was not performed) has a (substantial) detrimental effect on the quality of information processing and decisions made.

- *Lower information input quality:* When some of the information inputs lack in quality (e.g., are noisy, inaccurate, incomplete, incorrect, or lack clarity), it results in reduced quality of information outputs produced and decisions

made (compared to the quality of outputs produced and decisions made when all the inputs are at the required level of quality).
- *Expertise mismatch:* When the personnel of an organization lack the required expertise, it is likely to result in (substantially) reduced quality of information outputs produced and decisions made (compared to the quality of outputs produced and decisions made when personnel of the organizational unit have the required level of expertise).
- *Cumulative effect:* When several factors (such as overload, work deficit, lower information input quality, or expertise mismatch) combine, this is likely to result in a higher degradation of quality of information processing and decisions made than the sum of individual effects.

The reverse engineering methodology has applications in counteracting enemy organizations. To counteract an enemy organization, knowledge of the principles under which this organization operates is required. To successfully employ certain counteractions, additional knowledge about the enemy organization and its processes may be needed—ranging from the specifics of organizational command, control, communication, and information structures to the responsibility delegation and goals at the most salient enemy decision-making nodes [54].

Robust and Adaptive Designs of Attack-Resistant Organizations

The success of organizations can be measured along two dimensions: (1) how they can act to achieve their objectives, and (2) how they can react to deny the enemy his objectives. Any organization can potentially be attacked by its enemies in many ways. Enemies may seek substantially different and often asymmetric ways suited to their resource levels and objectives. Hence, it is important not only to be cognizant of organization's vulnerabilities, but to be prepared to compensate for these vulnerabilities and to defend organizations from possible attacks. This relates not only to overt attacks, but also to covert attacks aimed at undermining an organization's infrastructure and at inducing the longer term (invisible) adverse effects on the organization's ability to operate successfully.

Our objective is to defend an organization's infrastructure from different types of enemy attacks that may cause the organizational failures. Some of the ways to deter or to cope with attacks were discussed earlier. Here we explore how to engineer the design that would facilitate defensive actions and would limit the propagation of adverse effects. In addition to processes for monitoring and making sense of the environment to help detect attacks, the following design features have potential for increasing attack resistance in an organization:

- Robustness—derived from built-in alternative means to support critical processes and to pursue mission goals;
- Resilience—derived from means for blocking or diverting the propagation of adverse effects or for reversing such effects;
- Responsiveness—derived from means for timely preemption of attacks and for rapid reversing of the adverse effects of attacks;

- Adaptation—derived from means for timely detection of the need to adapt and for efficient implementation of adaptation options (both structure and process related);
- Flexibility—derived from a variety of feasible organizational forms or modes of operation to which an organization may switch in order to maintain superior performance under new conditions.

To devise attack-resistant organizations, we can utilize the congruent design methodologies for engineering robust and adaptive organizations [52]. We extend the concept of congruence from structural and functional congruence between the organization and its mission to congruence with the added goals of resisting certain attacks (Figure 9.14).

We first look at designing organizations that are robust to various kinds of attacks (i.e., at designing organizations that can maintain a high level of performance when facing potential attacks). The procedure for designing organizations that are robust to attacks is as follows (Figure 9.15):

1. *Augment mission and design objectives to include attack resistance.* We begin by augmenting the mission objectives with those for resisting specific attacks and for adhering to the corresponding robust design principles:

    (a) *Expect enemy attacks.* We first identify the expected (i.e., likely) attacks. To predict the likely attacks, we can use our knowledge of the enemy practices from the past, combined with examining our own organizational vulnerabilities.

    (b) *Design principles to minimize vulnerability to attacks.* We then determine the design principles (e.g., high connectivity of communication

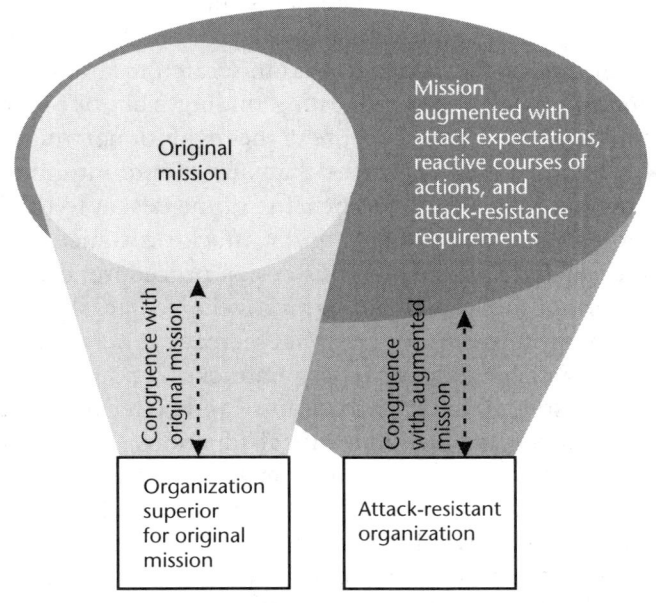

Figure 9.14 Notional visualization of congruence to combined performance and attack-resistance mission requirements.

networks, diverse supplier base, operations security, and so on) to explicitly account for (and minimize) vulnerability to the attacks identified and to increase the reliability of design components directly affected.

(c) *Determine courses of actions to cope with attacks.* We determine the required courses of action (sequences of tasks) contingent on each specific attack. These may include the efforts to replace or repair equipment and other resources, to replace or rotate personnel, specific means to counterattack the enemy if opportunity exists to prevent future attacks, and so on.

2. *Adjust mission expectations to account for attacks.* Next, we augment our mission expectations by adding the specific (future) events that represent the hypothesized enemy attacks on our infrastructure and processes (e.g., attacks targeted at our most critical vulnerabilities).

3. *Adjust mission requirements to include the need to resist attacks.* We augment our mission requirements by adding response actions devised to react to enemy attacks. These include actions such as restoring disrupted communications, repairing malfunctions, adapting decision processes to meet time-critical objectives, and so on.

4. *Engineer for organizational congruence with augmented mission.* Finally, we use the design methodology [36, 52] (outlined earlier and summarized in Figure 9.8) to engineer the organization optimized for this modified mission (which explicitly accounts for the likely attacks) to guarantee that our organization would be able to tolerate and withstand the attacks. The emphasis on congruence with the need to resist attacks (while maintaining acceptable performance) leads to the design that minimizes organizational vulnerabilities and failures (Figure 9.15).

This design procedure can be applied both in an evolutionary mode (i.e., augmenting the existing organizational design to enhance its attack resistance) and in a revolutionary mode (i.e., designing from scratch for attack resistance and superior mission performance). One interesting finding from our empirical research is that the organizational design procedure in the revolutionary mode may result in organizational structures and processes that are counterintuitive to humans and may require training to exert the full benefits of the design [44].

We note that, when optimizing for attack resistance, the step of designing for redundancies (step b in the previous design procedure) can make operations under normal conditions less efficient but more reliable in case of attack. For example, an organization optimized for operational secrecy may choose to sacrifice both performance and resistance to some types of attacks (e.g., on certain types of supplies or lower level personnel) in order to disguise and better protect the information about higher level core personnel and critical functions. This would be reflected in the choice of required courses of action (step c in the previous design procedure) driven by each specific attack (e.g., in case of attacks on noncritical supplies or lower level personnel, no actions would be taken).

While this approach can help make our organizations resistant to specific attacks, it does not guarantee that an intelligent enemy would not find unanticipated ways to attack our organizations. As we said before, no organizations are univer-

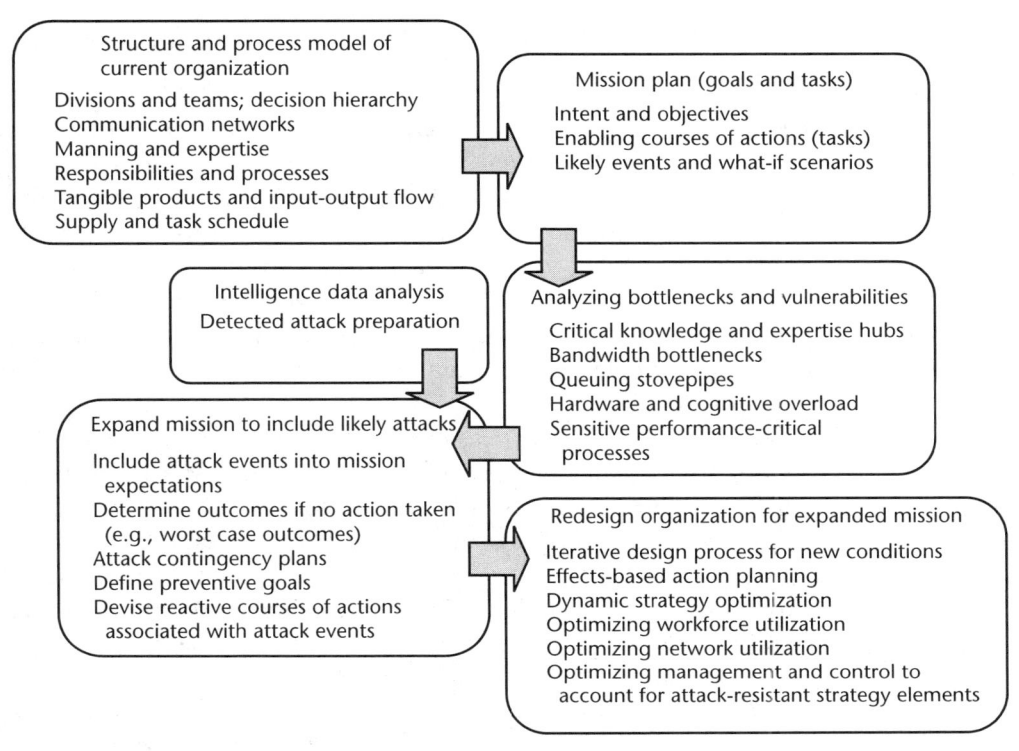

Figure 9.15 Designing organizations robust to potential attacks.

sally superior. Or, to restate this differently, every design has its bottlenecks (i.e., for every design, the situation may arise when its bottlenecks will be revealed).

To this end, we can apply our adaptive design methodology [52] to engineer organizations that will be able to adapt to new situations with relative ease. The design for adaptiveness defines alternative structures and modes of operations between which an organization needs to switch when situations warrant, in order to preserve the congruence with the mission and (as a result) to maintain adequate performance. In addition to specifying the alternative structures and modes of operation, an adaptive design needs to also: (1) explicate means and allocate responsibilities for detecting the need to adapt, and (2) specify the mechanisms for implementing (e.g., enforcing) the adaptation.

The decision as to which state to adapt to can be based on extrapolated knowledge of best practices combined with commander's intuition. As a more rigorous solution, commanders can use decision support tools, which would combine the executable process assessment methodology [35] and the team optimal design methodology [36] (briefly described earlier in this chapter) to detect the adaptation triggers. While monitoring and making sense of the environment assists in detecting attacks, monitoring the internal processes supports in detecting adaptation triggers that enable the organization to adapt in time to prevent performance breakdown. The adaptation triggers could thus include not only actual performance degradation, but also conditions that imply that such degradation is likely to occur in the near future (e.g., one trigger could be the detected process bottlenecks that slow down the response; another trigger could be the workload increase indicating imminent overload). The procedure for designing adaptive organizations to cope with attacks is as follows (Figure 9.16):

1. *Define feasible adaptation options.* Based on the expected attacks and their effects, the likely changes in the environment, or changes in the organization's objectives, we identify feasible adjustments to organizational structure or its processes to improve (or restore) the fit between the organization and its prospective mission scenarios:

    (a) *Expected enemy attacks and changes in the environment.* We identify the likely attacks (and their likely effects) from our knowledge of the enemy practices combined with examining the organization's vulnerabilities and environmental dynamics.

    (b) *Conditions that require adaptation.* We identify those likely changes (e.g., resulting from the likely attacks) for which the current organizational design would be unable to maintain adequate performance (i.e., performance above a tolerance threshold), thus implying the need to adapt.

    (c) *Design for adaptation options.* For each of the conditions that require adaptation, we then use the design methodology [36, 52] outlined earlier to engineer the optimized organizational structures and modes of operation

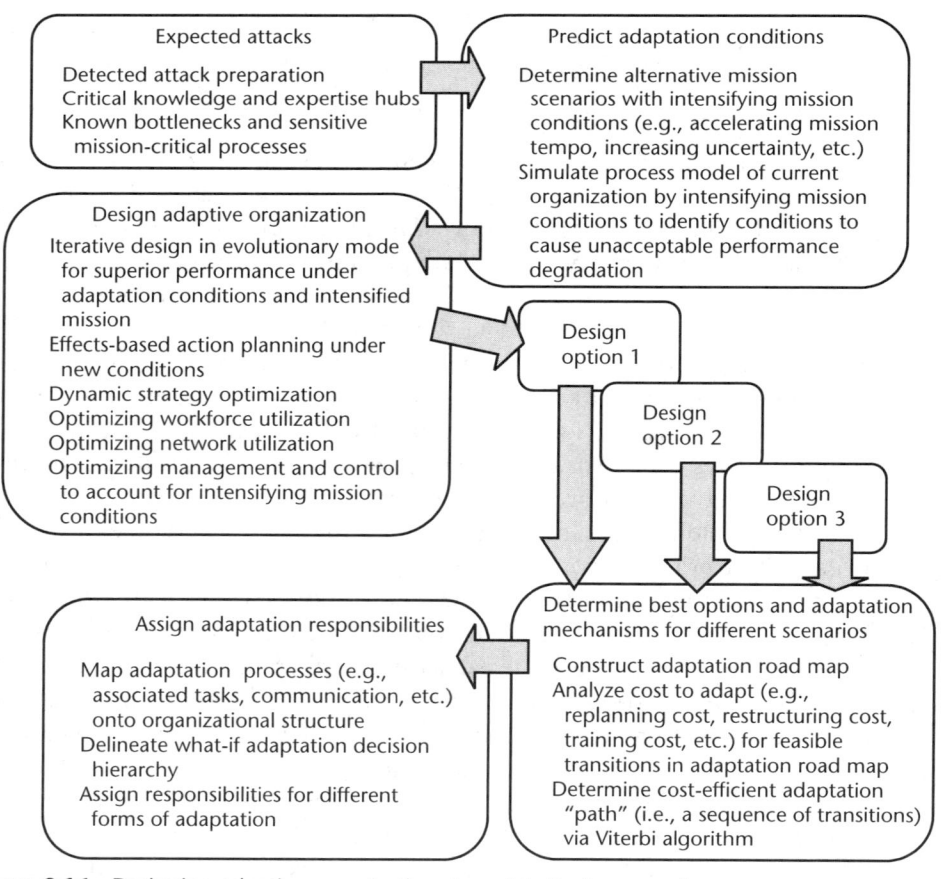

Figure 9.16 Designing adaptive organizations to resist attacks—overview.

that are congruent with the new mission scenarios stemming from the likely attacks. We use the evolutionary design mode (i.e., using the existing design as a starting point while assessing the cost of adaptation versus the corresponding benefits) to guarantee not only that the new organization would be able to tolerate changes and meet new challenges, but also that the adaptation options are feasible for the current organization.

2. *Define adaptation mechanisms.* We analyze the transition paths among adaptation options and the associated costs (Figure 9.16) to design cost-efficient transition mechanisms to implement adaptation and boost organizational performance.

3. *Define adaptation triggers and allocate monitoring responsibilities.* We use the process assessment methodology [35] to link observable organizational process measures (e.g., workload, response time, communication bandwidth, supplies and information queues, and so on) with the likely future performance measures, in order to identify triggers that would signal the need to adapt before a decrement in performance ensues. We also prescribe the structure and processes responsible for monitoring these triggers and informing the commander on the need to adapt.

4. *Designate responsibility and authority to initiate and manage adaptation.* Finally, we prescribe the appropriate authority to initiate and control the adaptation (e.g., the leader of a team, subgroup, or subdivision that needs to adapt; Figure 9.16).

When using automated tools to assist adaptation in real time, it is oftentimes not feasible to analyze all available adaptation options, as such an analysis can be time consuming and computationally expensive. When this is the case, the number of examined options will be restricted by the timeliness requirements and the processing speed of the supporting algorithms.

The concept of adaptive design is emphasized in [27], which underscores the importance of planning for adaptation and calls for agile organizations capable of coping with dynamic high tempo environments and increased complexity of missions. As one of the adaptive design principles, it describes the so-called empowerment of the *edge*. This principle refers to a centralized organization viewed as a sphere, with the command center notionally placed at the center of the sphere and with operational units notionally viewed as located at the edge of the sphere. Reference [27] argues that the agility of an organization (and its ability to adapt rapidly) could be facilitated by empowering the low-level managers (or local leadership) with more decision authority in order to allow parts of organizations to adapt as needed (as opposed to centralized adaptation of the whole organization). This would require accurate information and situation awareness within the operating units (teams), as well as the ability to self-synchronize both locally and horizontally, to ensure that, as teams achieve their local objectives, this builds toward achieving the overall mission objectives.

Agile organizations rapidly adapt (e.g., adapt the relationships of actions in time and space) when (1) objectives change, and (2) important developments occur in the environment. When the situation requires an organization to undergo structural adaptation in order to maintain superior performance, such adaptation is initiated and controlled by the appropriate leader (of a team, group, division, and so on). Only when the required structural adaptation stretches across the entire organization does it need to be initiated and controlled in a centralized manner by the strategic leadership of the organization.

Illustrative Example—Redesigning an Organization to Enhance Attack Resistance

We illustrate the process of engineering an attack-resistant organization via a hypothetical example next.

Example Scenario

We assume that a friendly military organization (FMO) is facing the mission of identifying and eliminating pop-up targets in a battle space (Figure 9.17). For simplicity, we assume that the FMO consists of a command center that has to dynamically manage a field force (FF). The FF contains the field units that move across the battle space and prosecute the assigned targets with various weapons (resources). The units are organized into a hierarchical divisional structure subordinate to the FF's command on the ground (FFCG).

The command center maintains communication with the FFCG, which in turn communicates with the unit leaders (Figure 9.18). The command center receives information on targets from the ISR assets (e.g., from satellites and unmanned aerial vehicles), associates priorities to targets, decides on the synchronization requirements, allocates targets to units, and communicates to the FFCG and units, the target locations, as well as the target priorities and synchronization requirements (when applicable). The units then search and prosecute targets according to the schedule defined by the unit leaders or FFCG. The FFCG coordinates the scheduling and prosecution of targets that require synchronization among units. The FFCG has the authority to reallocate targets to different units (from those originally assigned by the command center), if such a reallocation would significantly improve the overall target-processing schedule. However, in practice, because the command center utilizes decision support tools to assist its target-to-unit assignments, such reallocations by FFCG are infrequent.

We further assume that each of the pop-up targets in the battle theater could be assigned to one (or both) of the following two categories (Figure 9.18): (1) the so-called high-precision targets (HPTs), and (2) the time-critical targets (TCTs). The HPTs require planning and careful selection of weapons; hence, it typically takes longer for the command center to assign these targets to units. For the TCTs, the window of opportunity is relatively small and requires shortening of the decision cycle by the command center. This implies a potential decreased accuracy of the target-to-unit assignment decisions (e.g., potential reduced efficiency in selecting adequate resources to prosecute TCT). The command center addressed the need to maintain two different decision cycles for different target categories (time critical and not time critical) by dedicating two independent parts of the command center—the time-critical cell (TCC) and the high-precision cell (HPC)—to service these

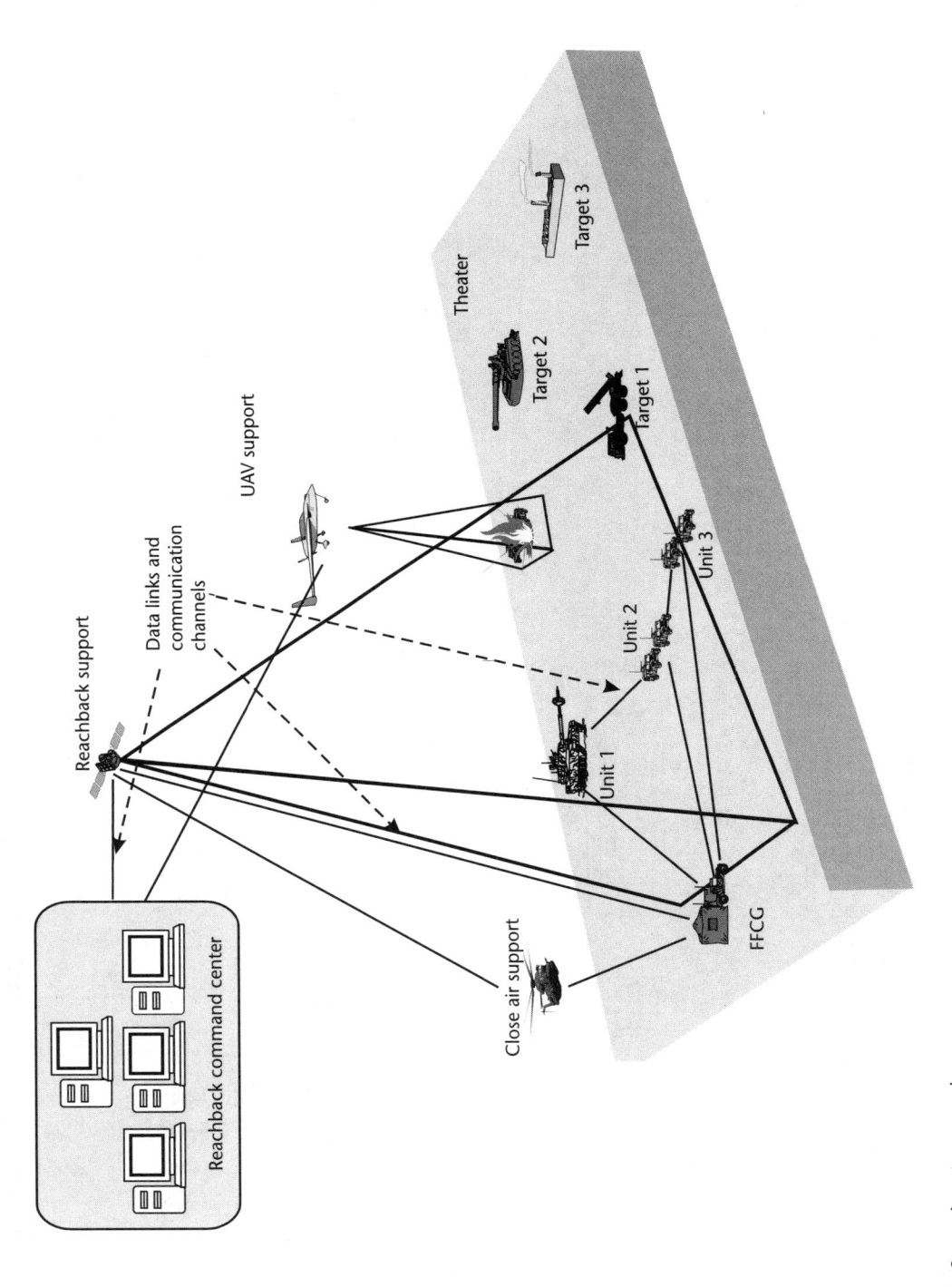

Illustrative Example—Redesigning an Organization to Enhance Attack Resistance 243

Figure 9.17 Scenario setup overview.

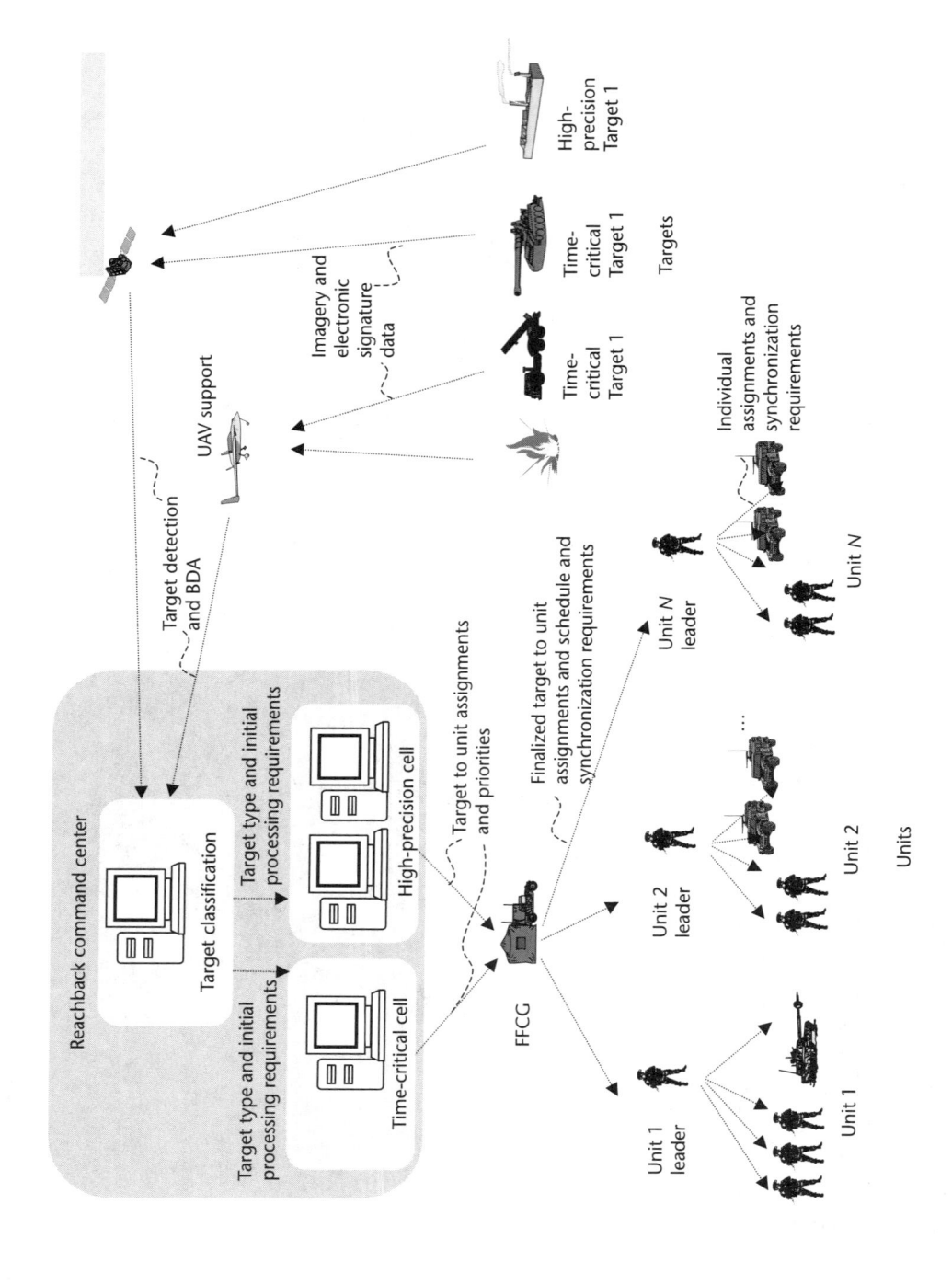

Figure 9.18 Information flow across a hypothetical FMO.

two target categories. The command center classifies targets (into either the TCT or the HPT category), before passing them for processing by the appropriate cell (Figure 9.18). Figure 9.19 compares the hypothetical target processing at TCC versus that at HPC.

Enemy Attacks

After several days of actions, the following patterns of enemy behavior have become apparent. First, in addition to direct attrition against the field units, the enemy attempted to jam the channels that linked the command center, the FFCG, and the units. In some cases, when the enemy had temporarily succeeded in doing this (until the communications were restored), it caused delays in target prosecution, which, in turn, resulted in hostile targets' inflicting damage on the friendly forces.

Second, the enemy introduced multiple decoys (false targets) into the battle space, which resulted in false alarms and overloaded the friendly organization, thereby causing delays in the command center's prioritization of real targets due to the command center's cognitive resources being diverted to dealing with the decoys. The command center's inability to process all the data caused occasional failure to address critical information on real targets. As a result, some of the targets were not prosecuted in a timely manner, and some of these targets also inflicted damage on the friendly forces.

Third, the enemy has periodically altered the arrival patterns of HPTs and TCTs. In particular, there were periods during in which there were a large number of TCTs in the battle space (causing overloads of the TCC, which resulted in offloading the processing of some of the TCT to the HPC). At other periods, most of

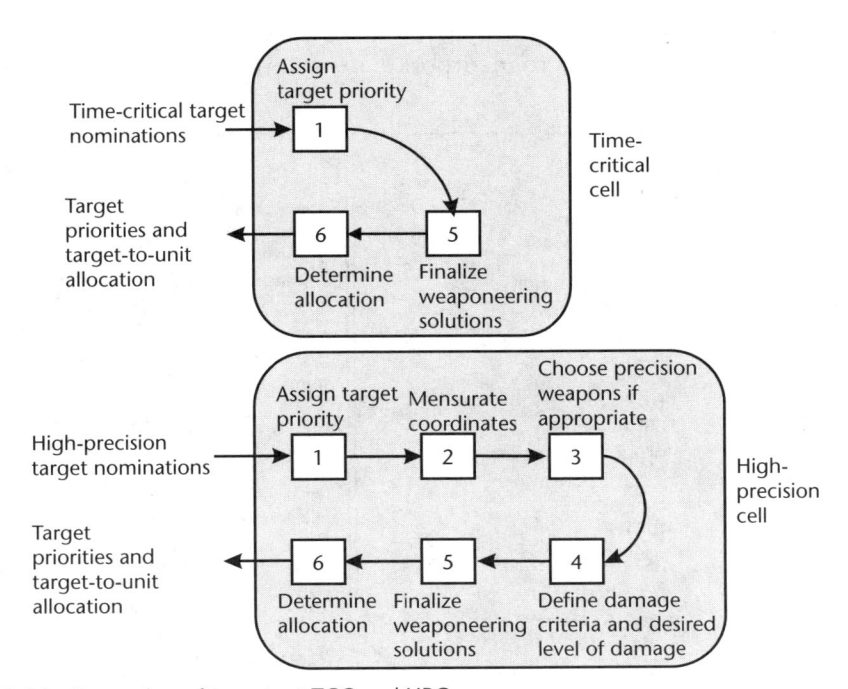

Figure 9.19 Processing of targets at TCC and HPC.

the targets were of the HPT type, causing overload of the HPC and the near-idleness of the TCC.

Finally, the enemy has mounted attempts (albeit unsuccessfully) to destroy the FFCG. Had the enemy been successful in doing this, it would have limited the friendly force's ability to conduct synchronized attacks on the enemy's targets.

In summary, the enemy's behavior revealed attempts to conduct five different types of attacks (Figure 9.20):

1. Direct attrition against individual field units (i.e., attempts to destroy the field units);
2. Jamming communication channels;
3. Varying TCTs and HPTs to reduce the FMO's ability to react in a timely fashion;
4. Misinformation (false alarms) to cause information overload;
5. Attempts to destroy the FFCG in order to destabilize the coordination of friendly units.

The effects of the attacks ranged from temporary loss of contact with the command center to the diversion of command center's time and resources, ultimately resulting in the command center's occasional inability to process critical information and in delays in target prosecution (Figure 9.20). Other potential threats included the destruction of the FFCG, which would jeopardize the ability of friendly units to synchronize their actions.

Redesign Principles

To enhance the FMO's resistance to these attacks, we will redesign the FMO's C2 structure and processes to incorporate the following design principles:

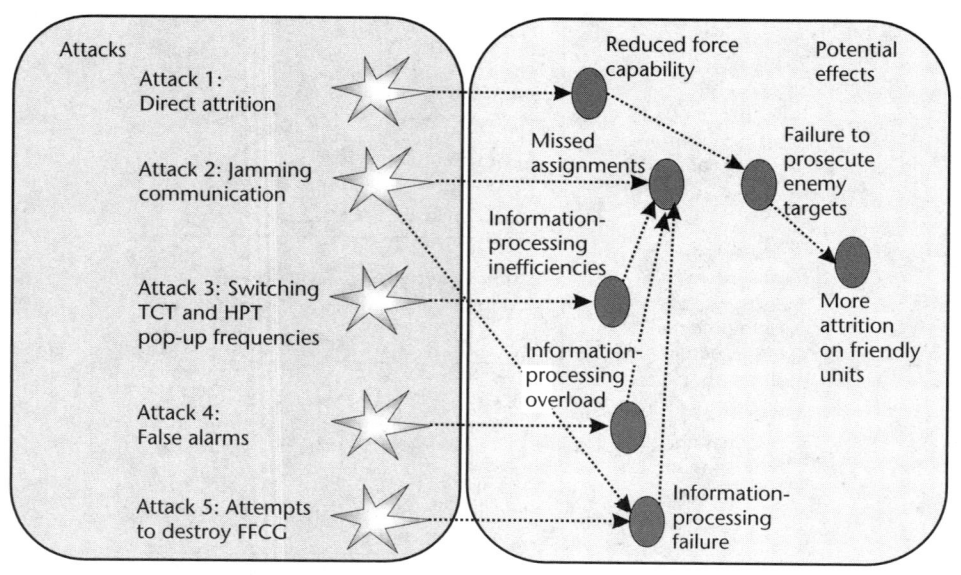

Figure 9.20 Types of enemy attacks and their effects.

1. Devise three modes of operation: (a) TCT-heavy mode; (b) TCT-HPT balanced mode; and (c) HPT-heavy mode. This would allow the command center to process targets faster for each of the target pop-up scenarios.
2. Include mechanisms that would allow the command center to dynamically switch, based on the mission demands, among these modes of operation. This would allow the command center to rapidly adapt to the enemy's attempts to mismatch the command center's mode of operation with its mission demands (by altering the balance between the TCT and HPT in the battle space).
3. Complement the selective control of the FMO units by the command center or FFCG with a process to allow units to operate with high autonomy to ensure that the units are not constrained by a single information hub designed around the command center and FFCG.
4. Allow for coordination by negation (in addition to coordination via communication) and prescribe default courses of action to deal with temporary loss of communication.
5. Enable direct communication between the command center and the unit leaders to increase robustness to communication jamming, compensate for potential disruption of communication between the FFCG and the unit leaders, and allow the command center to assume control over the mission in the event of a successful attack on the FFCG.

Attack-Specific Courses of Action

We augment the FMO's original mission requirements (for detecting and eliminating the pop-up enemy targets) by adding the need to conduct reactive control actions in response to the enemy attacks. These actions include:

1. Process adaptation:
   (a) Information flow rerouting (by engaging in series of predefined alternative command center–to–unit leader and unit leader–to–unit leader communications) in response to disrupted communications or FFCG failure to operate (attacks 2 and 5 in Figure 9.20).
   (b) Periodic situation-driven process adaptation among the TCT-heavy, TCT-HPT balanced, and HPT-heavy modes of operation in response to enemy-induced target load among high and low TCT and HPT pop-up frequencies (attack 3 in Figure 9.20).
   (c) Dynamic cognitive resource management (i.e., workload-driven reallocation of information tasks to command center operators) to more evenly balance the information processing load across the command center in response to information overload caused by increased false alarms (attack 4 in Figure 9.20).
2. Default courses of action (in response to potential communication blackout due to attacks 2 and 5 in Figure 9.20):
   (d) Switching to autonomous operations by blacked-out units to service predefined geographic areas (i.e., autonomously seek and destroy high-value enemy targets).

(e) Synchronized maneuvering to predefined rendezvous points along predefined emergency routes (know who is responsible for what geographic area of operations and where to retreat and from whom to seek coverage, if needed).

Engineering for Congruence with Mission in the Face of Attacks

For each of the three prospective mission scenarios—the TCT-heavy scenario, the TCT-HPT balanced scenario, and the HPT-heavy scenario—we can employ the congruency-driven organizational design methodology referenced earlier in this chapter to design the corresponding mode of operation for the command center (the outcomes of the design process are summarized in Figure 9.21).

The redesign of the FMO's structure and processes to cope with the attacks then proceeds (via an evolutionary design process) as follows:

Mission (re)definition:
  Offensive mission:
- Prosecute enemy targets based on priorities.

  Attack-driven tasks:
- Detect attempts to destroy the FFCG or the friendly units; prosecute the enemy resources involved in these attacks as high priority TCT.
- Monitor the workload of TCT and HPT to detect the need to adapt the command center's processes or to cognitively manage resources.
- Switch to alternative communication channels or use alternative message routing if possible in case of communication breakdown; otherwise, switch to default courses of action if unable to communicate with both the FFCG and the command center.

Structural modifications:
  Added communication links:
- Communication channels are added to allow direct communication between the command center and the unit commanders.
- The direct ISR link is provided to the FFCG and the unit commanders.

  Process assessment cell (PAC):
- The PAC, which is added to the command center structure (Figure 9.22), is designated to detect and dynamically assess changes in the mission mode (TCT-heavy mode, TCT-HPT balanced mode, or HPT-heavy mode).
- The PAC signals when the command center needs to change to an appropriate mode of operation: (1) the TCT-heavy mode, (2) TCT-HPT balanced mode, or (3) HPT-heavy mode. The modes of operation are illustrated in Figure 9.21.
- The PAC is responsible for dynamic cognitive resources management (as described later).

Process modifications:

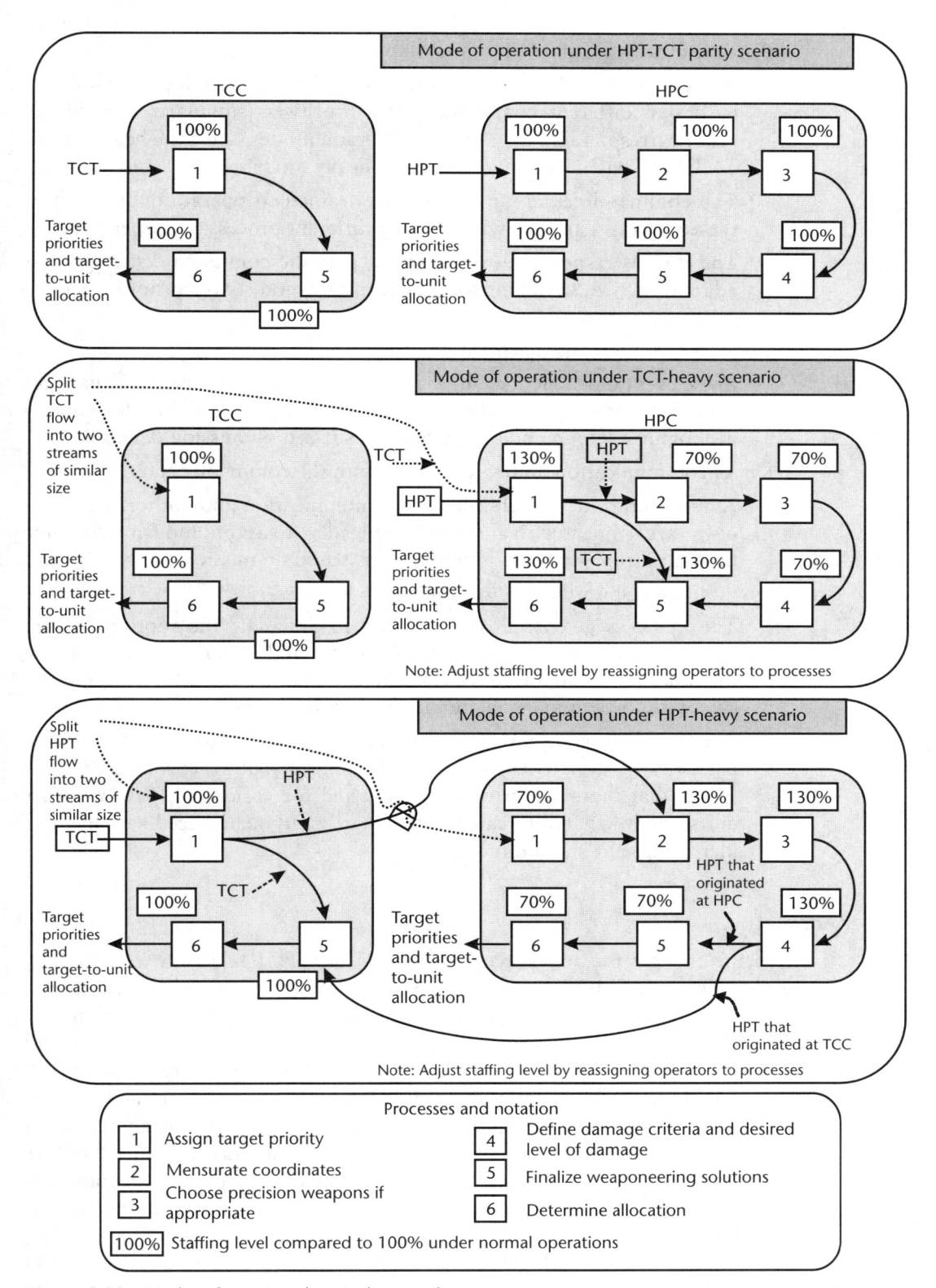

Illustrative Example—Redesigning an Organization to Enhance Attack Resistance 249

Figure 9.21 Modes of command center's operations.

Periodic process adaptation:
- The congruency-driven organizational design methodology is employed to design different command center processes (summarized in Figure 9.21) corresponding to each of the mission modes (the TCT-heavy mode, TCT-HPT balanced mode, and HPT-heavy mode).
- The command center's personnel are trained to operate under each of these processes and to switch among different processes of operations.
- The PAC is responsible for initiating periodic command center process adaptation (i.e., switching to an appropriate mode of operation) when the situation warrants.

Cognitive resource management:
- In the event of frequent changes among mission modes, the PAC manages cognitive resources dynamically by making recommendations for allocating cognitive command center resources based on mission demands.

Direct communication among units and with the command center:
- A multihub spider's web interunit communication and authority mechanisms are delineated to allow the unit leaders to autonomously determine the supporting-supported relationships based on mission needs.

Distributing responsibilities for attacks:
- The FFCG is given the responsibility to designate the supporting-supported relationships among units. The command center can also designate the supporting-supported relationships, if units are only able to communicate with the command center and not with the FFCG.
- A relatively small subset of all units (augmented with special resource packages) is designated to handle the majority of synchronized attacks (given that these are primarily conducted for various HPTs). All other units are designated to carry out primarily unsynchronized attacks, thus increasing unit's mobility and agility.

Adaptation

If the enemy changes the arrival patterns of HPT and TCT relatively infrequently, the newly redesigned FMO (a portion of which is shown in Figure 9.22) can adapt to one of the three modes of operation (shown in Figure 9.21): (1) TCT-heavy mode, (2) TCT-HPT balanced mode, and (3) HPT-heavy mode. This allows the command center to process targets faster for each of the different target pop-up scenarios. Alternatively, if the enemy begins to alter the patterns for popping-up the HPT and the TCT with relatively high frequency (so that the command center's switching among modes of operation becomes impractical), the command center then employs dynamic cognitive resource management (without switching between the modes of operation) and dynamically allocates each targeting cell (the TCC and the HPC) a concomitant share of the TCT and the HPT. The FMO can assess the attack frequencies by analyzing the intelligence data and the attack history in order to allocate the FMO's resources that are expected to deal with the attacks, as well as to determine processes for dealing with attacks that are likely to render superior results (e.g.,

Illustrative Example—Redesigning an Organization to Enhance Attack Resistance 251

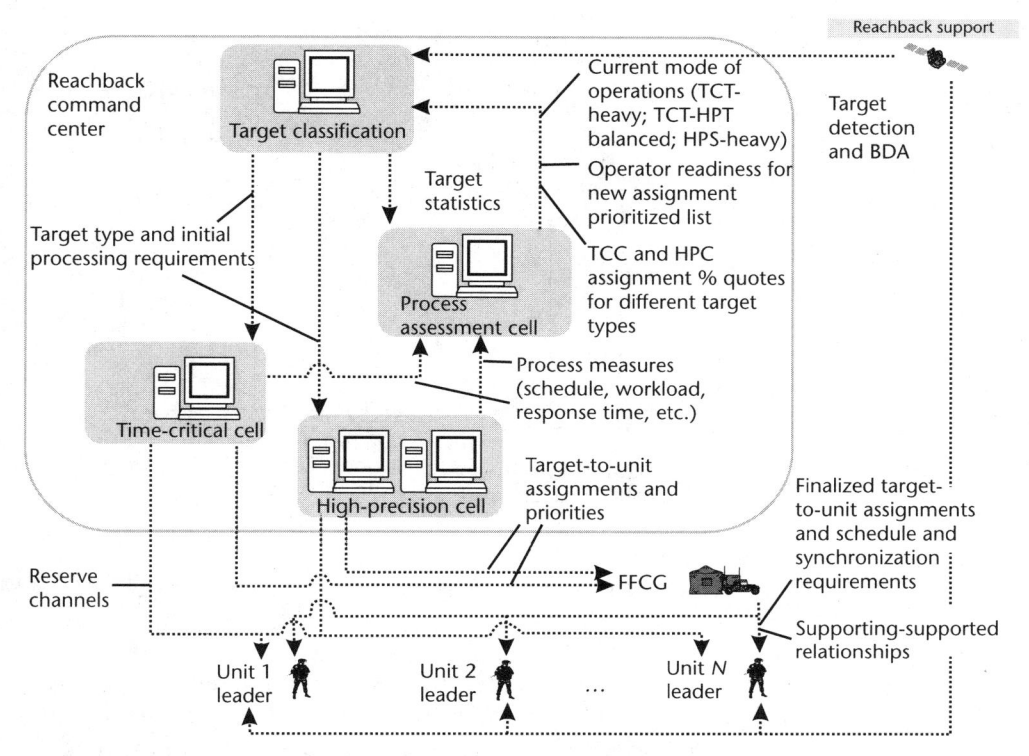

Figure 9.22 (Portion of) information flow across redesigned organization.

to determine whether to switch between modes of operation or to employ dynamic cognitive resource management at the command center).

In the event that a friendly unit finds itself cut off from the communication network, it employs the default courses of action (e.g., it switches to autonomous operations to seek and destroy high-value enemy targets in a predefined geographic area, or it conducts a synchronized maneuver to a predefined rendezvous point) until the communication is restored. If the enemy succeeds in destroying the FFCG, the command center then assumes control over the mission and directly manages the friendly units.

Analyze Organizational Design

The attack-resistance features (and mechanisms) of the FMO (and its command center) were designed to prevent performance degradation effects of attacks. For example, Figure 9.23 illustrates how the redesigned command center is able to maintain superior performance (as measured in terms of the number of targets serviced and the timeliness of servicing targets) despite the enemy's attempts to mismatch the command center's mode of operation with its mission demands by altering the balance between the TCT and the HPT in the theater (attack 3 in Figure 9.20).

The newly redesigned FMO combines features of robustness (alternative communication channels and default actions and coordination procedures; back-up

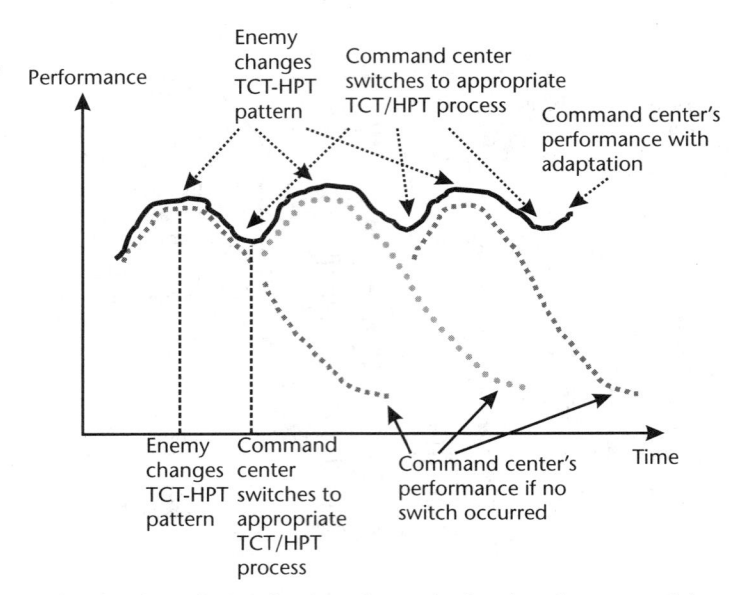

Figure 9.23 Notional (informal) visualization of organizational performance of the command center under new design.

command center's mission control option) and adaptation (switching among different modes of operations and workload-driven reallocation of information tasks to command center operators). The coordination by negation and the default courses of actions allow units to operate autonomously for extended periods of time. The redesigned communication and control structure (multihub spider's web interunit and unit-to-ISR communication links and interunit coordination mechanisms) ensures that the units are not constrained by a single information hub designed around the command center and the FFCG.

References

[1] Dalton, D. R., et al., "Organization Structure and Performance: A Critical Review," *Academy of Management Review*, Vol. 5, No. 1, January 1980, pp. 49–64.

[2] Naman, J. L., and D. P. Slevin, "Entrepreneurship and the Concept of Fit: A Model and Empirical Tests," *Strategic Management Journal*, February 1993.

[3] Ryan, A. M., and M. J. Schmit, "An Assessment of Organizational Climate and P-E Fit: A Tool for Organization Change," *International Journal of Organizational Analysis*, Vol. 4, No. 1, 1996, pp. 75–95.

[4] Tu, H., Y. Levchuk, and K. Pattipati, "Robust Action Strategies to Induce Desired Effects," *Proc. 2002 Command and Control Research and Technology Symposium*, Monterey, CA, June 2002.

[5] Roberts, F., *Measurement Theory, with Applications to Decision-Making, Utility and Social Sciences*, Reading, MA: Addison-Wesley, 1979.

[6] Pete, A., et al., "An Overview of Decision Networks and Organizations," *IEEE Trans. on Systems, Man and Cybernetics*, May 1998, pp. 172–192.

[7] Boutilier, C., R. I., Brafman, and C. Geib, "Prioritized Goal Decomposition of Markov Decision Processes: Toward a Synthesis of Classical and Decision Theoretic Planning," *Proc. 15th International Joint Conference on Artificial Intelligence*, Nagoya, Japan, August 1977, pp. 1156–1162.

[8] Carley, K. M., and M. J. Prietula, *Computational Organization Theory*, Hillsdale, NJ: Lawerence Erlbaum Associates, 1994.

[9] Falzon, L., L. Zhang, and M. Davies, "Hierarchical Probabilistic Models for Operational-Level Course of Action Development," *Proc. 6th International Command and Control Research and Technology Symposium*, Annapolis, MD, June 19–21, 2001.

[10] Donaldson, L., *Performance-Driven Organizational Change: The Organizational Portfolio*, London, U.K.: Sage Publications, 1999.

[11] Holly, A., H. Handley, and Alexander H. Levis, "A Model to Evaluate the Effect of Organizational Adaptation," *Computational & Mathematical Organization Theory*, Vol. 7, Kluwer Academic Publishers, 2001, pp. 5–44.

[12] Carley, K. M., and D. M. Svoboda, "Modeling Organizational Adaptation As a Simulated Annealing Process," *Sociological Methods and Research*, Vol. 25, No. 1, 1996, pp. 138–168.

[13] Burton, R. M., J. Lauridsen, and B. Obel, "Return on Assets Loss from Situational and Contingency Misfits," *Management Science*, November 2002.

[14] Hashemian, M., "Design for Adaptability," Ph.D. thesis, Department of Mechanical Engineering, University of Saskatchewan, Saskatoon, 2005.

[15] Morris, R., "Computers Under Attack: Intruders, Worms, and Viruses," in Denning, P. J., (ed.), *ACM Press*, Reading, MA: Addison-Wesley, 1990.

[16] Ellison, R. J., et al., *A Case Study in Survivable Network System Analysis*, Technical Report, CMU/SEI-98-TR-014, ESC-TR-98-014, September 1998, http://www.cert.org/archive/pdf/98tr014.pdf.

[17] Houle, K. J., and G. M. Weaver, in collaboration with N. Long and R. Thomas, "Trends in Denial of Service Attack Technology," CERT Coordination Center, October 2001, http://www.cert.org/archive/pdf/DoS_trends.pdf.

[18] Arbaugh, William A., W. L. Fithen, and J. McHugh, "Windows of Vulnerability: A Case Study Analysis," *IEEE Distributed Systems Online*, December 2000, computer.org/dsonline.

[19] Kendall, K., "A Database of Computer Attacks for the Evaluation of Intrusion Detection Systems," BS/MS thesis, Massachusetts Institute of Technology (MIT), June 1999.

[20] Government Accounting Office, *Information Security: Computer Attacks at Department of Defense Pose Increasing Risks*, Washington, D.C., 1996.

[21] "2004 eCrime Watch Survey: Summary of Findings," 2004 eCrime Watch Survey conducted by *CSO* magazine in cooperation with the U.S. Secret Service & CERT Coordination Center, http://www.cert.org/archive/pdf/2004eCrimeWatchSummary.pdf.

[22] Levchuk, G. M., et al., "Normative Design of Organizations—Part I: Mission Planning," *IEEE Trans. on Systems, Man and Cybernetics—Part A: Systems and Humans*, Vol. 32, No. 3, May 2002, pp. 346–359.

[23] Levchuk, G. M., et al., "Normative Design of Organizations—Part II: Organizational Structure," *IEEE Trans. on Systems, Man and Cybernetics—Part A: Systems and Humans*, Vol. 32, No. 3, May 2002, pp. 360–375.

[24] Jen, E., (ed.), *Robust Design: a Repertoire of Biology, Ecology and Engineering Case Studies*, Oxford, U.K.: Oxford University Press, December 2003.

[25] Eisenstat, R., et al., "Beyond the Business Unit," *McKinsey Quarterly*, No. 1, 2001.

[26] Renner, S., "Building Information Systems for Network-Centric Warfare," *Proceedings of the 8th International Command and Control Research and Technology Symposium*, Monterey, CA, 2000.

[27] Alberts, David, John Garstka, and Frederick Stein, *Network Centric Warfare*, DOD Command and Control Research Program, October 2003.

[28] Hutchins, S. G., et al., "Expeditionary Strike Group: Command Structure Design Support," *Proc. 2005 Command and Control Research & Technology Symposium*, Tyson's Corner, VA, June 14–16, 2005.

[29] Levchuk, G. M., et al., "Design and Analysis of Robust and Adaptive Organizations," *Proc. 6th International Command and Control Research and Technology Symp.*, Annapolis, MD, June 19–21, 2001.

[30] Serfaty, D., et al., "On the Performance of FORCEnet Command and Control Structure in Support of Strategic Studies Group XXI: Modeling and Simulation Analysis," Prepared for Strategic Studies Group XXI, Naval War College, Newport, RI, June 17, 2002.

[31] Diedrich, F. J., et al., "When Do Organizations Need to Change (Part I)? Coping with Incongruence," *Proc. 2003 Command and Control Research and Technology Symp.*, Washington, D.C., 2003.

[32] Entin, E. E., et al., "When Do Organizations Need to Change (Part II)? Incongruence in Action," *Proc. 2003 Command and Control Research and Technology Symp.*, Washington, D.C., 2003.

[33] Bower, M., "Company Philosophy: 'The Way We Do Things Around Here,'" *McKinsey Quarterly*, No. 2, 2003.

[34] U.S. General Accounting Office, *Human Capital: Taking Steps to Meet Current and Emerging Human Capital Challenges*, Washington, D.C.: GAO-01-965T, July 17, 2001.

[35] Levchuk, Y., et al., "Air Operations Center Process Assessment Tool," Final Technical Report for Phase II SBIR, *Integrating Tools, Organizations, and Information to Optimize Effect-Based C2*, Aptima for AFRL/IFSA/Contract F30602-01-C-0159, 2004.

[36] Levchuk, Y., "A Systematic Approach to Optimizing Organizations: Models, Design Methodology, and Applications," Ph.D. thesis, University of Connecticut, June 2003 (request an electronic copy at levchuk@aptima.com).

[37] Mintzberg, H., *Structure in Fives: Designing Effective Organizations*, Englewood Cliffs, NJ: Prentice-Hall, 1993.

[38] Mintzberg, H., *The Strategy Process: Concepts, Contexts, Cases*, Harlow, U.K.: Pearson Education, 2003.

[39] Robbins, S., *Organization Theory: Structure, Design and Applications*, Englewood Cliffs, NJ: Prentice-Hall, 1990.

[40] Lin, Z., and K. Carley, "Maydays and Murphies: A Study of the Effect of Organizational Design, Task and Stress on Organizational Performance," *Sociological Abstracts*, 1992.

[41] Burton, R. M., and B. Obel, *Strategic Organizational Diagnosis and Design: Developing Theory for Application*, 2nd ed., Boston, MA: Kluwer Academic Publishers, 1998.

[42] Levis, A. H., and W. S. Vaughan, "Model Driven Experimentation," *Systems Engineering*, Vol. 2, No. 2, August 23, 1999, pp. 62–68.

[43] Kleinman, D. L., P. Young, and G. S. Higgins, "The DDD-III: A Tool For Empirical Research in Adaptive Organizations," *Proc. 1996 Command and Control Research and Technology Symposium*, Monterey, CA, June 1996.

[44] Entin, E. E., "Optimized Command and Control Architectures for Improved Process and Performance," *Proc. 1999 Command & Control Research & Technology Symposium*, NWC, Newport, RI, June 1999.

[45] Levchuk, G. M., et al., "Congruence of Human Organizations and Missions: Theory Versus Data," *Proc. 2003 International Command and Control Research and Technology Symp.*, Washington, D.C., June 2003.

[46] Hocevar, S. P., et al., "Assessments of Simulated Performance of Alternative Architectures for Command and Control: The Role of Coordination," *Proc. 1999 Command & Control Research & Technology Symposium*, NWC, Newport, RI, June 1999.

[47] Levchuk, G. M., et al., "Networks of Decision-Making and Communicating Agents: A New Methodology for Design and Evaluation of Organizational Strategies and Heterarchical Structures," *Proc. 2004 International Command and Control Research and Technology Symposium*, San Diego, CA, June 2004.

[48] Levchuk, G. M., et al., "Model-Based Organization Manning, Strategy, and Structure Design via Team Optimal Design (TOD) Methodology," *Proc. 10th International Command and Control Research and Technology Symposium*, McLean, VA, June 2005.

[49] Entin, E. E., et al., "Inducing Adaptation in Organizations: Concept and Experiment Design," *Proc. 2004 Command and Control Research and Technology Symposium*, San Diego, CA, 2004.

[50] Levchuk, Y. N., K. R. Pattipati, and M. L. Curry, "Normative Design of Organizations to Solve a Complex Mission: Theory and Algorithms," *Proc. 1997 Command and Control Research and Technology Symp.*, Washington, D.C., June 1997.

[51] Meirina, C., et al., "Goal Management in Organizations: A Markov Decision Process (MDP) Approach," *Proc. 2002 Command and Control Research and Technology Symp.*, Monterey, CA, June 2002.

[52] Georgiy, M., et al., "Normative Design of Project-Based Organizations—Part III: Modeling Congruent, Robust and Adaptive Organizations," *IEEE Trans. on Systems, Man and Cybernetics*, Part A: Systems and Humans, Vol. 34, No. 3, May 2004, pp. 337–350.

[53] Kleinman, D. L., et al., "Scenario Design for the Empirical Testing of Organizational Congruence," *Proc. 2003 International Command and Control Research and Technology Symposium*, Washington, D.C., June 2003.

[54] Levchuk, G., D. Serfaty, and K. Pattipati, "Normative Design of Project-Based Adaptive Organizations," in E. Salsa, (ed.), *Advances in Human Performance and Cognitive Engineering Research*, New York: Elsevier, 2005.

About the Authors

Kathleen M. Carley is a professor of computer science in the Institute of Software Research and the director of the Center for Computational Analysis of Social and Organizational Systems (CASOS) at Carnegie Mellon University. She received a Ph.D. from Harvard University. Her research combines cognitive science, social networks, and computer science to address complex social and organizational problems. Her specific research areas are computational social and organization theory; group, organizational, and social adaptation and evolution; dynamic network analysis; computational text analysis; and the impact of telecommunication technologies and policy on communication, information diffusion, disease contagion, and response within and among groups particularly in disaster or crisis situations. Her models meld multiagent technology with network dynamics and empirical data. Three of the large-scale multiagent network models that she and the CASOS group have developed in the counterterrorism area are BioWar, a city-scale model of the impact of, and response to, weaponized biological and chemical attacks; DyNet, a model of the change in covert networks, naturally and in response to attacks, under varying levels of uncertainty; and VISTA, a model for informing officials (e.g., military and police) of possible hostile and nonhostile events (e.g., riots and suicide bombings) in urban settings as changes occur within and among red, blue, and green forces. One of her tools, ORA, produces intelligence reports identifying vulnerabilities in groups and organizations.

Katya Drozdova focuses her work on the role of technology in U.S. national security and policy issues with particular emphasis on asymmetric threats in the post-9/11 environment. This includes problems of counterterrorism, intelligence analysis, nuclear nonproliferation, critical infrastructure protection, cybersecurity, and privacy. She has also published articles on questions ranging from balancing national security with individual liberty (*Hoover Digest*) to pursuing international cooperation against cybercrime and terrorism (*Hoover Press*). She was also a member of the NSA-sponsored Consortium for Research on Information Security and Policy and was awarded the Science Fellowship as well as MacArthur Affiliate status at the Center for International Security and Cooperation (CISAC) at Stanford University. Dr. Drozdova earned a Ph.D. from New York University, Stern School of Business Information, the Operations and Management Sciences Department—on the subject of organizational fault tolerance and technology risks with national security implications—and a B.S. and an M.S. in international relations and international policy studies from Stanford Universtiy.

D. Andrew "Disco" Gerdes holds a B.A. in philosophy from the University of Colorado, Colorado Springs, and an M.S. in logic and computation from Carnegie-Mellon University. Clark Glymour advised his master's thesis, entitled

"An Algorithm for Command Structure Identification from Communication Logs." The WDL algorithm described in his contribution to this volume is based on this master's thesis work. Currently, he is a post–master's research associate in the computational science and engineering division at the Oak Ridge National Laboratory. His work at ORNL is focused on creating analyst tools to detect rare events. His academic interests include social network analysis, data mining, knowledge discovery, network searches, and information retrieval.

Clark Glymour is an alumni university professor at Carnegie Mellon University and a senior research scientist at the Florida Institute for Human and Machine Cognition. He is a former Phi Beta Kappa lecturer, a former Guggenheim Fellow, and a Fellow of the Statistics Section of the American Association for the Advancement of Science. His principal statistical work is on algorithms for inferring causal relations. His joint books include *Causation, Prediction, and Search* (MIT, 2000), *Computation, Causation, and Discovery* (MIT, 1999) and *The Mind's Arrows: Bayes Nets and Graphical Causal Models in Psychology* (MIT, 2003).

Paul Hubbard is a defense scientist at Defence R&D Canada (DRDC) in Ottawa, where he leads a group studying the use of synthetic environments for emerging technologies, such as autonomous vehicles and new advanced sensors. Dr. Hubbard has worked previously on control systems applications in various industries, including working at General Electric, Inc., Carnegie Mellon University, and KVH Industries. Immediately prior to joining DRDC, Dr. Hubbard worked on a DARPA program on the advanced command and control of teams of autonomous systems. Dr. Hubbard holds a Ph.D. in electrical engineering with specialization in discrete-event systems from McGill University in Montreal but has focused in his career on modeling and simulation for defense applications as well as the development of algorithms for autonomous systems. His current research interests include discrete-event and hybrid control systems, information theory, and human-system interaction.

Kari Kelton is an organizational engineer and the lead of the organizational networks team at Aptima. Dr. Kelton has served as the principal investigator in R&D projects to support intelligence analysis for counterterrorism, with a particular focus on information fusion and agent-based simulation of adversarial behavior. Key sponsors of this stream of research include the Army Research Laboratory, the Office of Naval Research, and the Defense Advanced Research Projects Agency. Additional areas of research include simulation of organizational safety dynamics for NASA, modeling of cultural factors in multinational coalitions, and graph-theoretic methods for automatically integrating information and decision models. Dr. Kelton holds a Ph.D. in decision sciences and engineering systems and an M.S. in operations research and statistics from Rensselaer Polytechnic Institute, as well as a B.A. in mathematics from Western Kentucky University.

David L. Kleinman is a professor at the Naval Postgraduate School in Monterey, California. He received a B.B.E. from Copper Union, New York, in 1962 and an M.S. and a Ph.D., both in electrical engineering from MIT in 1963 and 1967, respectively. From 1967 to 1971 he worked at Bolt Beranek and Newman, Inc., in Cambridge, Massachusetts, where he pioneered the application of modern control and estimation theory to develop and validate an analytical model for describing human control and information processing performance in manned vehicle systems. From

1971 to 1973, he established and directed the Cambridge office of Systems Control, Inc., where he led applied research projects in both manual and automatic control. From 1973 to 1991, Dr. Kleinman was a professor with the Electrical and Systems Engineering Department at the University of Connecticut; he was also the director of the CYBERLAB, a laboratory for empirical research in cybernetic systems. Dr. Kleinman was a technical committee chair (Human Decision Making) and the vice president of finance and long-range planning for the IEEE-SMC Society. He was the program chairman for the 1989 IEEE International Conference on Systems, Man, and Cybernetics, and for the 1990 JDL Symposium on Command and Control Research. In 1980, he was the program chairman for the 19th IEEE Conference on Decision and Control. In 1993, Dr. Kleinman was elected Fellow of the IEEE. He is also a cofounder and director of ALPHATECH, Inc., in Burlington, Massachusetts, and of QUALTECH Systems, Inc., in Storrs, Connecticut.

Alexander Kott earned a Ph.D. from the University of Pittsburgh, Pennsylvania, where his research focused on applications of artificial intelligence for innovative engineering design. Later he directed R&D organizations at technology companies, including Carnegie Group, Honeywell, and BBN. Dr. Kott's affiliation with DARPA included serving as the chief architect of DARPA's Joint Forces Air Component Commander (JFACC) program and managing the Advanced ISR Management program as well as the Mixed Initiative Control of Automa-teams program. He initiated the Real-Time Adversarial Intelligence and Decision-Making (RAID) program and also manages the DARPA program called Multicell and Dismounted Command and Control. Dr. Kott's research interests include dynamic planning in resource-, time-, and space-constrained problems, in dynamically changing, uncertain, and adversarial environments; and dynamic, unstable, and "pathological" phenomena in distributed decision-making systems. He has published over 60 technical papers and has served as the editor and coauthor of several books.

John Kunz is the executive director and chief scientist of the Center for Integrated Facility Engineering at Stanford University. Current and recent projects focus on virtual design and construction, including developing models of engineering products and processes, organizational modeling, automated construction planning, visualization, use of business objectives, and automated building code checking. His teaching includes classes in virtual design and construction, an engineering seminar series, professional short courses in information technology in design and construction, advising the Ph.D. thesis research of many Stanford doctoral students, and supervising the research of professionals visiting the center. Research interests cross the engineering life cycle from preproject planning through design-build, retrofit, and decommissioning. Prior to joining Stanford, he was the chief knowledge systems engineer at IntelliCorp, which developed commercial applications of artificial intelligence methods.

Georgiy M. Levchuk is a simulation and optimization engineer at Aptima, Inc. He received a B.S. and an M.S. in mathematics with highest honors from the National Taras Shevchenko University, Kiev, Ukraine, in 1995 and a Ph.D. in electrical engineering from the University of Connecticut, Storrs, Connecticut, in 2003. His research interests include global, multiobjective optimization and its applications in the areas of organizational design and adaptation, and network optimization. Prior to joining Aptima, he held a research assistant position at the Institute of

Mathematics (Kiev, Ukraine), a teaching assistantship at Northeastern University (Boston, Massachusetts), and a research assistantship at University of Connecticut (Storrs, Connecticut) working on projects sponsored by the Office of Naval Research. Dr. Levchuk received the Best Student Paper Awards at the 2002 and 2003 Command and Control Research and Technology Symposia and the Best Paper Award at the 2004 Command and Control Research and Technology Symposium.

Yuri N. Levchuk is a senior scientist and senior engineer at Aptima, Inc. Dr. Levchuk's interests include multiobjective optimization and its applications in the areas of human information processing and decision-making, organizational design, and adaptation. Dr. Levchuk developed optimized organizational structures for a variety of military command and control teams, including Joint Task Forces; Combat Information Center teams for the new generation of Navy surface combatants; JFACC Time Critical Targeting teams; and AWACS surveillance and Weapons Director crews. Dr. Levchuk received the Best Paper Award at the 2004 and 1997 International Command and Control Research and Technology Symposia. Some of his other awards include Best Student Paper Awards at the 2001 and 2002 Command and Control Symposia and an Outstanding Performance Award at the Mathematics Olympiad of Ukraine in 1982. Dr. Levchuk received a Ph.D. in electrical and computer engineering, an M.S. in mathematics from the University of Connecticut, and a B.S./M.S.M. in mathematics from the National Taras Shevchenko University of Kiev, Ukraine.

Michael Martin has a formal education that combines postdoctoral training in computational cognitive modeling at Carnegie Mellon University with a Ph.D. in experimental cognitive psychology from the University of Kansas and an M.S. in human factors psychology from the University of Arkansas at Little Rock. His research experience includes work performed in government, industrial, and academic environments, first as a research psychologist at the Naval Air Warfare Center Training Systems Division and later as a cognitive systems engineer at Logica Carnegie Group, BBN, and ManTech. Dr. Martin is currently a cognitive scientist in the Dynamic Decision Making Laboratory at Carnegie Mellon University, where he combines behavioral studies with computational cognitive modeling techniques to explore how people learn to control slow-responding dynamic systems. His current line of research concerns practical methods for providing feedforward to improve anticipatory control.

Candra Meirina is pursuing a Ph.D. at the University of Connecticut. She received a B.S. and an M.S. in electrical engineering from Purdue University, West Lafayette, Indiana, in 1995 and 2000, respectively. She currently holds a research assistantship position at the University of Connecticut. Prior to that, she held a researcher position at the Calibration, Instrumentation, and Metrology Department of the Indonesia Institute of Sciences, Jakarta, Indonesia, from 1995 to 1997. She was also a lecturer at the Academy of Instrumentations and Metrology, Jakarta, from 1995 to 1997. Her current research interests include the area of adaptive organizations for dynamic and uncertain environments, in particular in utilizing optimization-based multiagent systems for the organizational designs, analyses, simulations, and decision-support systems. Ms. Meirina was a recipient of Science, Technology, and Industrial Development Scholarships in 1991 and 1997.

Krishna R. Pattipati is a professor of electrical and computer engineering at the University of Connecticut, Storrs, Connecticut. He has published more than 300 articles, primarily in the application of systems theory and optimization techniques to large-scale systems. Professor Pattipati received the Centennial Key to the Future award in 1984 from the IEEE Systems, Man and Cybernetics (SMC) Society and was elected a Fellow of the IEEE in 1995. He received the Andrew P. Sage award for the Best SMC Transactions Paper for 1999, the Barry Carlton award for the Best AES Transactions Paper for 2000, the 2002 NASA Space Act Award, the 2003 AAUP Research Excellence Award, and the 2005 School of Engineering Teaching Excellence Award at the University of Connecticut. He also won the best technical paper awards at the 1985, 1990, 1994, 2002, 2004, and 2005 IEEE AUTOTEST Conferences, as well as at the 1997 and 2004 Command and Control Conferences. Professor Pattipati served as editor-in-chief of the *IEEE Transactions on SMC-Cybernetics* (Part B) from 1998 to 2001.

Michael Prietula is a professor of decision and information science in the Goizueta Business School at Emory University, holds an adjunct appointment in the Department of Psychology, and holds a courtesy appointment of research scholar/scientist at the Institute for Human and Machine Cognition in Pensacola, Florida. Professor Prietula holds a Ph.D. in information systems, with minors in computer science and psychology from the University of Minnesota. He worked as a research scientist at Honeywell's Systems Research and Development Center, has taught at Dartmouth College and Carnegie Mellon University, and was the department chair at the Johns Hopkins University, where also held an adjunct appointment in the Johns Hopkins University School of Medicine. Professor Prietula has published in such journals as *Organizational Science*, *Human Factors*, *Cognitive Science*, *Management Science*, *Information Systems Research*, *MIS Quarterly*, *Journal of Personality and Social Psychology*, *Journal of Experimental Social Psychology*, the *ORSA Journal on Computing*, and the *Harvard Business Review*. He has coedited two books: *Computational Organization Theory* (Erlbaum, 1994), coedited with Kathleen Carley, and *Simulating Organizations: Computational Models of Institutions and Groups* (MIT Press, 1998), coedited with Kathleen Carley and Les Gasser. His primary research areas are computational models of individual and group behaviors (such as social algorithms), cognitive models of negotiation, and the analysis and representation of expert knowledge. Professor Prietula has worked on such simulation systems as CM2 (market simulation based on A Behavioral Theory of the Firm), TrustMe, the ACT Emotion Engine and Affect (emotional agents), Tides (group emergence), Norm (norm emergence), and xChng (knowledge exchange).

Joseph Ramsey is special faculty and the director of research computing in the Department of Philosophy at Carnegie Mellon University. He received a Ph.D. in 2001 from the University of California, San Diego, for work on computer algorithms for mineral identification from reflectance spectra. Since then, he has worked on a variety of computational problems mostly (but not exclusively) related to the development and implementation of computerized search methods for graphical causal models. He is the lead developer of the Tetrad IV project, a project that implements a number of discrete and continuous procedures for causal modeling. Dr. Ramsey has also worked on the development of a number of novel causal search algorithms, including a scalable algorithm for Markov blanket search and a refinement of the PC algorithm that improves orientation accuracy. He has also done extensive work in the development of online courseware for causal reasoning and sentential and first order logic and in the development of algorithms for the detection of graphical command structures in social networks. His work has been published in a variety of venues, including *Data Mining and Knowledge Discovery* and *Uncertainty in Artificial Intelligence*.

Sui Ruan is a Ph.D. student in the Electrical and Computer Engineering Department at the University of Connecticut. She received a B.S and an M.S. in electrical engineering from the Information and Electronic Systems Department of Zhejiang University, Hangzhou, China, in 1996 and 1999, respectively. From 1999 to 2002, she was with Bell Labs, Lucent Technologies, Beijing, China. Her research interests include adaptive organizational design and optimization algorithms for fault diagnosis.

Satnam Singh is currently a Ph.D. student in electrical and computer engineering at the University of Connecticut. He received a B.S. in chemical engineering from the Indian Institute of Technology (IIT), Rookree, India, in 1996 and an M.S. in electrical engineering from the University of Wyoming in 2003. Prior to his graduate studies, he worked as a design engineer in Engineers India Limited, Delhi, India. His research interests include signal processing, pattern recognition, fault detection and diagnosis, wireless communications, and optimization theory.

Ed Waltz is the chief scientist in the Intelligence Innovation Division of BAE Systems Advanced Information Technologies, where he leads intelligence analysis and information operations planning research for the intelligence community and DoD. For the past decade, his research has focused on modeling human systems, including foreign leadership, organizations, and social populations. He holds a B.S.E.E. from the Case Institute of Technology and an M.S. in computer, information, and control engineering from the University of Michigan. He has over 35 years of experience in developing and deploying signal processing, data fusion, and intelligence analysis capabilities. He is the author of *Knowledge Management in the Intelligence Enterprise* (Artech House, 2003) and *Information and Warfare Principles and Operations* (Artech House, 1998), and the coauthor of *Counterdeception Principles and Applications for National Security* (Artech House, 2007) and *Multisensor Data Fusion* (Artech House, 1990). He was the cochair of the 2000 NATO summer study on data fusion and is an editor of the results in *Multisensor Data Fusion* (Kluwer, 2001). He was a recipient of the DoD Joseph Mignona Data Fusion Award in 2004 and became a Veridian Technology Fellow in 2002.

Peter Willett received a B.A.Sc. (engineering science) from the University of Toronto in 1982 and a Ph.D. from Princeton University in 1986. He has been a faculty member at the University of Connecticut ever since, and since 1998 he has been a professor. He has published 76 journal articles (11 more under review), 189 conference papers, and 7 book chapters. He was awarded IEEE Fellow status in 2003. His primary areas of research have been statistical signal processing, detection, machine learning, data fusion, and tracking. He has interests in and has published in the areas of change and abnormality detection, optical pattern recognition, communications, and industrial/security condition monitoring. He is presently the editor or associate editor for three active journals—editor-in-chief of *IEEE Transactions on*

Aerospace and Electronic Systems (for data fusion and target tracking), and associate editor of *IEEE Transactions on Systems, Man and Cybernetics*, parts A and B, and of the *IEEE AES Magazine*, editor of the *AES Magazine*'s periodic *Tutorial* issues, associate editor for ISIF's electronic *Journal of Advances in Information Fusion*, and is a member of the editorial board of IEEE's *Signal Processing Magazine*.

Feili Yu received a B.S. in optical engineering and an M.S. in information and electronic systems from Zhejiang University, Hangzhou, China, in 1994 and 1999, respectively, and an M.S. in electrical engineering from University of Connecticut, Storrs, Connecticut, in 2004. He is currently a research assistant in the System & Cyber Optimization Lab in the Department of Electrical & Computer Engineering at the University of Connecticut. His primary research interests include optimization using evolutionary algorithms, modeling and optimization of command control organizations, and fault diagnosis.

Index

A

Access
 influence versus, 23–24
 organization data, 23
Accuracy-workload tradeoff curve, 138
Actionable implications, 194
Actions, 95
Active compensation, 144–46
Activities
 discovery, 56
 enemy, simulating, 53–55
 HMMs of, 53
 planning, 89–90
Activity patterns
 defined, 39
 discovery, 48–53
 example, 41
 identifying, 44
Actors
 centrality, 70
 defined, 15
 index measures, 17
 as known perpetrators, 83
Adaptability, 217
 cost, 218
 promotion of, 218
Adaptation mechanisms, 241
Adaptation triggers, 241
Adaptive Architectures for Command and Control program, 223–24
 experiment, 224
 experiment results, 233
 human-team-in-the-loop research, 230
 organizational performance, 224
 organizational processes, 225
 research, 224
Adaptive organizations, 230
 adaptation mechanisms, 241
 design process, 231
 triggers, 241
Adaptive safety analysis and monitoring (ASAM), 57
Adversaries
 decision-making, impacting, 89–111
 identification research, 33–36
 relationships, 33
Adversary identification research, 33–36
 decision aids, 33
 DM, 33, 34
 SAN, 33, 34–35
 summary, 34
Advice
 decisions based on, 169
 in TAG model, 167
Agent networks, 54
Ali Baba datasets
 actors, 72
 with inclusion of topics, 85
 scenarios, 66–67
All-to-all network, 179
Analytic workflow, 19–23
 all-source analysis, 22
 collection support, 19–21
 effect analysis, 23
 entity-relationship extraction, 21
 exploratory simulation, 23
 group detection, 21
 hypothesis generation, 22
 illustrated, 20
 link discovery, 21
 network visualization, 22
 query and visualize, 21
 refinement, 21
 SNA, 22
 unsupervised data mining, 21
Anonymous information exchange networks, 163–64
 defined, 163
 individual trust mechanisms, 164
 miscreant markets, 165–67
 online chat rooms, 164–65
 participant properties, 163–64
 trust mechanisms, 164
The Art of War, 2
Attack-resistant organizations
 congruent design methodologies, 237
 design, 213–52
 design features, 236–37
 devising, 237

Attack-resistant organizations (continued)
   key components, 219
   robust and adaptive designs, 236–42
   solutions, 218–21
Attacks
   courses of action, 247–48
   direct effects from, 220
   distributing responsibilities for, 250
   effects, 246
   flow of information, 220
   mission expectations and, 238
   organization design and, 239–40
   resistance, optimizing for, 238
   success inhibition, 223
   types, 246
   vulnerability, minimizing, 237–38
Atypical Signal Analysis and Processing (ASAP) Tool, 33
Authority, misallocation of, 127–28
Available intelligence, 110

B

Baited gambit, 105
Behavior
   analysis, 109
   consistent with optimizing for social network resiliency, 194
   learning, 53–55
   model, 38–39
Belief-desire-and-intention (BDI) architectures, 55
Betweenness
   centrality, 70, 71
   high degree, 15
Blind trust, 178

C

C2 organizations
   behavior characterization, 38–42
   command structure, 37
   communication structure, 37–38
   design, 36
   examples, 37
   identifying, 55–57
   nodes, 36–37, 48
   principles, 36
   task structure, 38
Cascading collapse, 125–26, 143
   avoiding, 126
   example, 125
   See also Decision-making
Centrality, 15
   betweenness, 70, 71
   closeness, 70, 71
   degree, 69
   eigenvector, 70, 71
   measurement, 69–71
Channels, 95–96
Chattering, 132
Chunk, 172
Closeness centrality, 70, 71
Clustering coefficient, 18
Cognition, 92
Cognitive-emotive defeat, 101
Command
   center, operation modes, 249
   by negation, 154–55
Command, control, and reporting (CCR)
   structures, 64–66
   actor connection to, 65, 66
   graphical representation, 64–66
   groups, 64
Command and control organizations. *See* C2 organizations
Communications
   categories, 45
   human capability, 159
   links, 102
   peer-to-peer, 130
   for synchronization, 130
Compensating component, 144–45
Complex adaptive systems, 99
Computational Analysis of Social and Organizational Systems (CASOS), 207, 208
Computational organization theory, 55
Confirmation bias, xi
Congruent design, 229, 237
Coordination
   lack of, 128–31
   minimizing need for, 155
Credentialing, 178
CrimeNet Explorer, 35
Critical chain (CC), 208
Critical path method (CPM), 208
Cumulative effect, 236
Cyber warfare, 165

D

Data mining (DM), 33
   in adversary identification research, 34–35
   defined, 33–34
Deadlock, 131–32
Deception, 90, 104–5
   matrix, 104

Index 267

in uncertainty, 117
Decision-accuracy relationship, 149
Decision-making
    accuracy-workload tradeoff curve, 138
    adversary, 89–111
    cascading collapse, 125–26
    deadlock, 131–32
    dynamic (DDD), 57–58
    excess timidity/aggressiveness, 121–22
    lack of synchronization/coordination, 128–31
    low and high threshold, 120
    malfunctions, 115–33
    misallocation of authority, 127–28
    overload, 124–25
    performance measurement, 150–51
    prioritizing, 154
    rational elements, 94
    representation, 137
    saturation, 124–25
    self-reinforcing error, 122–24
    sequence, 91–92
    tardy decision, 117–20
    target, 93–94
    team, 147–50
    thrashing and livelock, 132–33
    Vroom-Yetton, 14
Decision Making: A Psychological Analysis of Conflict, Choice, and Commitment, 1
Decision-responsibility structure, 149
Decisions
    based on advice and gossip, 169
    dependency criterion, 150
    in dynamic environments, 138
    performance levels, 94
    pull mode, 149
    push mode, 149
    requirement forecasting, 151
    responsibilities, dynamic allocation of, 147
    self-reinforcing overload, 136–41
    tardy, 117–20
Decision-sharing structure, 149
Defeat
    cognitive-emotive, 101
    effects, inducing, 100–107
    information, 101
    military mechanisms, 101
    physical, 101
    propagation of, 135–56
Defection, 90
Degradation, avoiding, 154

Degree centrality, 69
Degree distribution, 18
Denial operations, 102
Diagnosis, 155
Difference equations, 145
Dimensionality, 11
Disinformation probe, 204
Disruption(s)
    baselines, 172–73
    coping methods, 216
    defined, 103
    experiment results, 174–75
    gossip and, 172–77
    guidelines, 103–4
    interpretation, 175–77
    low sensitivity to, 215
    organizations coping with, 214–18
    propagation, 141–44
    timing, 103
    virtual experiment, 173–74
    withstanding, 215
Distributed decision-making agent (DDA) architecture, 55
Diversion, 90
Division, 90
Domains
    breadth, 11
    types, 31
Dynamic decision-making (DDD), 57–58
Dynamic reorganization
    compensation by, 155
    to mitigate malfunctions, 146–47
DyNet, 109, 208

E

Effects, 95
    analysis, 23
    attack, 246
    categories, 98, 108
    deception, 104–5
    for defeat, 100–107
    denial or destruction, 101–2
    direction and reflexion, 105–7
    disruption, 103–4
    fourth-order, 98
    indirect, 98
    influence, 5
    targeting for, 107–10
Effects-based planning, 98
Effects-based targeting, 95–100
Eigenvector centrality, 70, 71
Emotion, 92

Enemy organizations
    counteracting, 30
    counteraction strategies, 193–94
    defeat implications, 211
    detection implications, 211
    dynamics, 193–94
    identifying, 31–33
    processes, 29
    strategy development, 195–97
    structural discovery, 30
    structure, 29
    transactions, 30
    See also Organizations
Enumeration, 10, 24
Error
    percentage, 141
    self-reinforcing, 122–24
Essence of Decision, 12
Executable organizational model, 226
Expertise mismatch, 236
Explicit modeling, 111
Exploitation, 4
    information, 5
    physical, 5
    social, 5
Exploratory simulation, 23

F

Feedback
    channels, modeling, 140
    as information control device, 177
Field force command on the ground (FFCG), 242
First in, first out (FIFO), 172
Fourth-order effects, 98
Friendly military organization (FMO), 242
    attack-specific courses of action, 247–48
    cognitive resource management, 250
    direct communication, 250
    mission redefinition, 248
    periodic process adaptation, 250
    process modifications, 248–50
    redesign, 246–47
    responsibilities distribution, 250
    structural modifications, 248

G

Geospatial intelligence (GEOINT), 7
Gossip
    anonymous information exchange networks, 163–64
    as cultural learning mechanism, 178
    decisions based on, 169
    defined, 162–63
    disruption and, 172–77
    as dysfunctional, 178
    good/bad views, 160–62
    impact, 159, 179–80
    importance, 159–81
    key properties, 162
    model, 171–72
    mutation into rumor, 180
    norms, 176, 177
    reputations and, 160
    socially focused, 178
    social need, 161
    TAG model, 166, 167–69
    in TrustMe, 168
    trust mechanisms, 169–72
    See also Suspicion
Graphical network structures, 16
Groups, 18–19

H

Hackers, 166
Heterogeneity, 18
Hidden Markov model (HMM), 41–42, 48
    activities, 53
    automated construction, 54
    continuous-state, 50
    defined, 41
    detection, 50, 51
    hierarchical (HHMM), 51–52
    to identify activity patterns, 44
    learning, 53
    in modeling dynamic behavior, 49
    multinomial assumption, 52
    observation-space, 50
    parameters, 42
    premise, 41
    representation, 49
    superimposed, 51
Hierarchical HMMs (HHMMs), 51–52
High-level decision-makers (HLDMs), 116, 121, 124, 125, 127
High-precision cell (HPC), 242, 245
    overload, 246
    processing of targets, 245
High-precision targets (HPTs), 242
Holistic analysis, 9
Honest models, 171
Human intelligence (HUMINT), 7
Human-in-the-loop (HIL), 57
Hypothesis generation, 22
Hypothesis-testing process, 56

Hypothesized organization, 47

I

Identification
    adversary, research, 33–36
    domains of, 31–33
    probe, 196
    process validation, 57–59
Impact
    gossip, 179–80
    means of attack for, 91
    planning, effects-based approach, 97
    planning operations for, 90–100
Individuals
    knowledge of connections, 33
    tracking, 42
    trust mechanisms, 164
Influence
    access versus, 23–24
    effects, 5
    information, 5
    operations planning, 5
    physical, 5
    planning approaches, 100
Information
    defeat, 101
    exploitation, 5
    influence, 5
    networks, 5
    spread, 180
    structure, 149
Informational operations (IO), 3
Interactions, pattern discovery, 48–53
Intermediate component, 142
Internet, 180, 181
Intersecting connectivity, 180
Intervention planning, 196–97
Isolated reputational networks, 164

J

Joint task force (JTF), 58

K

Knowledge, skills, abilities, and other characteristics (KSAOs), 221

L

Latency, 18
Leadership analysis, 12–14
    defined, 9
    detailed level, 13–14
    focus, 12
    high-level, 12–13
    intermediate level, 13
    profiles, 12
Linearization, 139
Livelock, 133
Lower-level decision-makers (LLDMs), 116, 121, 123, 124, 127

M

Malfunctions, decision-making, 115–33
    cascading collapse, 125–26
    deadlock, 131–32
    dynamic reorganization to mitigate, 146–47
    excess timidity/aggressiveness, 121–22
    inducing, 153–56
    lack of synchronization/coordination, 128–31
    low and high threshold, 120
    misallocation of authority, 127–28
    mitigating, 153–56
    overload, 124–25
    self-reinforcing error, 122–24
    tardy decision, 117–20
    thrashing and livelock, 132–33
    types of, 116–17
Markov transition model, 39–40
Matlab, 140
Matlab Simulink, 151–53
Membership, 178
Message accuracy, 143
Military decision-making process (MDMP), 32
Mission complexity
    defined, 192
    opposing pressures, 193
    options for dealing with, 192–93
    test-case scenario, 200–203
    total, 192, 200
Mission expectations, 238
Mission model
    defined, 39
    example parameters, 40
    generic example, 209
Mission requirements, 238
Model learning, 56
Model networks, 47
Monte Carlo, 209

N

NetSTAR validation process, 58
Networks
    analysis, 68
    anonymous information exchange, 163–65
    graphical structures, 16
    information, 5

Networks (continued)
    model, 47
    physical, 5
    scaled, 205–6
    scale-free, 192, 206
    social, 71–73, 179, 194
    stability, 145
    star, 179–80
    targeting, 95
    visualization, 22
Networks and Netwars, 90
Nodal analysis, 9
Node mapping, 55
    problem, 46
    solution, 47
Normative design, 230–33
Numerical computations, 140

O

Observations
    collected, 45
    definition of, 42–44
    relating to structural links/activities, 44
    sequences, 53
    structure, 149, 150
Online chat rooms, 164–65, 169
Open-source information (OSINT), 7
Operational planning
    focus, 91
    for organization impact, 90–100
Operational security (OPSEC), 7
Operations
    defined, 3
    denial, 102
    informational (IO), 3
    planning for organization impact, 90–100
Organization, this book, xii–xvi
Organizational design
    algorithms, robust, 230
    analysis, 251–52
    attack resistance and, 239–40
    congruent, 229, 237
    features, 218
    formalism, 221–22
    iterative, 229
    key components, 219
    mission-driven, 222
    normative, 228–30, 230–33
    redesign, 246–51
Organizational dysfunction, 25
Organizational intelligence, 6–12
    activities, 6
    available, 110
    collection/utilization challenges, 42
    defined, 6
    fidelity, levels, 10–11
    needs, 8
    purpose, 6
    sources, 7
Organizational learning, 178
Organizational performance
    computational approach, 225–28
    model, 226
    precursors, 222–25
Organizational vulnerabilities
    cumulative effect, 236
    expertise mismatch, 236
    lower information input quality, 235–36
    overload, 235
    work deficit, 235
Organization complexity, 192
Organization discovery, 38
Organization research, 2
Organizations
    access, 4
    adaptive, 230
    agile, 241
    attack-resistant, 213–52
    C2, 36–42
    contextual knowledge, 15
    coping with disruptions, 214–18
    defined, 1, 31
    domains of identification, 31–33
    dynamics, 25
    exploitation, 4
    field components, 138
    as fragile structures, ix
    function, 18
    groups within, 18–19
    headquarters components, 138
    heterogeneity, 18
    hypothesized, 47
    influence, 4
    latency, 18
    redundancy, 18
    situation dependencies, 25
    size, 31
    strengths/vulnerabilities, 19
    structural properties, 15, 19
    structure, 18
    vulnerability to attacks, x–xii
OrgCon, 207
Overload, 124–25, 235

Index

P

Perception, 92
Perception and Misperception in International Politics, 1
Performance
    decision-making, 150–51
    levels, 94
    organizational, computational approach, 225–28
    organizational, precursors, 222–25
Personnel, 102
Person-to-person (P2P) architectures, 163
Physical defeat, 101
Physical exploitation, 5
Physical influence, 5
Physical networks, 5
Plan development, 108
Portfolio theory, 214
Predicted decision requirements, 151, 152
Predictive control, 147
Probability
    calculation, 80
    threshold, 81
Probe identification, 196
Process assessment methodology, 234

R

Ramsey Spy Finder (RSF) algorithm, 86–87
    defined, 86
    parameter variance, 87
    procedure count-word-frequencies, 87
Redesign principles, 246–47
Redundancy, 18
Reflexive control, 105–7
    deceptive matrix methods, 106
    defined, 105
    loop elements, 106
Relation, 10, 24, 92
Relationships
    adversary, 33
    decision-accuracy, 149
    defined, 15
    SNA, 15
    trust, 180
Reputations
    gossip about, 160
    as information control device, 177
Reverse engineering, 230
    applications, 236
    organizational vulnerabilities, 233–36
Rumors
    gossip mutation into, 180
    key properties, 162
    three Cs, 162

S

Saturation, 124–25
Scaled networks, 205–6
Scale-free networks, 192, 206
Scheduling algorithms, 231
Self-reinforcing decision overload, 136–41
Self-reinforcing degradation, 135–56
Self-reinforcing error, 122–24
Sensors, 102
Services, 102
Signal models, 48
Signals intelligence (SIGINT), 7
Simulation(s), 10–11, 24
    defined, 109
    firefighting example, 151–53
    mission, 56
    in system behavior prediction, 54
    tests, 81–82
    tools, 110
SimVision, 195
    defined, 208
    inputs, altering, 210
    Monte Carlo, 209
    organization modeling with, 208–10
    strategy development with, 208–10
    task/project schedule delay prediction, 210
Social exploitation, 5
Social network analysis (SNA), 14–19, 67–71
    actors, 15
    in adversary identification research, 33, 34–35
    in analytic workflow, 22
    defined, 9
    network characterization, 17
    patterns, 67
    premise, 14
    purpose, 15
    relationships, 15
    static, 9
Social networks
    flat, 179
    resiliency, 194
    work exploiting time in, 71–73
The Social Psychology of Organizing, 1–2
Stability
    envelope of, 143, 146
    network, 145
    Star networks, 179–80
Static SNA, 9

Strategem, 105
Strategic analysis, 109
Strategies against enemy organizations
    developing, 195–97
    intervention planning, 196–97
    probe identification, 196
Streaming, 73
Subordinates, empowerment, 154
Suspicion, injecting, 159–81
Synchronization
    of actions, 129
    communication importance, 130
    failures, 130, 143
    lack of, 128–31
Synthetic agent-based model, 227

T

Tardy decision, 117–20
Targeting
    for effects, 107–10
    effects-based, 95–100
    entire organization, 94–95
    networks, 95
    nodes, 95
    perspectives, 92
    planning, 107
    system mechanisms, 102
Targets
    categories, 108
    communication links, 102
    decision-making, 93–94
    modeling, 110
    organization, 96
    personnel, 102
    processing at TCC/HPC, 245
    sensors, 102
    services, 102
    time-critical (TCTs), 242
    types, 102
Tasks
    characteristics, 148
    completion duration, 206
    complexity, 192
    delay risks, 210
    instance project model, 210
Team decision-making, modeling, 147–50
Team optimal design (TOD), 226
Test-case scenario (torpedo attack)
    attack prevention, 198
    defined, 197–98
    enemy mission complexity characteristics, 200–203

    enemy organization model, 199–200
    interventions impact estimation, 205–6
    outline, 200
    potential damage, 198
    probe impact estimation, 203–5
    simulation results, 204
    VDT-based computational analysis, 199
Thrashing, 132–33
    chattering, 132
    livelock, 133
Ties. *See* Relationships
Time-critical cell (TCC), 242, 245
Time-critical targets (TCTs), 242
Time delays, 117–20
    examples, 119
    sources, 117
    See also Decision-making
Time overlapping, 77–79
    illustrated, 78
    results, 79
    testing, 77
    triad calculation, 79
    window data separation, 78
Time windowing, 75–77
    communication, 76
    defined, 75–76
    drawback, 77
    start time, 76
    using, 77
Transactions, 30
Trust
    blind, 178
    changes, 176
    defined, 167
    as enduring relationship component, 180
    gossip model, 171–72
    honesty models, 171
    mechanisms, 164, 169–72
    norms, 175
    ratio, 173
    state of, 169
    trust models, 169–71
Trust, advice, and gossip (TAG), 166
    advice, 167
    experiment results, 174–75
    gossip, 168
    models without exogenous contact, 179
    overview, 167–68
    search, 167
    trust, 167
    viability, 169
TrustMe, 167, 168

Index

Trust models, 169–71
    conflict resolution options, 170–71
    defined, 169–70
    levels of tolerance, 170
Typical solutions, 220

V

Veiled groups, 159
Virtual design team (VDT), 191, 197
    analytical foundations, 195, 207
    baseline model graphical representation, 201
    computational approach, 196
    model parameter, 203
    organization modeling, 193
    parametric models, 196–97
    probe analysis capability, 196
    research programs, 195
    test-case scenario analysis, 199
Vroom-Yetton organizational decision-making, 14

Vulnerability analysis, 5, 108

W

Will, 92
Windowing down the lines (WDL)
    algorithm, 79–81
    data, 74
    data formatting, 74–75
    evaluation, 82–86
    output graphical structure, 84
    probability calculation, 80
    scenario, 74
    time ordering, 75
    time overlapping, 77–79
    time windowing, 75–77
    with topics, 84–86
    without topics, 82–84
Work deficit, 235

The Artech House Information Warfare Library

Electronic Intelligence: The Analysis of Radar Signals, Second Edition, Richard G. Wiley

Electronic Warfare for the Digitized Battlefield, Michael R. Frater and Michael Ryan

Electronic Warfare in the Information Age, D. Curtis Schleher

Electronic Warfare Target Location Methods, Richard A. Poisel

EW 101: A First Course in Electronic Warfare, David Adamy

Information Warfare and Organizational Decision-Making, Alexander Kott, editor

Information Warfare Principles and Operations, Edward Waltz

Introduction to Communication Electronic Warfare Systems, Richard A. Poisel

Knowledge Management in the Intelligence Enterprise, Edward Waltz

Mathematical Techniques in Multisensor Data Fusion, Second Edition, David L. Hall and Sonya A. H. McMullen

Modern Communications Jamming Principles and Techniques, Richard A. Poisel

Principles of Data Fusion Automation, Richard T. Antony

Tactical Communications for the Digitized Battlefield, Michael Ryan and Michael R. Frater

Target Acquisition in Communication Electronic Warfare Systems, Richard A. Poisel

For further information on these and other Artech House titles, including previously considered out-of-print books now available through our In-Print-Forever® (IPF®) program, contact:

Artech House
685 Canton Street
Norwood, MA 02062
Phone: 781-769-9750
Fax: 781-769-6334
e-mail: artech@artechhouse.com

Artech House
46 Gillingham Street
London SW1V 1AH UK
Phone: +44 (0)20-7596-8750
Fax: +44 (0)20-7630-0166
e-mail: artech-uk@artechhouse.com

Find us on the World Wide Web at: www.artechhouse.com